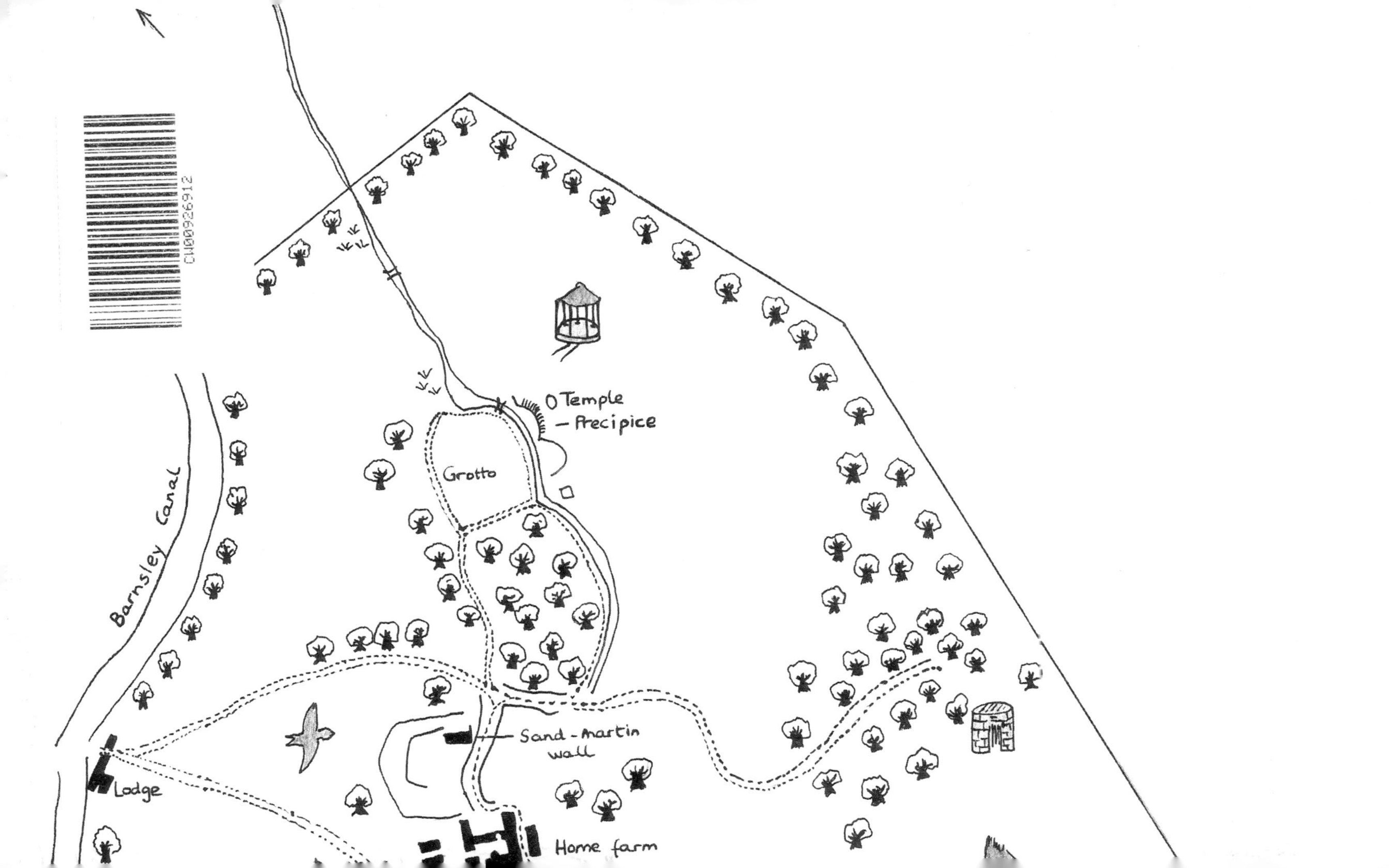
Temple
Precipice
Grotto
Barnsley Canal
Sand-Martin wall
Lodge
Home farm

CHARLES WATERTON
A BIOGRAPHY

Charles Waterton
A Biography

Brian W. Edginton

The Lutterworth Press
Cambridge

The Lutterworth Press
P.O. Box 60
Cambridge
CB1 2NT

British Library Cataloguing in Publication Data:
A catalogue record is available from the British Library.

ISBN 0 7188 2924 7

Printed in Great Britain by Redwood Books, Trowbridge

Contents

List of Illustrations

Dedication

To John, Mark, Celia, Debbie, Pamela, Michael, Simon and Jane.

Acknowledgements

Completion of a book like this depends to a great extent on the goodwill of others. Special thanks to:

Richard Bell, Wakefield Naturalists' Society; Dr J A Bennett, consultant Anaesthetist and Honorary Secretary of the History of Anaesthesia Society; Richard Ellis, Professor of Anaesthesia, St James's University Hospital, Leeds; Alan R Fulton, Head of Operations, City of Aberdeen Libraries; Jennifer Gill, Durham County Archivist; John Goodchild, Local Studies Officer and Keeper of the Archives, Wakefield Library; Gordon Hand, Group Librarian, North Yorkshire County Library; Graham Hopner, Local Studies Librarian, Dumbarton District Libraries; Martin L. Levitt, Manuscripts Librarian, American Philosophical Society, Philadelphia; Dr John Lines of Wisbech; Bari M Logan, Prosector in Anatomy at Cambridge University and formerly Chief Prosector to the Royal College of Surgeons; Lutterworth Press, Editorial Department; Roger Maltby, Professor of Anaesthesia, Foothills Hospital, Calgary; Lillian B Miller, Historian of American Culture and Editor of the Peale Family Papers, Smithsonian Institution, Washington DC; Anthony Smith, Curatorial Officer, Royal Commission on Historical Manuscripts; Dr W D A (Denis) Smith, Senior Lecturer (retd.), Department of Anaesthesia, University of Leeds; Doreen Stobo, Editor, *Spaghetti*; the late Howard K Swann and family, Wheldon & Wesley Ltd; Michael Taylor, Perth Museum and Art Gallery; John Thackery, Secretary, Society for the History of Natural History; Miss A Travers, Research Assistant, Royal Commission on Historical Manuscripts; The Reverend F J Turner, Librarian, Stonyhurst College; Waterton Natural History Association, Alberta; David Waterton-Anderson; The Waterton Park Hotel, Walton Hall; Ellen B Wells, Head of Special Collections Branch, Smithsonian Institution libraries; and last but most of all – Gordon Watson, Museums and Arts Manager, Wakefield Metropolitan District Council.

Chapter 1
Bless the Squire . . . and his Relations

Here's what you need for a self-catering holiday in the South American rain forest: umbrella obviously, top hat, braces, shoes perhaps, check shirts, dispensable unmentionables and flannel underwear ('a great preserver of health in the hot countries.')[1] You'll need a hammock, mosquito net, lancet, purgatives, taxidermy tools, specimen bags, coarse towels and plenty of tea – best warm and weak. You'll also need to have a suitable little speech prepared, to introduce yourself to the assembled natives. Something like:

> Ladies and Gentlemen, I am come to see you, and to admire your beautiful country, but not to eat you, nor to be eaten by you.[2]

The top hat should double as a boxing glove, the braces serve as a convenient muzzle for dangerous snakes, and the shoes come in handy as a head-rest when necessary. (Otherwise they're redundant.) Apart from all that – and, of course, the Latin classics (required tree-top reading) – all you need is a boundless appetite for accidental travel and implicit faith in Almighty God and the Catholic church. Assuming, that is, that you have cause to consider yourself, as Charles Waterton considered himself, 'amongst the most commonplace of men.'

He had good cause to consider himself one of the most commonplace of men, as vindication of the facts which proved otherwise. Herewith, the common facts. For the time being, let's persevere with the South American theme: he rode an alligator, boiled a toucan, talked to insects, fought with snakes, apostrophised woodpeckers, phlebotomised himself, offered his toe to the vampire bats and indulged in all manner of scientific monkey business with the primates, such as remoulding the skin from a howler monkey's backside into the likeness of a man, which he then represented as a newly-discovered species, or Nondescript, to take revenge on a customs official. All which and much more, advanced as evidence of eccentricity, makes a monkey out of any man, or woman, who claims that Charles Waterton was commonplace.

There are, of course, eccentrics and eccentrics – genial clowns and self-indulgent idiots. They're easy to confuse. They both flourish, and sometimes amalgamate, where social mores and common sense are sufficiently stupid to merit lunacy. Add a pinch of historical defeat, a few hundred years of second-class status to inhibit the otherwise privileged, and you see why the English Catholic aristocracy find so much to subvert in the Puritan ethic, so much to gain from fooling around. Throw in a few extra laughs for old time's sake, and you get, in practice, something like Charles Waterton, clown prince imperial of the optimal experience. He was a very lucky man; you can get eccentricity on prescription now, from the NHS – foolproof cure for facing the facts and killing yourself.

Charles Waterton, then, Squire of Walton Hall, Yorkshire; known to his world-wide parish as 'the Squire'. He was a great pioneer of the optimal experience, the Squire, his species of eccentricity almost always genial, though sometimes less genial than eccentric. He was occasionally a self-indulgent idiot – very occasionally – and often assumed to be when he wasn't. Objections to his 'odd and offensive peculiarities' by one peculiarly offensive oddity called James Simson,[3] aspired to give Charles Waterton and eccentricity a bad name, or at least to reinforce the bad name they'd already got. But his legacy transcends all ill-will and struggles still against persistent, sanctimonious efforts to compensate for his reputation by taking him too seriously. Eccentricity as a term of abuse – a disgraceful imputation on the good name of infamy.

The best way to demonstrate the inoffensiveness of Charles Waterton's eccentricity, and the incongruity of assuming that the tag itself either does not apply or somehow demeans him if it does, is to compare him with the sort of contemporary screwball with whom he's most likely to be confused. How about John Mytton, squire of Halston Hall, Shropshire? 'Mad Mytton'. 'Neck-or-nothing Mytton'.

Both men were country squires and both connoisseurs of that most aristocratic of all forms of humour, the practical joke. There, the similarity starts. But Mytton was a little Caligula by comparison, a spoiled aristocratic brat with an undeserved reputation as a folk hero in the local pubs. Charles Waterton, by contrast, was obviously the prototype for Lewis Carrol's Mad Hatter – or if he wasn't, he should have been.

Mytton dressed up as a tramp, to pick fights with his butler and beat up his footman; the Squire passed himself off as the butler, to tickle his visitors' ribs with the coal brush. Mytton stocked up with pheasants until they were 'thick as sparrows at a barn door', so that he could blast their brains out like genocide, 800 at a time; Charles Waterton protected what pheasants he'd got and stocked up with wooden dummies to draw the enemy's fire. Mytton locked up his 16-year-old wife in the dog kennel, threw her pet dog on the fire, then pushed the girl into the lake to cure her of 'melancholy'; Waterton revered his 17-year-old wife, unjustly blamed himself for her death in childbirth, and inflicted a lifelong penance on himself by never afterwards sleeping in a bed. Mytton released some imported rats onto his frozen lake and hunted them on ice skates; the Squire preserved his lake strictly for the birds and enlisted the help of owls, weasels and a South American tiger cat, to control the indigenous rats which threatened to over-run his house and grounds. Mytton climbed the tallest trees in his heronry, to take the eggs; Charles Waterton climbed the tallest trees in his heronry, to replace the nestlings blown out by a storm. Mytton disguised himself as a highwayman and shot two of his guests; Charles Waterton disguised himself as a dog and bit his guests' legs as they walked through his front door. Mytton set his own shirt-tail on fire, to cure a hiccup; Charles Waterton cupped his own blood, to cure anything and everything, from backache to malaria. Mytton rode a live bear in his own drawing-room; the Squire dissected a dead gorilla in his. And so on.

But all peculiarities are still assumed to be uniformly offensive by reason of being odd. And so we get the old, reiterated rigmarole about Charles Waterton's reputation as an eccentric obscuring his true merit as a serious naturalist. The truth is, he was a very indifferent naturalist but one of the truly great eccentrics of all time. Neither fact obscures the other, though both illuminate a good deal else.

Take the cuckoo, jug-jug, pu-we, to-witta-woo. The Squire of Walton Hall was a great authority on the cuckoo; on 4 July, 1822, at about half past one o'clock in the afternoon, he heard one call, 'repeatedly'.[4] The observation was verified several times in later life. So when fools bore false witness with regard to the cuckoo, as in the case of that persistent old wives' tale about a recently-hatched bird which allegedly evicted a young hedge sparrow from the nest, Charles Waterton was quick to assume truth's fiery cross:

> No bird in the creation could perform such an astounding feat under such embarrassing circumstances . . . I had much rather believe the story of baby Hercules throttling two snakes in his cradle.[5]

Or take the dipper, *Cinclus cinclus*, another speciality of Charles Waterton. There were those who claimed that it could defy the laws of physics and walk on the bed of a fast-flowing stream. The Squire was 'not a convert to the doctrine of a subaquatic promenade.'[6] He would much rather have believed, and certainly did believe, the much better authenticated story of Jesus Christ's observed ability to walk on the surface.

Or take the ornithological oil gland, or parson's nose. . . . But enough! Suffice to say that in matters of great import, on subjects of assumed competence, he was roughly as likely to be wrong as right the sole judge of truth, in endless error hurled. With so many pert jackanapes and conceited young coxcombs around – market zoologists and closet naturalists – natural history was a strictly confrontational sport. He managed to flay alive two generations of academic quacks, not least on the subject of ducks, and where the vulture's power of smell was concerned – a subject of great international passion at the time – he certainly got the better of that other doubtful expert, John James Audubon. But as often as not, as with the cuckoo, the dipper and the parson's nose, he was wrong – usually pig-headedly.

No stronger evidence of his glorious insignificance exists than his own humble protestations of greatness. One of the 'market naturalists' – William Swainson – had the temerity to call him an 'amateur naturalist' – clearly a pejorative. He was hurt by 'amateur' as badly as he was stung by 'eccentric.' So he set about proving he was a level-headed professional. The result was a curious medley of borrowed plumes and useless cock nests:

He'd succeeded in getting birds of inimical species to nest 'at a stone-throw of each other'; he'd noticed the eclipse phase in waterfowl plumage; he saw that rooks flew away to feed each morning and came back again later; he got a bit worked up about which birds cover their eggs and which don't; he noticed that kestrels generally don't eat birds but that wigeons do eat grass; he traced the introduction of the brown rat into England back to the same ship which had introduced the House of Hanover; he showed how to preserve birds' eggs and insects for cabinets of natural history; and he pointed out that, whilst cats are generally to be discouraged if you want birds (except at Walton Hall), if you're unlucky enough to be attacked by a dog, you may as well kick him. These were his own justifications for his own idea of his place in history, in response to a charge of being merely an amateur.[7] He also, be it remembered, heard the cuckoo call repeatedly.

But if you think all this sounds dismissive – slightly derogatory – for goodness sake don't let it interfere with your better judgement. It should not be allowed to obscure Charles Waterton's true merit as a first-rate eccentric. The opposite claim is ludicrous.

It's often claimed that the first 'biography' of Waterton – a book of 'Reminiscences of an Intimate and most Confiding Personal Association for Nearly Thirty Years' – purposely exaggerated his odd peculiarities, and served as the model on which his implicitly over-stated reputation as an eccentric is based. As the Squire himself might have said of almost anything else – Rubbish! He got funnier and dafter as he got older and wiser, but the most commonplace of men was already 'queer as Dick's hatband', by his own testimony, long before he met the man he accused of agreeing with him.

We're talking about Richard Hobson, his friend and physician – almost as daft and ten times as stupid – the most hilariously pompous lump of Victorian lardy cake you could possibly wish to meet. Exaggeration from him puts his own bumptious airs in context.

The Squire met him in the 1830s and the two men were good friends until they fell out in 1862. They exchanged visits, corresponded, and sat together in the Walton Hall 'grotto', to 'beguile many an hour in natural history pursuits, for upwards of five and twenty years.' A cherished image is of Doctor Hobson watching, 'in painful suspense and much against my own inclinations,' as his 77-year-old patient hopped along the brink of a 'precipice' in the grotto, whilst his other leg dangled over the ten foot 'chasm' below. The same high spirits found many another outlet, viewed with similar pained reproach by the doctor. But most of the episodes were pale reflections of much more daring exploits carried out long before the two men met.

Hobson's book – *Charles Waterton, His Home, Habits and Handiwork* – was first published in 1866, shortly after the subject's death. As Doctor Johnson said of somebody's edition of Shakespeare, 'its pomp recommends it more than its accuracy.' Which isn't to say that the book is inaccurate.

It talks of many things – of shoes and ships and sealing wax, of cabbages and kings. Its table of contents alone is a delicious hors d'oeuvre for the indigestible main course to follow, a seductive compendium of the joyful inconsequentiality of life in the World's First Nature Reserve. Here are just a few items from the menu, to whet the appetite:

'Mental and Physical Characteristics create a Diversity in the Choice of Amusements. Two singularly-constructed Rappers affixed to the folding Doors of the front Entrance. Singular Representation of the Nightmare in the Vestibule. Special Immunity in the female Sex from Death by Lightning. Visits of the Lunatics from Wakefield Asylum. Unusual Progeny of a Waterhen. Mr Waterton's Opinion on the Accusation against the Carrion Crow of sucking the Eggs of other Birds, with Evidence somewhat mitigating the Charge. Promiscuous Charity alluded to, which leads to singular Encounters with a lazy Beggar. Fearless Risks undertaken by Mr Waterton in connection with the Leeds daring Snake Adventures. Discriminating Courage of the Squire with an Orang-Outang from Borneo. Mr Waterton "fairly floored" by Mr Salvin's clever Imitation of a Pig. Mr Waterton cautioned as to the dangerous Bungling of operative Practice in Surgery. His jocular Retort. An Allusion to a Stench from a dead Herring near the Grotto induces the Squire to relate an Incident regarding dead Letters. The numerous good Deeds of Mr Waterton attest his truly estimable Character. The Squire attempts to navigate the Atmosphere. Mr Waterton becomes afterward convinced that he was "out of his Element" in attempting an atmospheric Trip. Mr Waterton faces the Snowstorm without his Hat, and throws his Slippers over his Head when approaching his 80th Year. Letter from Mr Waterton describing a Fall from a Ladder. Explanation why the Squire was able to make his Elbows meet. Cats, and his Opinion as regards their being fed. The Squire's Distinction of Criminality incurred between the upper Ranks and the Poor. The Squire expatiates on the Beauties of the Toad.' Etcetera.

But, for all the rich diversity of life at Walton Hall, and the many singularities associated with it, it's the style of presentation which stamps the *Home, Habits and Handiwork* as a great masterpiece of nineteenth-century bombast. The present author doffs his cap and salutes in mute acknowledgement of his debt to the great prime mover of Watertoniana; all attempts to follow are necessarily unworthy.

For many years, and despite religious differences, the two men were very close friends. Hobson was obviously schooled by Waterton as a possible Boswell. An attempt was made to stop the book, when it was learned that the publisher had it in his hand, but that was after the two men had quarrelled. In the name of all that's magnificently turgid and vacuous, posterity has good cause to be grateful that publication went ahead. The book is a mine of information and a triumph of fine feeling and artistic integrity.

We learn, for instance, that Charles Waterton liked trees:

> The Squire has several times, when speaking of his early pedestrianism and what he called vagabondising expeditions in South America, insinuated to me, that in admiration of nature in all her varied profusion of modified phases and alluring aspects, with extreme pleasure and with a deep and fervid interest, which none but an ardent admirer of nature could realise - that he had, he observed, in numerous instances, been delighted and influentially animated beyond measure, during a long succession of years, by closely watching simple arborescent vegetation during its gradually advancing growth, from the tiny twig to its umbrageous and approaching ornamental termination of its growth.[8]

He definitely liked trees.

As for that quarrel, the Squire, said the doctor, was liable to support his own opinions by very 'energetic' language, and occasionally by 'a tartness and an asperity of expression somewhat ungracious, and unfortunately, also, by a very decided insuavity of manner.' No-one was exempt from a 'somewhat irascible punition.'[9] He'd got plenty of spunk in his balls.

But Hobson's reminiscences, for all their apparent praise by faint damnation, were not intended as revenge. His eccentric language did not exaggerate his friend's eccentricity. That was left to Charles Waterton himself, with his top hat, braces, flannel underwear and assorted barrel-loads of funny monkeys. He needed no mud-slingers; he dived in head-first.

About that hat – a 'shocking bad hat', with the crown flapping loose and the sides pierced with holes to keep his head cool. It was a hat fit for a head fit for a hat like that. All it lacked was the price tag. He never allowed it to be brushed but soaked it in the same solution of bichloride of mercury – corrosive sublimate – which he used for preserving his zoological specimens and the lining of his carriage. The result satisfied him, and horrified Hobson, which was gratifying.

Figure 1. Frontspiece of 1838 edition of Watertons Essays, *1st series.(Wakefield MDC. Museums and Arts)*

It was an age of hats – still the sign of some degree for all the fashionable Tom Fools and young coxcombs of high society – the Miss Nancies and man milliners who were sartorial equivalent of those arch-knaves of fashionable science, the closet naturalists and market zoologists – pert jackanapes all. Charles Waterton wore his hat as a sign of his total disregard for social degree. He often wore no hat at all, for the same reason.

It was also an age of whiskers, so he shaved. It was an age of long hair, so he cropped his hair short. Doctor Harley of Harley Street said that he looked like a man recently discharged from prison. That would have pleased him.

Many paintings and sketches of Charles Waterton were made during his lifetime, and one or two after his death. In reference to all of these, the perfect complement, if uncomplimentary, is the autobiographical word-picture, painted in middle-age as an introduction to his first volume of *Essays on Natural History*, published in 1838.

> I stand six feet high, all but half an inch. On looking at myself in the glass, I can see that my face is anything but comely: continual exposure to the sun, and to the rains of the tropics, has furrowed it in places, and given it a tint which neither Rowland's Kalydor, nor all the cosmetics on Belinda's toilette, would ever be able to remove. My hair, which I wear very short, was once of a shade betwixt brown and black: it has now the appearance as though it had passed the night exposed to a November hoar frost. I can-not boast of any great strength of arm; but my legs, probably by much walking, and by frequently ascending trees, have acquired vast muscular power: so that, on taking a view of me from top to toe, you would say that the upper part of Tithonus has been placed upon the lower part of Ajax. Or to speak zoologically, were I exhibited for show at a horse fair, some learned jockey would exclaim, he is half Rosinante, half Bucephalus.[10]

Each November, as an old man, this zoological hybrid – half Rosinante, half Bucephalus, and the other half soaked – took a constitutional dip in the sea at Scarborough, in full-fig, calf-length, Victorian bathing costume and goose pimples. Its safe to assume that he also wore his top hat for the occasion, just to amuse the assembled seagulls. As sure as pigs have wings, the sea was boiling hot.

One of the main grounds for an eccentric reputation is prophetic insight. After leaving Scarborough each year, Charles Waterton would return with his sisters-in-law, by train and Whitechapel cart, 'to those gloomy regions where volumes of Stygian smoke poison a once wholesome atmosphere, and where filthy drainage from hells upon earth is allowed by law, for the sacred rights of modern trade, to pollute the waters in every river far and near.'[11] The country around Walton Hall was plagued with the 'long chimneys' of incipient progress. Even the rain was poisonous – it was killing the trees, he said. Poisonous rain! The man was obviously stark, staring mad.

That was the general view at the time and his physical appearance served to reinforce it. But we now call the rain acid and it's mad men who make it – 'Men my brothers, men the workers, / Ever reaping something new.'[12] Tennyson was thrilled to bits with the sacred rights of modern trade – 'fairy tales of science', he called it, and barely an Englishman alive disagreed. It was only a few selfish cranks like Charles Waterton who objected. Like he objected to the felling of forests in New England. 'A century will not replace their loss', he said. 'They can-not, they must not fall.' So they fell, like the credibility of all those selfish cranks like Charles Waterton who objected.

He objected, too, to the persecution of rooks and crows and magpies and jays and hawks and snakes and owls and herons and kingfishers. 'A keeper who can massacre the greatest number of these interesting denizens of earth and air, is sure to rise the highest in his employer's estimation.' The keeper at Walton Hall was made to promise, as he valued his place, to protect all such and all else besides except the 'Hanoverian rat.'

His neighbour's keeper handed him a bittern which he'd just shot and clearly expected effusive thanks for the gift. 'Your father murdered the last raven in England', muttered the Squire, indignantly, and the keeper went away with his tail between his legs and murder in his heart for all 'eccentric crows' like Charles Waterton – a title first bestowed on him by James Rennie, Regius Professor of natural history at King's College, London.[13]

Objection to shooting genuinely was considered extremely eccentric, even after Waterton's death. Someone saw an osprey near Tamworth, on the edge of the Black Country, and was roundly chastised by the scientific gentlemen because he 'let it go.' Yet who knows – that last raven in England could have been King Arthur himself; for was not the flower of kings transformed into a raven when he croaked? One day, for sure, he'll reappear to reclaim his throne and then, once more, the browsing herds will pass from vale to vale, the swains will sing from the bluebell-teeming groves, and nymphs with eglantine and roses in their neatly-braided hair will walk hand in hand to the flowery mead to weave garlands for their lambkins.[14] Depend on it. Charles Waterton did.

Meanwhile, we just have to do the best we can with materials to hand – succour the beast and spare the tree, in anticipation of that great second-coming when the lion shall eat straw like the ox and no-one shall hurt nor destroy in all God's holy mountain. That's what Charles Waterton did – he prepared the ground for the second-coming – and for all his limitations as an expert, in the modern sense of the word, he was thereby a great, pioneering high priest of that repudiated gift-horse of the next century, 'conservation of nature'.

You can cultivate expertise – in natural history, as in almost anything else – on the strength of your own prior assumption of it. Charles Waterton told the world he was an expert and most of the world believed him. The rest was better educated, less well-informed – they stuck to the closet.

Their classification gave him jaw-ache – 'Gampsonyx Swainsonii, Lophophorus, Tachipetes, Pachycephela, Thamnophilinae, Dendrocolaptes, Myiagra rubiculoides, Ceblepyrinae, Ptilonorynchus, Opistholophus, Palaeornis, Meliphagidae, Eurylaimus, Phalacrocorax and a host of others.'[15] He much preferred the cosy ambivalence of local names; they covered any embarrassing doubts he might have about identification.

But his sense of the 'economy' of nature, his understanding of habitat and interdependence, his contempt for pollution and for all the shallow, pestilential notions of what constitutes a 'pest', and why, was years ahead of his time and faintly derivative of Pliny the Elder, 1800 years before his time. Reminiscent, too, of Sir Thomas More, which isn't surprising – the two men were related. The founder of the World's First Nature Reserve was 8th in direct line of descent from the author of *Utopia* – the wonderful island of Utopia, where men have singular delight in fools but take no pleasure of foolishness, and where the contemplation of nature and the praise thereof is to God a very acceptable honour.[16]

A lifetime of intimacy with nature – the common or garden variety – gave him a certain folksy authority which he often used to discredit everyone else's. It sets him head and shoulders above all the meticulous classifiers of his own day, and several heads and shoulders above all the woolly-hatted twits and twitchers, the ornithological pachycephela, of today's small outdoors. He knew which trees it was safe to climb and strongly advised the importance of climbing them. Excessive caution was 'melancholy doctrine.'[17] He was still reading his Latin classics in the tree-tops – aetatis suae 82 – a few weeks before he died. And many a family of commonplace crows had an occasional visit from an eccentric relative. He just failed to discover a suitably lofty subject for his intimate investigations; that's all.

If the spleen is the seat of all humour, as claimed by the classical Greeks, it comes as no surprise that Charles Waterton probably died of a ruptured spleen. But the best of all belly laughs undoubtedly proceeds from the region of the bowels – inferior apertures and caudal extremities; medieval adventures in Cloacina's temple, the true seat of all humour. The theme recurs in Charles Waterton's work and generally points an instructive little moral, as in the tales from royal courts of old. It also subserved his religious belief.

The Protestant cathedral of St Paul's, in London, was a veritable 'pandemonium of pollution' – the pews turned into *cabinets d'aisance*, and the whole filthy den of thieves rank with the stench of 'smoking ordure', in lieu of incense.[18] But the Catholic church of the Gesù, in Rome, was spotless, except, of course, when there were thieves about; that's when the church treasures were hidden under a few bucketfuls of 'imported guano' from the street outside – much better than a padlock. Such stories, he sought to relate 'without trenching on the very nervous sensibility of modern decorum.'[19] Others on the same lines were less moral.

In the Waterton literature, you meet a diarrhoeic Scotsman plagued by a swarm of coushie ants, you hear about an eel which was defecated twelve times, you hold your nose in the lavatories of gorgeous palaces – 'O ye nasty people of Bologna'[20]– and you get presented with the admeasured components of a jackass's caudal deposits, adduced as evidence to acquit two innocent horses of disrespect. All without trenching on the very nervous sensibility of modern decorum.

He didn't much care for modern decorum, special preserve of 'those pharisaical parsons who think it damnation to whistle on a Sunday.'[21] He was raised under the Jesuits, at Stonyhurst, and cared not a fig leaf for the Pharisees of 'Mother Damnable's church'[22]– products, for the most part, of those unprincipled dens of dunces at Oxford and Cambridge. His religion and politics best conformed to what would now be called liberation theology, with a strong undertow of Catholic credulity and fundamentalism. His views of Creation made no more concession to science than a tentative acceptance of the long-discredited theories of geological catastrophism and the Great Chain of Being. His religious conviction drew no distinction between belief in God and unquestioning belief in all the most unbelievable miracles and holy relics. His response to a few hundred years of religious oppression, by 'a creed infected with the rot', was to act the silly jackass and rub all puritanical noses in the products of caudal extremities.

Yet some of his best friends were Protestants – in fact most of his best friends were Protestants. Bishop Longley of Ripon was a frequent visitor to Walton Hall, and he became Archbishop of Canterbury. They remained good friends.

You can take it as an article of faith that Longley cast more than a casual glance at 'The Reformation Zoologically Illustrated' – the Walton Hall rogues gallery of flies and beetles and frogs and toads and lizards and newts and apes and snails and puppy dogs' tails, preserved for all time and eternity as John Knox, Martin Luther, Titus Oates, Bishop Cranmer, Bishop Burnett, Old Nick, Queen Bess at Lunch and sundry other 'dissenting fry' of the Church by Law Established. It was all harmless fun, of course, though a Protestant satire on Catholic equivalents might not have been. He meant no harm – just vengeance. Where his own faith's repression was concerned, another close Protestant friend – J. G .Wood – put it all in a neat little ecumenical nutshell: 'He felt the wounds but he could jest at the scars.'[23]

He could also fight back, and usually did – whenever he felt that his faith was impugned by a creed infected with the rot. His polemical pamphlets against the manic pope-bashers of the Reformation Society were as close as he came to the provocative last word – the trenchant indecorum of great literature. His family's history demanded no less.

The Watertons took their name, 800 years ago, from land called Watretone, on the banks of the river Trent, in the Isle of Axholme, Lincolnshire. The farm is still there and still called Waterton Hall. It was Reiner, son of Norman de Normanbi, who first adopted the surname Waterton.[24] That was in the 12th century.

By the 14th century, the family was illustrious. Sir Hugh, Sir Robert and Sir John Waterton were all in the service of the House of Lancaster. When Henry Bolingbroke

Figure 2. Pontefract Castle. Remains of stone keep. (Wakefield MDC. Museums and Arts)

returned home from exile to claim the throne of England, Sir Robert Waterton was the first man to greet him at the quayside.[25] And when King Richard II was taken prisoner at Flint, he was sent to Pontefract Castle – 'Bloody Pomfret'[26] – where he was duly murdered whilst in the custody of its warden, the same Sir Robert Waterton. When the Duke of Orleans was captured at Agincourt, he, too, was sent to Pontefract Castle[27] but enjoyed such honeurs de guerre as allowed him to visit his gaoler's home for the jeux de chasse. King Henry got nervous and desired that the duke should no longer go to 'Robert's place . . . for it is better he lack his desport than we were deceived.'[28]

Robert's place was Methley Hall, a magnificent new manor house between Pontefract and Leeds. It was demolished in 1964. Robert died in 1425 and his younger brother, John, enhanced the family scutcheon by marrying a daughter of Sir John de Burgh, Lord of Borough Green, Cambridgeshire, and Walton, Yorkshire. John Waterton died in 1430, and his son, Richard, also married into the de Burgh family and acquired Walton Hall, jure uxoris, in 1435.[29] It remained in the Waterton name until 1878, shortly after Charles Waterton's death.

Another Sir Robert Waterton was Henry VIII's Master of Horse. He refused to acknowledge the king as 'supreme head of the church in England' and therefore received an internal memo from the royal goat himself:

Waterton, I will take thy estate, but I will save thy life. HENRY REX.[30]

The king's chancellor, Sir Thomas More, was not so lucky. He was detained in the Tower during the king's pleasure, declined to give up eternity for a house in Chelsea, and finally lost his head but came to no harm. His descendant, Charles Waterton, followed that example many times.

Sir Robert Waterton kept his head but lost his rent roll. (In Methley and the Isle of

Axholme.) One of his descendants married Charles Waterton's Aunt Mary and thereby remerged the Methley and Walton lines of the family.[31]

The Watertons of Walton Hall kept their estate, though their fortunes declined steadily. Penal laws against Catholics were by varying degrees severe under the Tudors, with a brief pause to burn Protestants in the 1550s. Elizabeth did a stretch inside whilst her Catholic sister ruled the country and her gaoler at Woodstock was Mary Tudor's vice chamberlain and governor of the Tower of London,[32] Sir Henry Bedingfeld, great great great great great grandfather of Charles Waterton's mother, Anne Bedingfeld. 'If we have any prisoner whom we would have sharply and straitly kept, we will send for you', said the queen,[33] afterwards, and Bedingfeld was clearly forgiven.

But not many Catholics were as lucky as that. She was not her father's daughter for nothing. Doubt her spiritual supremacy, and you lost your estate or your life or both. Stay away from church, and you were fined £20 a month.[34] Successive generations of Watertons paid £20 a month like clockwork during the sixteenth and seventeenth centuries.

During the Civil War, the family fought for God and king in that order, against parliament, in the name of free worship and the right to whistle on a Sunday. At Marston Moor, in 1644, the 5th generation of Thomas Waterton was one of the 4,000 Royalist dead. Shortly after the battle, so the story goes, Oliver Cromwell took a small force of men to Walton Hall, to pass the time of day with Thomas's widow, Alice. The siege troops set themselves up on a little hill about 100 yards from the fortified watergate tower which in those days was three storeys high and connected to the mainland by a wooden drawbridge. The drawbridge was destroyed in the siege but the gatehouse ruins still stand and figure large, not only in the Waterton family history, but in the life of Charles Waterton himself, lord protector of barn owls and starlings. As one of the ruins that Cromwell knocked about a bit, its claim to authenticity is a good deal stronger than most.

In the eighteenth century, two members of the British fauna began to play a significant role in British history, to the edification of all readers of Charles Waterton. William of Orange was killed, in 1702, by 'a little gentleman in black velvet', and the same Jacobite tradition which toasted the mole who made the hill that tripped the horse which threw Dutch William to his death, insisted that his successor, German George, introduced the brown rat into England, in 1714, to all but evict the Catholic black rat from its native country. Charles Waterton declared war on the 'Hanoverian rat' but the Jacobite mole was as clever a little drainage engineer as ever a squire could wish to have on his land.

His grandfather – 25th Lord of Walton Hall – was a Jacobite bold and true. It was he who married Anne Cresacre More, of Barnborough, near Doncaster - 6th in line of descent from Sir Thomas More – and who, shortly before the Battle of Culloden, 'had the honour of being sent prisoner to York . . . on account of his well-known attachment to the hereditary rights of kings.'[35] When he was released from gaol, he found that his home had been ransacked for arms by the redcoats. He died in 1767 and left at least seven children. The first-born son, Thomas, became 26th Lord and married Anne Bedingfeld, daughter and heir of Edward Bedingfeld, the second son of Sir Henry Bedingfeld, third baronet of Oxburgh Hall, Norfolk.[36] On 3 June, 1782, to a glorious fanfare of birds in the park, Anne gave birth to her first son, Charles – 27th Lord apparent of Walton Hall.

Bless the Squire. . . and his relations.

Charles Waterton often claimed that after Henry VIII's vandals had done their worst, his family was the first to disclose a real catholic icon to public view a stone crucifix attatched to the ornamental precipice in the grotto. A similar cross still stands on the derelict watergate tower in front of the Hall.

Graven images were the just perquisite of uninfected worship, to Charles Waterton, as the rudest simplicity was its practical application. His bed-less bedroom on the top floor of Walton Hall was to all intents and practical purposes a monastic cell, bodily comforts sacrificed to spiritual hygiene and 'delicate manipulations' – the preservation of zoological specimens. Bed was a mat, pillow a wooden block, wardrobe a nail. Dressing table and taxidermy bench were one and the same. The eviscerated baboon hanging from the bare rafters served as inducement to profitable contemplation and sweet dreams.

All the wealth and colour absent from his bedroom was found in the private chapel on the ground floor. But to fully appreciate the true splendour of Catholic worship, and to put the lavish delights of religious asceticism in context, he obviously had to take the Grand Tour. And the Grand Tour to Charles Waterton was grander than most – a Catholic pilgrimage from church to church, from holy relic to holy relic, from ecstatic virgin to authenticated 'supernatural occurrence'. The occurrences used to occur in Britain, before the Reformation – history assured us of it 'in the most positive terms.'[37]

Mecca, of course, was Rome – the quietest of cities in his day, as New York was the most law-abiding. Apart from great Christian festivals and glorious shrines, there was also the taxidermical treasure trove of the open-air bird market, outside the Pantheon; there he could pick up the odd 'Italian heron' or 'chesnut drake', for delicate manipulation, or a few live 'civettas' for naturalisation abroad. (After a nice hot bath en route.) By Italian heron, he meant night heron, *Nycticorax nycticorax*, and the chesnut drake (sic) is a red-crested pochard, *Netta rufina*; they're now at home in Wakefield Museum. The civettas were little owls, *Athene noctua*, which Charles Waterton was first to introduce into Britain. They're now at home on your local telegraph poles and television aerials.

Italy had its drawbacks, of course. The frog soup, goat flesh and snail cream – not to mention roast little owls and heron pie – was tough cheese to chew after a diet of dry bread and weak tea. It affected the Squire's health to such an extent that for some time after he got home to Yorkshire, he was disinclined to swarm up the pillars of his Greek portico, as was his wont. A more damning indictment of Italian food, it bids scorn to suggest.

He also took the Italians to task over disposal of the aforementioned 'fertilising matter' – the products of caudal extremities. They should put more of it on their fields and less in the streets, he said, though it did undeniably come in handy for hiding church valuables from Napoleon's troops. In this, as in other environmental issues, his comments now seem like the words of the wise, much matter of great issue decocted therein. In 1989, at last, a book was published in the U.S.A. which finally recognised the ecological value of properly disposed fertilising matter: *How to Shit in the Woods - An Environmentally Sound Approach to a Lost Art.*[38]

Charles Waterton never lost the art; there's some corner of a foreign field which is, to this day, forever England . . . In that rich earth a richer dust concealed. But he never thought to write a treatise solely devoted to the matter, though he dealt with it at some length in various places. Instead, he applied himself to other aspects of field natural history and general disdain for the products of the closet. Unless, of course, the closet was his own bedroom and the products were those of the more delicate manipulations associated with his own, far-famed Waterton Method.

His war on the closet naturalists ran through his life like a running sore. It was 'deficiencies in bog education'[39] which marked them out for eternal damnation. They might invoke big words or fancy titles, honorary memberships or complicated theories – anything to conceal their deficiencies – but there was no fooling a dedicated bog scholar like Charles Waterton. He saw them for what they were – scientific reprobates – 'morbid and presumptuous men', weighed down by the hopeless, melancholy doctrine of keeping feet firmly planted on the ground. Like flightless birds of a feather, they stuck together.

He clearly felt threatened by the closet naturalists, despite their lack of popularity. They were rival 'experts', competitors for the same ecological niche of ornithological fame, wholesale dealers in whatever branch of market zoology best served their shameful need to earn a living. Served them right if they failed to earn one.

William Swainson produced two series of *Zoological Illustrations*, the second volume of the first series of which sold like stale cakes: '100 – no. Twenty-five, no. Fifteen, no. Ten? Yes. Positively ten copies and no more.'[40] What was he to think of the 'generally diffused taste', as the phrase is, for natural history?'

But Waterton's *Wanderings in South America* was already 'in every hand.' From its very first edition, in 1825, the author was a very famous man; the taste for his brand of natural history was generally diffused throughout the English speaking world, and, ultimately, a few other worlds besides. Zoology was his property; he claimed copyright.

The fame lasted until almost the end of the 19th century. In *Walford's County Families*, for 1890, William Erasmus Darwin was the son of Charles Robert Darwin Esq., F.R.S., of Down, Kent. Edmund Waterton was the son of Charles Waterton, 'author of *Wanderings in South America*.'

But as early as the 1840s, the generally diffused taste was more for the trivial issues of political economy than for even the most trivial manifestations of the much more important economy of nature. (Economy, as being the old word for ecology.) 'The history of political parties has put natural history out of countenance', said the Dublin Review, 'and political leaders have so far outstripped tomtits in the noble art of feathering their nests, as, in parliamentary phrase, to have kept politicians 'in' and pastorals 'out'.'[41] They remain out still, except insofar as they become political, which, by due process of urbanisation, puts them in again.

Charles Waterton remained a sentimental pastoral, against the patriotic grain. His attitude to nature was old-fashioned to a degree, and child-like withal –jongleur of the jungle, mountebank of the precipice, the glory, jest and riddle of the world. 'The strangest mixture of kindness, harshness and bigotry' that ever Charles Darwin saw.[42] 'By no means a wise man and certainly not a witty one', as one obituarist mourned.[43] So much the worse for wit and wisdom.

In the modern sophisticate's noisy little world of pea-brained reaction against nature – wilful agoraphobia, cultivated philistinism, misanthropic 'people'–preference, blind insensitivity masquerading as sense, or even wit – his example offers no very powerful antidote or corrective at all. He was too entertaining, too bigoted, too suggestible, too old-fashioned, too sentimental, too easy to laugh at. He possibly serves to add fuel to the prejudice. But viewed through more friendly eyes, the life, and, in particular, the Home, Habits and Handiwork of Charles Waterton, Esquire, 27th Lord of Walton Hall, represents a species of bizarre sanity by which civilized affections persevere despite facile reaction. You won't be sorry if you get to know him. And here's your chance.

Gentle Reader, Charles Waterton. Charles Waterton, Gentle Reader.

Chapter 2
1782-1801
Organ Blower and Football Maker

The old Walton Hall was demolished 15 years before Charles Waterton was born. The new house was built on the same site, on an island at the north-western end of a 17-acre lake, connected to the mainland by a cast-iron footbridge. The bridge was also new and is said to have been the first of its kind outside Ironbridge itself. All that remains of the old, fortified manor house – apart from the ruined gatehouse – is some oak panelling which was taken from the old, 90-foot long dining hall and put into the entrance hall of the new house. Some of the panels were later used to build a pigeon cot.

The park surrounding the Hall was about 250 acres in extent – mixed woodland and grassland, with a small orchard and a new landscaped grotto, complete with sunken flower garden and classical temples. For the first ten years of his life, Charles was educated at home and much of his time was spent in the park, with his sister Helen. He began to learn the basics of what Gilbert White always called 'the life and conversation of animals.' He also talked to the trees.

He started climbing trees at roughly the same time that he began talking to them. But generally speaking, they offered little in the way of a meaningful challenge – you may as well climb the stairs. For a real demand on his climbing skill, the true indefeasible seeker after truth and bird nests should tackle something more prejudicial to life and limb – a sheer rock face if possible. Unfortunately, there were no suitable mountains or sea cliffs on the Walton Hall estate, so Charles had to make do with man-made alternatives – buildings and precipices. He'd have liked no better fun than to have had an occasional spree up the chimney with 'the little black fellow' from Barnsley[1] – a healthy profession, chimney sweep. But melancholy doctrine and social etiquette forbad it, so he looked elsewhere.

His accredited career as a steeplejack began when he was 'not quite eight years old.' Under one of the roof slates of an outhouse at Walton, he'd spotted what looked like a starling's nest and decided to investigate – all in the interests of science. 'Had my foot slipped,' he said, 'I should have been in as bad a plight as was poor Ophelia in the willow tree, when the 'envious sliver broke'.' But Mrs. Bennett was on hand to 'cast her rambling eye' on the lad, and seeing the danger he was in, rushed indoors for something to lure him down – a piece of gingerbread. When he finally did get down to the ground – a paved courtyard – she seized him – 'as though I had been a malefactor' – and thereby hurt his sense of justice, for lack of mere physical injury. Melancholy

Figure 3. Walton Hall and the Watergate c. 1988. (Wakefield MDC. Museums and Arts.)

doctrine and no doubt about it.

Another recorded incident from this pre-prep-school period of life also involved female betrayal and also, inevitably, a bird's nest. It happened during one of those 'childish but very natural' days spent with his sister. Somewhere in the park, they'd found a skylark's nest with a full clutch of eggs; the father of the man thought he'd like to try a little experiment – he picked up one of the eggs, popped it in his mouth and swallowed it, shell and all, just for a bit of a lark (presumably). His sister ran home to tell Mama about it and Charles was given a fashionably drastic home remedy with what came to hand from the household effects – hot mustard. One outcome was a line of pure poetry in his memoirs: 'I could never afterwards endure the taste of mustard.' Another outcome was the egg.

His mother had a penchant for turned stomachs and usually chose the most violent recipes for turning them. 'Ormskirk medicine' for instance – taken with warm beer and generally thrown back again with bell, book and candle. It was 'bulky' and 'the colour of brick dust', said Charles Waterton[2] – a more nauseous antidote than Don Quixote's balsam of Fierabras, which made Sancho Panza so sick at stomach. Good stuff, Ormskirk medicine.

But all this female solicitude suddenly came to an end when Charles was ten years old – nine, according to himself (a very unreliable source). He was packed off to a small preparatory school at Tudhoe, near Durham, and left in the care of two 'holy and benevolent' clergymen, both of whom had a wholly benevolent affection for the rod. During the four years he spent at Tudhoe, he had a taste of the rod many times; it served to concentrate his mind.

It was all the fault of birds, of course, of which there were plenty in the neighbourhood to corrupt him. In a letter to one of his Catholic friends, Bryan Salvin of Croxdale Hall – written over 20 years later[3] – he mentioned passing through the area on the mail coach and looking for some old oak and ash trees whose tops, as a schoolboy, he had 'many a time visited in search of nests.' 'Our boyish days are happiest of all,' says the letter, informatively, 'and the sight of Croxdale Woods brought back to my mind happy moments not to be described on paper.'

Those same woods, as avouched in the autobiographical preface to his essays, had ensured 'a sufficient supply of carrion crows, jackdaws, jays, magpies, brown owls, kestrels, merlins and sparrowhawks, for the benefit of natural history and my own instruction and amusement.' Much of it was subsequently described on paper.

But there was too much amusement and not enough instruction in it for the teachers' liking. Literature had 'scarcely any effect' on the boy, and his 'vast proficiency in the art of finding birds' nests' was considered a hindrance to his progress. As often as not, he was punished for his interest by the 'unwelcome application of the birch rod.'

But the warm application of it, in lieu of effacing my ruling passion, did but tend to render it more distinct and clear. Thus are bright colours in crockery-ware made permanent by the action of fire.[4]

It was during one of these benevolent floggings, at the hands of the Reverend Joseph Shepherd – 'a very correct disciplinarian' – that he learned how to retaliate in kind. 'I flew at the calf of his leg, and made him remember the sharpness of my teeth.' 70 years later, he was still biting people's legs, purely for the entertainment of friends.

So the rod proved accidentally auspicious to the boy. It enlarged his repertoire and encouraged his natural inclinations. It helped to produce one of the great green gurus of the nineteenth century. But all saints have to sin first or there's no virtue in sainthood; the ruling passion was not always protective, as a series of recollections written for his cousin, George Waterton, illustrate.[5]

Soon after arriving at Tudhoe, some of his school-friends told him that a flock of geese were trespassing on school property. They offered him a wooden stake, with instructions to defend the school's honour, and little Lord Waterton obligingly attacked the gander, 'whilst he was hissing defiance,' and killed him. Whenever, after that, he had occasion to pass the farmyard where the geese belonged, the farmer's children would call after him – in his own translation – 'Yew killed aur guise.' He didn't seem much concerned about it, even in retrospect.

It was the boys who put him up to it who got punished for the offence but Charles was not always so lucky in sin. He was on a 'roving expedition in search of hens' nests' when he chose to throw a half end brick at the headmaster's King Charles spaniel, devoid of all thought for retribution or Restoration. How was he to know that the headmaster himself was watching from a nearby window? The rod was not spared.

He was also caught in the act on another occasion at prep school:

> There was a large horse pond, separated by a hedge from the field which was allotted to the scholars for recreation ground. An oblong tub, used for holding dough before it is baked, had just been placed on the side of the pond. I thought that I could like to have an excursion on the deep; so taking a couple of stakes out of the hedge, to serve as oars, I got into the tub and pushed off . . . I had got above half way over, when, behold, the master, and the late Sir John Lawson of Brough Hall, [Catterick, Yorks.] suddenly rounded a corner and hove into sight.

> Terrified at their appearance, I first lost a stake, and then my balance: this caused the tub to roll like a man-of-war in a calm. Down I went to the bottom, and rose again covered with mud and dirt. 'Terribili squalore Charon.' My good old master looked grave, and I read my destiny in his countenance: but Sir John said it was a brave adventure, and he saved me from being brought to a court martial for disobedience of orders, and for having lost my vessel.[6]

His friends and peers were more than happy to exploit his brave adventurousness. A group of them persuaded him to ride on the back of a cow and were treated to a free rodeo for his pains. The cowboy flew over the bronco's head and the noble peers adjourned. And when someone was needed to raid the school larder at dead of night, Charles Waterton was yer lad.

He also began to show signs of more original sin. He was said to have feigned a bad arm in school so that he could wear a makeshift sling and attempt to hatch out an egg in his arm-pit. He was within four days of being a mother when someone pushed him and broke the egg.[7]

When he needed to blow an egg, he used a fine darning needle which he apparently carried around with him wherever he went. The gift of this darning needle, to an old lady beggar near the school, went some way to compensate for his early life of crime. He later considered it one of his most meritorious acts of charity.[8]

After four years at Tudhoe, Charles learned that he was to have a few months holiday at home, before starting at his next school in the autumn of 1796. One episode from this period earned a place in the preface to his *Essays*. The family chaplain, Monsieur Raquedal, whose room was next to the boy's, heard something go bump in the night and rushed next door to investigate. He was just in time to catch a long-leggity beasty, in bare feet and night-gown, in the act of lifting up the sash of his bedroom window, three storeys above the ground, and stepping out. The 14-year-old schoolboy was sleep-walking. He claimed that he thought he was on his way to a neighbouring wood, where he knew of a crow's nest, but another version of the same story says that he was edging along the window sill to get to a nest in the eaves. The exact truth, as always, is immaterial; what scarcely needs saying is that he was, by this time, completely obsessed with birds.

At his next school, he could 'attend to birds, without much risk of neglecting books.' For that school was Stonyhurst, and the good fathers who taught there soon found a way of dealing with his 'predominant propensity'.

> By a mutual understanding, I was considered rat-catcher to the establishment, and also fox-taker, foumart-killer, and crossbow-charger at the time when the young rooks were fledged. Moreover, I fulfilled the duties of organ-blower and football-maker with entire satisfaction to the public.[9]

Now you might think, as Edmund Selous thought, that to turn a budding young naturalist into a rat-catcher and polecat-killer constitutes what ought to be thought a 'derogation'.[10] But natural history in those days was generally thought of as no more than a harmless pastime – a *studium inutile* – to be humoured rather than encouraged. That was the view of Charles Waterton's own father, himself a naturalist,[11] and not many people seemed to think otherwise.

There are those now who decline even to humour natural history – majestically indoor men, too intellectually sophisticated for anything as feeble-minded as Nature. The breed was barely born in Waterton's youth, when there was much more nature about, but a faint suspicion, even then, of anything which smacked of bumpkinism, was reflected in the attitude of his father and of his first school at Tudhoe. The Stonyhurst policy was rash.

Charles Waterton's appointment as rat-catcher, fox-taker, foumart-killer and crossbow-charger, in forbearance of *studium inutile*, suited him well.

> I followed up my calling with great success. The vermin disappeared by the dozen; the books were moderately well-thumbed; and, according to my notion of things, all went on perfectly right.[12]

He was sure he had good reason to thank Stonyhurst and his loyalty to the school lasted to his grave. To understand the origins of that loyalty, and the impact of his Stonyhurst education on the rest of this story, its helpful to know a little of the school's history:

Queen Elizabeth's penal code prohibited all teaching of the 'old religion' in England; so an outlaw race of Catholic schools and colleges was established abroad, and came to be known as the 'seminaries beyond the seas'. One of these – the English college of St Omer's, in France, founded in 1592 by a Jesuit priest, effectively served as progenitor of Stonyhurst itself.

The English government had declared it high treason for parents to send their children to St Omer's and were not above capturing boys at sea, when they were on their way to school, and putting their parents in prison. Most of those that were not in prison were themselves in exile abroad, their English property sequestrated. In the early days of the Grand Tour, Englishmen had to apply for a travel licence from the Privy Council, which expressly forbad them to visit St Omer's.

In 1762, the French parliament officially declared war on the Society of Jesus and resolved to disband the teachers and to deny them the opportunity to teach anywhere in France. So the priests of St Omer's conceived a plan to transplant the whole establishment – masters and boys – to Bruges, then in the Austrian Netherlands. But when they got to Bruges, their troubles doubled. The kings of France and Spain united against the Society and eventually persuaded Pope Clement XIV, in 1773, to issue a brief designed to scupper the Order once and for all. The Society was put down, its members released from their vows, and its property confiscated. The fathers of Bruges were taken into custody.

But one English college – at Liège – suffered comparatively little persecution. English students from Bruges fled to Liège, where the Academy for English Jesuits remained for the next 21 years. Eventually, this refuge also became insecure and by 1794 the boys and staff were on the run again, this time from the advancing French Revolutionary Armies, bent on 'de-Christianisation'.

The safest retreat this time was obviously back home in England, from where the Order had escaped 200 years earlier. Penal laws had recently eased in England and a certain Thomas Weld, of Lulworth Castle, in Dorset, who had himself been educated at Bruges, and who owned a large mansion in Lancashire, offered that property to the exiles as their new home.[13]

So it was that the English Jesuits established Stonyhurst College in a vast and dilapidated stately home, close to the banks of the river Hodder, near Clitheroe. Charles

Figure 4. West Front, Stonyhurst College, c.1830. Engraving. J M W Turner. (Stonyhurst College Library.)

Waterton himself entered Stonyhurst two years later, in 1796. His father had been educated at St Omer's and some of the Bedingfelds also went to one or other of the seminaries beyond the seas. Charles was always a staunchly proud advocate of Jesuit education and his affection for Stonyhurst never faltered.

> Under Almighty God and my parents, I owe everything to the Fathers of the Order of St Ignatius. . . . During the whole of my stay with them (and I remained at their college till I was nearly twenty years old), I never heard one single expression come from their lips that was not suited to the ear of a gentleman and a Christian. Their watchfulness over the morals of their pupils was so intense, that I am ready to declare, were I on my death-bed, I never once had it in my power to open a book in which there was to be found a single paragraph of an immoral tendency.[14]

How he managed to establish whether a book was free from paragraphs of an immoral tendency before opening it, he never disclosed. But Stonyhurst did give him a good grounding in literature – Latin especially – at a time when the likes of Cowper, Cobbett and Byron were complaining about the 'profane jesters'[15], 'dens of dunces'[16] and 'stagnant intellects'[17] of Oxbridge. Stonyhurst was not typical. Nor was Charles Waterton.

There seems to have been a slight difficulty at Stonyhurst in keeping the boy within bounds – his affection for the place knew none. He went to examine a magpie's nest in a tree which was strictly out of bounds, and a group of holy fathers appeared, with Bryan Salvin, and passed by, beneath him, unaware that he was there.

Another time, he escaped out of bounds through a thicket of yew and holly trees – inspiration for subsequent planting policy at Walton Hall – to a neighbouring wood where he knew of a carrion crow's nest he felt he should visit. (He wasn't dreaming this time.) A prefect spotted him and gave chase but the intrepid wanderer managed to reach the back door of a pig sty, where he found old Joe Bowren, the brewer, taking straw in for the pigs. Charles knew he'd found a pal: 'He was more attached to me than to any other boy.' As with Mr Salvin, they'd known one another at Tudhoe, where Charles had 'made him a present of a very fine terrier.' One good turn deserves another so the general dog's body put him in with the pigs. The prefect lost track and gave up the chase. When he'd gone, the fugitive stole out of cover, 'as strongly perfumed as was old Falstaff when they turned him out of the buck basket.'[18]

It was much the same within the school bounds. A jackdaw chose to nest near the top of one of the college's quadrangle towers, and Charles Waterton obviously couldn't rest until he'd taken a look. He was about three parts of the way up the tower when the rector spotted him and ordered him down, deaf to his pleas to be allowed to continue. The stone parapet which avowedly fell to the ground that night – and which the boy would have had to negotiate if he had carried on – is still there, propped up against the quadrangle wall, as a curio for visiting parents and guests.

His own father occasionally visited the college during his stay there and by this time, accepted news of his little adventures with resignation. He was probably present in 1799, a few days before the start of the new century, when his son 'bore a part' in the school play – a hastily written drama called *Sidney*, which commemorated Sir Sidney Smith's defence of the Syrian port of Acre, earlier that year, against Napoleon Bonaparte.[19] The play was one of the Christmas activities which were a feature of life at Stonyhurst during the first 100 years, when very few of the boys went home for Christmas. During the 19th century, old boys would descend on the college *en masse* at Christmas time, and Charles Waterton regularly spent his Christmas holidays at Stonyhurst, even as a very old man. He always dressed for these, as for other important occasions, in a fair copy of his old school uniform – bright blue swallow-tail coat, brass or gold buttons, check waistcoat and grey trousers.

One item of uniform which he declined to wear as an adult was the head gear – a curious-looking leather cap, with ear flaps, which could be worn in any of seven different ways. The cap was lined with the fur of polecats – supplied by the official Stonyhurst foumart-killer, Charles Waterton. He seems to have rejoiced in his alleged success in controlling the polecats (though they were still common in the area long after his death) just as he saw fit to congratulate himself on the alleged elimination of rats from Walton Hall. With a bit of luck, the polecat will soon return to Lancashire and be spared the depredations of incipient conservationists.

His attitude towards rats was one of the characteristics partly inherited from his Stonyhurst education. A not entirely playful belief in the Protestant affinities of the brown rat was definitely encouraged there, and Charles Waterton was not the first to call them Hanoverian. But the college's attitude to natural history in general was best reflected in its appointment of fox-taker and foumart-killer, with official blessing to pursue predominant propensities. Had Stonyhurst chosen to repress natural history as diligently as Tudhoe had done, it might have been flushed out of his system for good – his trust in the Jesuit fathers was complete and unchangeable.

But it seems to be generally agreed that the greatest benefit of all from his Stonyhurst education bore no immediate relation either to natural history or to the more conventional aspects of education:

> My master was Father Clifford, a first cousin of the noble Lord of that name [and possibly related to Charles Waterton]. . . . One day, when I was in a class of poetry, and which was about two years before I left the college for good and all, he called me to his room. 'Charles,' he said to me, in a tone of voice perfectly irresistable, ' I have long been studying your disposition, and I clearly foresee that nothing will keep you at home. You will journey into far distant countries, where you will be exposed to many dangers. There is only one way for you to escape them. Promise me that, from this day forward you will never put your lips to wine, or to spiritous liquors.

Gloriously sober, he obeyed th'important call. For a short time after leaving Stonyhurst, he occasionally took a little beer with his dinner but he soon gave that up and seems to have kept the vow with regard to wines and spirits to his dying day. He certainly did journey into far distant countries, where he certainly was exposed to many dangers, mostly self-inflicted. But the only claret he ever tapped was his own blood. The wine of life could have been the death of him.

With the help of Father Clifford's advice, Charles Waterton survived plague, tempest, crime and warfare, snakes, vampires, alligators and buffalos, purgatives, venesection, bone-setting and surgery, perilous ascents and descents of trees, ladders, domes, towers, sea-cliffs and precipices, a ship-wreck, a sea battle, an earthquake, a road accident, a shooting accident, a climbing accident, another road accident, an attempt to fly and a fall into the sea at dead of night, until, at the ripe old prime of 82, with a touch of the Desperate Dan, he tripped over a bramble and died in bed. He owed everything to the Fathers of the Order of St. Ignatius. They taught him all he needed to know and gave him leave to learn the rest for himself.

> To the latest hour of my life I shall acknowledge, with feelings of sincerest gratitude, the many acts of paternal kindness which I so often received at the hands of the learned and generous Fathers of Stonyhurst College.[21]

The day he left there was a day of 'heartfelt sorrow'.

Chapter 3
1801-1804
Black Vomit

To the young beneficiary of Walton Hall, nature was made of fun and fresh air. When he left Stonyhurst, in 1801, his father presented him with the obligatory pair of thoroughbred hunters and gave him leave to ride to hounds with the far-famed pack of Lord Darlington, ingloriously known as the Raby.

Lord Darlington, later Duke of Cleveland, was a power in the land amongst the right sort, on account of his 'elegant seat on horseback and cool intrepidity in charging fences' – the very summit of achievement in the English aristocracy. He was Master of the Hunt for 53 years and was said to have spent nearly every day of his life in the saddle. Whenever possible, he spent six days a week chasing foxes.[1]

But where cool intrepidity was concerned, the new young member from Walton Hall soon proved a match for the MFH himself, according to the new young member from Walton Hall. There was no curb rein to his bridle.

'Zounds! Mr Waterton, what a jump!' cried an eloquent baronet, as the said Mr Waterton leaped over a tall hedge and landed safely in a little 'quarry' on the other side. It must have been a good jump because the admirer was head of a neighbouring family which had not been on speaking terms with the Watertons for years. The family feud was over and the two men rode home friends.

Mr Waterton Senior was less impressed. He'd heard stories of his son jumping five-barred gates and such, and of Lord Darlington himself congratulating him for his boldness. But praise from such a quarter 'glibly glides into the brains of young men', warned Pater, 'and, now and then, turns them upside down.' (Several members of the Hunt were in serious distress) 'You obliged me by commencing to hunt – will you still further oblige me by giving it up.'

So, after a year spent 'amid the pleasures of the field', Charles Waterton grudgingly agreed never again to appear in crimson, a vow which later words of his own suggest that he broke but which, making allowance for tricks of memory, he possibly kept. (As he claimed.)

Anyway, after a year of dissipation with the right sort – most of whom were obviously the wrong sort – Charles was now considered ready for the traditional finish to every gentleman's education – foreign travel. The Peace of Amiens had suspended the war against Napoleon Bonaparte and European travel was therefore feasible for the first time in several years. It was decided that Charles and his young brother, Christopher, should visit two of their maternal uncles who, rather than apostatise their faith, had left England and settled in Andalucia, where they'd taken out Spanish citizenship.

> Perhaps there was no happier lad in all the world than Charles Waterton, with his youth, his position, his wealth, his intelligence, his activity and his complete freedom, starting to visit the lands of his dreams.

That was the slightly damp speculation of one of his more effusive admirers, Mrs. W. Pitt-Byrne.[2] Certainly, as he set sail from Hull, in November 1802, aboard the glamorous ship *Industry,* he and his teenage brother must have thought that the world lay at their feet. So many worlds, so much to do. But they found more than either of them bargained for in the lands of their dreams.

First port of call was Margate. Then Cadiz. In Margate, they got wind of a plot to kill the captain of a Scottish ship which was anchored nearby. Bound by the code of a gentleman and an old salty dog, Charles Waterton decided to do something suitably nautical about it: when they left port, next day, he put a message in a bottle and threw it onto the quarter-deck of the Scottish ship as the Industry ran alongside. The captain took the message below, read it, then returned on deck to give a low bow of gratitude to his unknown young pen-friend.

In Cadiz, the two brothers trod on foreign soil for the first time. The British consul took them to the bullfight ('in honour of royal nuptials') and Charles managed to get lost as the crowd left the stadium, his roving eye enthralled by 'a thousand objects', all unspecified. He walked the streets until close on midnight and was finally led to the consul's house by a passing Frenchman.

They spent a fortnight in Cadiz and no account is given of any bird watching excursion during that time, not even to the Coto Donana, now Europe's largest nature reserve. But at Malaga, the final destination, Charles Waterton made up for time unquestionably lost on the Maids of Cadiz.

His uncles lived in a pleasant home in the foothills of the sierra, and many were the days of 'rural amusement' spent there. Red-legged partridges 'abounded', vultures were 'remarkably large', goldfinches were more common than sparrows in England, and quails and bee-eaters arrived 'in vast numbers' during spring migration from Africa. 'Once, when I was rambling on the sea-shore, a flock of a dozen red flamingoes passed nearly within gun-shot of me'.

During his stay in Malaga, Charles made friends with an English commercial traveller whose firm was 'most respectable.' He was clearly the 'rich merchant' later mentioned in correspondence with Father Clifford. Together, they embarked on an expedition overland, across the Sierra Bermeja and along what was then a quiet, almost inaccessible stretch of coast which less respectable firms have since turned into the Costa del Sol. Torremolinos was then a 'delightful village'.[3] Marbella was a small, Moorish fishing port.

The place they were aiming for was the Rock of Gibraltar, previously passed at sea, and the main reason for going there was to see the small colony of Barbary apes which thrives there to this day – the only primates, other than Man, to be found wild in Europe.

The two men arrived in Gibraltar just in time to witness a vertical migration of the apes – 50 or 60 of them – from the top of the rock to the bottom. Charles Waterton reflected on how the animals had got to Europe in the first place. To one reviewer, many years later, his explanation recalled 'those extreme catastrophist doctrines in geology which are now happily extinct.'[4] But in fact, Waterton was anticipating Cuvier's theory, or at least its most explicit statement, by about ten years. With biblical stories of the flood in mind, he believed that, 'in times long gone by', Gibraltar was connected to Apes Hill, on the Barbary coast, and that they were separated by 'some tremendous convulsion of nature' which formed the Mediterranean Sea.[5]

It is now generally agreed that Barbary apes were probably introduced into Europe as pets, but Charles Waterton's geology was not all that unsound, except insofar as 'times long gone by', to him, as to the likes of Baron Georges Cuvier, were not so very long ago.

Whether or not the travelling salesman was interested in geological doctrines, its not possible to judge – merchants write memos, not memoirs. Later in life, the Squire discovered that his friend had 'fallen in love, had wooed in vain, and hanged himself in despair', as behoves all faithful employees of most respectable firms.

When they returned to Malaga, Charles prepared himself for another little excursion, this time to Malta, recently rescued from Napoleon. He was clearly enjoying himself immensely. No hint of despair in his life.

> More than a year of my life had now passed away in Malaga and its vicinity without misfortune, without care, and without annoyance of any kind.[6]

But all of this was to change, very suddenly, when rumours began to circulate of a violent plague of fever which rejoiced in the popular local name of 'il vomito negro' – black vomit. The trip to Malta was cancelled.

Black vomit was an apparently virulent strain of yellow fever. It originated in Africa and was spread by the slave trade, via the female sex of a species of mosquito called *Aedes Egyptis*. The plague in Malaga was one of the worst outbreaks ever to occur in Europe.

> I myself, in an alley near my uncles' house, saw a mattress of most suspicious appearance hung out to dry. A Maltese captain, who had dined with us in good health at one o'clock, lay dead in his cabin before sunrise the next morning. A few days after this I was seized with vomiting and fever during the night. I had the most dreadful spasms, and it was supposed that I could not last out till noon the next day. However, strength of constitution got me through it.[7]

Disease is no respecter of persons, and noblemen and peasants alike were dying like flies. The older of the two uncles went to visit a dying priest and returned home to die himself, 'a martyr to his charity.'

We got him a kind of coffin made, in which he was conveyed at midnight to the outskirts of the town, there to be put into one of the pits which the galley-slaves had dug, during the day, for the reception of the dead. But they could not spare room for the coffin; so the body was taken out of it, and thrown upon the heap which already occupied the pit. A Spanish marquis lay just below him.[8]

Many of the bodies were left in the street at night, to be collected by the dead carts before morning, whilst others, buried at sea, were washed ashore by the east wind, to be picked at on the beach by the vultures. Everyone with any sense was leaving the city, so Charles Waterton stayed on to perfect his Spanish – God's language. He was soon to be witness to another great gift of God, such as made yellow fever seem like gentle release: everyone was now subjected to the most extreme of all catastrophes – tremendous convulsions of nature.

> The pestilence killed you by degrees; and its approaches were sufficiently slow, in general, to enable you to submit to it with firmness and resignation. But the idea of being swallowed up alive by the yawning earth, at a moment's notice, made you sick at heart, and rendered you almost fearful of your own shadow.[9]

The first shock came at 6 p.m., 'with a noise as though a thousand carriages had dashed against each other.' Everyone rushed outside and walked up and down the Alameda rather than risk being buried alive. Charles eventually went to bed a little after midnight (very late for him) but was roused at 5 a.m. by another quake, more frightening than the first. 'I sprang up, and, having put on my unmentionables (we wore no trousers in those days), I ran out, in all haste, to the Alameda.' But the tremors gradually became weaker and finally stopped altogether.

At last, the accidental traveller considered it 'high time to fly' from Malaga. He'd had just about enough of tremendous convulsions, seismic or otherwise. But it was becoming more and more difficult to get away. The whole city was in quarantine and no ships were allowed to leave port until the Admiralty gave permission.

During his stay in Malaga, he'd become acquainted with a Swedish captain named Bolin, who was luckily possessed of those two important virtues essential to all the most dedicated practitioners of accidental travel – 'coolness and intrepidity'. Both qualities were soon to be needed.

Charles managed to extort papers from the British consul, Mr Laird, declaring that the city was now free from sickness (which was almost true) and set about conspiring with Captain Bolin to break the embargo on shipping. The consul feared the worst:

> 'My young friend,' said he, in a very feeling tone, 'I shall either have to see you sunk by the cannon of the fort, or hear of your being sent prisoner for life to the fortress of Ceuta, on the coast of Africa.'[10]

Malaga, at that time, was full of 'presidarios' from Ceuta, which gave the place, in a subsequent consul's words, 'a good claim to be entitled the sink of Spain in respect of vice, and the almost total demoralisation of the lower classes.'[11] As the blackmailing of Laird appears to confirm, the consulate itself was 'not upon a respectable footing.'[12] The upper classes, too, could be led astray.

Whilst Charles was getting a clean bill of health in Malaga, Captain Bolin was making out false papers for the ship, and for its valuable cargo – a necessary precaution since Britain and France were now back at war. Christopher Waterton was entered as a passenger, and Charles as a Swedish carpenter. They tried to persuade their surviving uncle to join them but he chose to stay behind and face plague and earthquake, rather than submit to anything as pestilential as Thirty-nine Articles. He died the following year, in a fresh outbreak of black vomit which reportedly killed another 36,000 people in the area.

For several nights, as the Swedish ship lay in harbour, Charles and Christopher slept on board, waiting for a favourable wind to blow up and make the escape possible. A strong east wind at last arrived, the harbour master made his daily round of inspection, the governor went for an airing in his carriage and the Spanish naval officers went ashore for their customary afternoon entertainments. Craftily, his face 'a portrait of cool intrepidity', the noble Captain Bolin worked his ship clear of the 40 or so vessels in harbour and instantly put up 'a great cloud of canvas.'

Long before the brigs of war had got their officers on board, and weighed in chase of us, we were far at sea; and when night had set in, we lost sight of them forever – our vessel passing Gibraltar at the rate of nearly eleven knots an hour. (sic).[13]

Greatly impressed, Charles Waterton declared that if he'd had it in his power, he'd have made the captain an admiral on the spot; it was just the sort of escapade he loved, just the sort of spirit he admired. The Battle of Trafalgar, 12 months later, involved some of the same brigs of war.

Once in the Atlantic, with winter fast approaching, the Swedish carpenter and his intrepid shipmates came across cold and stormy conditions which they must have known they could expect at that time of year and in those latitudes. They were obliged to endure these conditions for almost another month. Still in his thin Mediterranean clothes, and still convalescing from his bout of fever, Charles now went down with a severe cold which, in his own words, reduced him 'to the brink of the grave.' Fears of being sent back to Spain for a regular bill of health, served to make him feel worse. During the period of quarantine in Malaga, the ship had been laden with fresh fruit which, by now, must have been slightly over-ripe; Charles had serious doubts about permission being granted to dock. But with just the informal certificate provided by the British consul, the vessel was allowed to enter the Thames and at last unloaded its cargo in the Port of London.

It had been a long and dispiriting journey, after the initial excitement, and Charles went home to Walton Hall, and the rigours of a Yorkshire winter, in a sad state of health. 'I must have sunk', he said, perhaps melodramatically, 'had it not been for the skill [probably blood-letting] of the late celebrated surgeon, Mr Hay of Leeds: he set me on my legs again; and I again hunted with Lord Darlington.'[14] (Despite his vow to the contrary.)

Christopher was not so lucky. He stayed indoors and eventually succumbed to a fresh attack of yellow fever, apparently latent in his bloodstream. Charles Waterton's 'iron constitution' – an all-purpose compound of good luck and auto-suggestion – saved him from a similar fate.

But his thin young frame was not well suited to a cold climate, and his family had money. War had again ruled out European travel but, as good luck would have it, his paternal uncles owned some sugar and coffee plantations in Demerara, and Charles asked his parents if he could go out to superintend the estates, on behalf of his younger brothers, Thomas, Edward and William, whose patrimony they were eventually to become. In November 1804, after a year spent with his family at Walton Hall, he set sail from Portsmouth, on board the bad ship Fame, for the first of many subsequent journeys to the strange new world of South America. By the time he got back for the first of several trips home, Christopher Waterton was dead and buried.

Chapter 4
1804-1812
Outlaws and In-Laws

Colonial prosperity, in the first years of the nineteenth century, was based very largely on the exploitation of African slaves. In British Guiana, something like 100,000 slaves were employed on about 700 estates. Coffee, cotton, sugar and rum were booming industries – a bull-run sustained by the bull-whip. The population of Georgetown (then called Stabroek) increased tenfold in less than twenty years. Land values multiplied proportionally.

Charles Waterton experienced this mercantile golden-age at first hand. His own family had acquired property in Demerara through his paternal aunt Anne, who was out walking in Wakefield one day when 'a gentleman, by name Daly, from Demerara, met her accidentally and fell desperately in love with her.' A real live sugar-daddy with a flair for happy accident – just the man to exalt the family succession.

The two sweethearts got married, went out to Demerara and were duly followed by Anne's brother, Christopher – Charles Waterton's uncle – who, like the maternal uncles in Andalucia, 'had no hopes at home on account of the penal laws.'

Uncle Christopher married an English woman in Demerara and so acquired a half-share in a couple more plantations, with the appropriate complement of African slaves. He then returned to England with his wife and not long afterwards, his brother, Thomas – father of Charles – in turn acquired a plantation in Demerara which inevitably came to be known as Walton Hall. It was these three estates – Walton Hall and the two part-owned by his uncle Christopher – which Charles went out to to superintend in 1804.

Port of embarkation was Portsmouth, but first there was some family business to attend to in London. At a coaching inn, on his way to London, he met the great Lord Darlington himself, who tried to persuade him to return home for another season's hunting with the Badsworth, near Walton, but though one source claims that he did steal back to follow the hounds a few times,[1] Charles Waterton makes no mention of it. According to his own words, he spent at least some of that pre-embarkation period at the London home of another of his maternal uncles, 'the late intrepid Sir John Bedingfeld, who had saved the king's life in 1796.'

The king went in fear of his life, every time he went anywhere near his subjects – too many people, not enough bread. A stone had been thrown through his carriage window, after a state opening of parliament, and in 1794, someone had fired a shot at him as he left a performance at the Theatre Royal in Drury Lane. In 1796, he was back at Drury Lane and in danger of being lynched by an angry mob as he left, but this time the intrepid Lord Bedingfeld was on hand, to avert all danger. As the lower orders became disorderly, Sir John jumped onto the step of the king's carriage and pointed his spectacle case menacingly at the crowd. 'I'll shoot the first man to approach nearer', he said, but no-one called his bluff.

To show his gratitude for the rescue, King George later bestowed on Sir John the ribbon and badge of the papal Guelphic order. A portrait of Lord Bedingfeld, thus bedecked and bedizened, at one time hung in the dining-room at Walton Hall. It was said to bear a close resemblance to Charles Waterton himself.

So this was the uncle with whom Charles Waterton stayed before leaving for Demerara. One great advantage in possessing relations like that is that they tend to know all the right people, mad kings notwithstanding. Since Charles was going to the colonies, and since he was obviously a bit of a nature boy, it would be nothing less than criminal negligence to leave London without making the acquaintance of Sir Joseph Banks, president of the Royal Society, circumnavigator of the globe (with Captain Cook), and 'munificent patron of science'.

For some years, the house in Soho Square where Banks lived when the Royal Society was in session (November to June) had been a great meeting place for scientists of all kinds, particularly at his breakfast table, where many a great project had been planned, and official scientific information passed round with the sliced ham and kidneys. It was to Soho Square, one Friday morning in November, that Sir John Bedingfeld took his nephew, Charles, for the obligatory levee and hobnob with Sir Joseph. They received a bonus – an invitation to dinner:

> 'I am a Catholic, Sir Joseph,' said I, 'and I am not allowed to eat meat on Fridays.' – 'I am a Catholic too,' said he, 'as far as abstinence from meat is concerned; for the doctors have lately put me on a pudding diet: so you shall sit by my side, and we shall have our eggs and pudding together.'[2]

It was the beginning of a firm friendship between Banks and Waterton which lasted until the older man's death, in 1820. At this first meeting, Sir Joseph was at pains to impress upon the young man the extreme health hazard of low, swampy countries. In the early nineteenth century, the dangers were acute, not only from yellow fever and malaria but from other such traveller's delights as smallpox, typhoid, dysentery and various psychological disorders. Banks suggested that the best way to avoid that sort of thing was to return home once every three years or so.

> I followed this admirable advice with great success: still, I used to think that I ran less risk of perishing in those unwholesome swamps than most other Europeans, as I never found the weather too hot, and I could go bareheaded under a nearly equatorial sun, without experiencing any inconvenience.[3]

He sailed from Portsmouth on 29 November, 1804, and arrived in Stabroek, in what had recently been Dutch Guiana, about six weeks later.

> I liked the country uncommonly and administered to the estates till 1812, coming home at intervals agreeable to the excellent and necessary advice which I had received from Sir Joseph Banks.[4]

The traditional urban activities of colonial life left Charles Waterton cold:

> In Demerara he was looked upon by the rollocking boys of that age as not quite right in his mind. For, first of all, he was a teetotaller, when custom never

> excused a man from having a drink, and we may also safely state that he would be out of place among the cock-fighting and card-playing gamblers, as well as those votaries of Venus then prominent in Georgetown.[5]

Whether because or in spite of his personal habits, he survived in Demerara and managed the three plantations, off and on, for eight years. The venomous James Simson, many years later, tried to claim that he left the estates in disgrace, on account of his mismanagement of them,[6] but that particular slander was based on extremely flimsy evidence and pathological hatred. To fully understand Charles Waterton's role during those eight years, a bit more history is necessary.

Plantations in the Guianas were first developed by Dutch settlers during the seventeenth and eighteenth centuries, but the British ruled what is now called Guyana (Berbice, Demerara and Essequibo) from 1796 to 1802 and again from 1803 until 1814, when the region was formally ceded as a crown colony. It remained British Guiana until independence, in 1966, when it became the republic of Guyana.

So the country was in British administration during the whole of Charles Waterton's association with it, though the Dutch colonial influence was evident then, as it still is, and many of the Dutch administrators remained.

It was not until the Abolition of Slavery began to take effect that the prosperity of the country began to wane. The Abolition Bill was passed in 1807 but an illegal traffic continued until complete emancipation of existing slaves, in 1834, by which time most of the plantations had gone out of cultivation. But during Charles Waterton's period of estate management, slavery was still the order of things, plantations thrived, and fortunes, generally, remained intact. His own attitude towards slavery was almost unreservedly hostile:

> Slavery can never be defended; he whose heart is not of iron can never wish to be able to defend it: while he heaves a sigh for the poor negro in captivity, he wishes from his soul that the traffic had been stifled in its birth.[7]

The sentiment gradually became fashionable in the world at large but amongst the West Indian owners themselves, with a vested interest of about £50 million in plantations alone, it was possibly an unusual point of view. What remained of the Waterton family's vested interest found expression in the claim that the conditions of slaves on the plantations was not as wretched as was generally supposed. The Briton's heart, he said, was 'proverbially kind and generous. . . . He cheers his negroes in labour, comforts them in sickness, is kind to them in old age, and never forgets that they are his fellow creatures.' As the price of slaves rose, after 1807, it became policy to take care of them anyway, as James Rodway pointed out: 'There is always a desire to keep a valuable horse in good condition, especially when it cannot be replaced.'[8] That was the brute reality.

Charles Waterton was probably in charge of something like four or five hundred slaves on the three estates that he managed, though he claimed never to have 'owned' a slave in his life. Responsibility was obviously delegated and demands on his own time would not have been great. Much of his time was spent at his Aunt Anne's estate – Bellevue – or at the up-country home of a Scottish planter named Charles Edmonstone – Warrows Place. It was this Charles Edmonstone – 'the most valued friend I ever had in the world' – who was later to become Charles Waterton's father-in-law.

Waterton's first term of duty in the tropics lasted little more than 12 months. In

1805, his father died and the oldest son went home to succeed to the title of 27th Lord of Walton Hall – the Squire. He spent about a year at home, clearing up details of inheritance, and by 1807, he was back in Demerara, performing the kind of duties James Simson was sure he was not qualified for, either by training or temperament. A better man would have had seven years training in West India House, in London, and then have gone through a suitable initiation period in Guiana, 'with every advantage over a man like Charles Waterton, who went out fresh from Stonyhurst, a visit to Spain, fox-hunting, and the 'hedgerow mousing' of an embryo naturalist.'[9] He never showed the slightest talent for business, said Simson, with obvious disgust, and his 'clients' should have paid him to stay at home, as long as he undertook to stay away from the Jesuits.

It was all very unfair criticism, and the 27th Lord did his level best to earn it whenever possible.

As entrepreneur – attorney for the family business – the new young massa from Walton Hall cut a very fine figure as a jockey. First duty of any self-respecting overseer is naturally to oversee, and that must obviously involve the basic task of any man in authority – showing himself off to the workers. He chose to do this on a 'cream-coloured and beautiful animal', imported from the Orinoco region to work in the sugar mills. (Steam hadn't arrived yet.) He took great pride in his choice, 'because no other person seemed inclined to engage him.' This was not so very surprising, since the animal in question was a mule, and no-one else had the necessary sense of majestic humility and general horsing around.

He called his steed Philip and clearly took delight in his unpredictability. 'At times, he went quietly enough; but every now and then, he would shew who had been his father, and then you would fancy that the devil of stubbornness had got entire possession of him.' On one occasion, Philip was stung in the face by 'a large brown wasp of the country' which temporarily changed his personality from pig-headed mule to kangaroo; Charles went 'neck and crop' over Philip's head, and landed imprudently on his own. When he finally came to his senses – in the broadest sense of the word – he did what a man in full command of his senses must needs do in such extreme circumstances: he took out his lancet and drew 20 ounces of blood from his arm. 'This prevented bad consequences; – and put all to rights.' It was never known to fail.

On another occasion, the mule came to a dead halt on one of the plantation roads close to the coast – land reclaimed from the sea by the Dutch polder system.

> 'Philip,' said I, 'I can't afford to stop just now, as I have an appointment; so pray thee, my lad, go on.' 'I won't', said he. 'Now do, my dear fellow', said I, patting him on the shoulder as I spoke the words; 'we must not remain here, a laughing stock to every passing nigger.' Philip declared that he would not move a peg. 'Then, master obstinacy,' said I, 'take that for your pains;' and I instantly assailed his ears with a stick which I carried in lieu of a whip. 'It won't do,' said Philip, 'I'm determined not to go on;' and then he laid him down, I keeping my seat on the saddle, only moving in it sufficiently to maintain an upright position; so that, whilst he lay on the ground, I appeared like a man astride a barrel.

The sluice-keeper stood at the door of his little hut, grinning from ear to ear. 'Daddy,' said the Squire, 'bring me a fire-stick.' So Daddy brought a blazing stick from his fire and put it under Philip's tail. 'Up he started, the hair of his tail smoking and crackling like a mutton chop on a grid-iron.' Charles kept his seat – trained neck and crop for just such an emergency – and Philip galloped off down the road, never again to lie down in a public place with a distinguished passenger on board. So ends the story of Philip the mule, recorded, naturally, in an essay on dogs.[10]

Whilst business responsibilities were thus discharged on a beautiful cream charger, explorations of the great primeval forest were apparently limited to a few social visits up-river, by tent-boat and woodskin. One of these visits was made, ostensibly, to check on the story of a hybridised cow, the results of which inquiry were recorded, similarly, in an essay on monkeys.[11]

It was the year 1807, the year that the slave trade was abolished and cheap labour became expensive. Big business began to shrink and human values soared to more reasonable proportions. 30 miles up the river Demerara, in a little forest clearing, lived an old Dutch settler with just enough room for a little dairy, some pasture for his cattle, and a small garden to provide vegetables for his family. The man's name was Laing, a 'farming-looking gentleman' who served excellent coffee without a trace of chicory in it.

His house stood on top of a small hill, surrounded on three sides by the un-tamed forest, whilst the fourth was rich 'greensward', sloping down to the water's edge. On viewing it from the river, said Charles, 'You would have said that it was as lovely a place, for a man of moderate desires, as could be found on this terrestrial globe.'

But Nature is a brazen little hussy, even in Paradise. One of the cows from Mynheer Laing's pedigree herd in due course dropped calf which instantly aroused suspicions as to family honour; it was 'misshapen from birth.' Scandal spreads like wildfire in the primeval forest, and word soon got around that the mother had been playing away with a rogue bull of the bush, suitably equipped with a flexible proboscis and a rigid phallus – a wild tapir.

Charles always professed a profound distaste for freaks and hybrids, though he kept a few himself for the strictly scientific purpose of idle amusement. It was a thoroughly conventional distaste for an almost universal fashion, but he was always first in the queue for the peep shows. When he'd heard that an albino negro lived in Georgetown, he'd hied him along to the man's house, knocked on his door, and announced that he'd come to make his acquaintance forthwith; he'd never seen a black man with fair skin before.[12]

So when news arrived of a cross between a dairy cow and a wild bush cow, he was off with his worthy friend, Charles Edmonstone, to check out the story for himself. He was already convinced that it was all a load of bull – 'a union of animals so opposite in their nature could not be.' A thorough veterinary examination confirmed his belief and the scandal soon died down.

A few months later, in September – in recognition of services rendered – he was made a lieutenant in the 2nd regiment of the Demerara Militia, an appointment of some dignity which ultimately led to a chance to explore the Orinoco. On 2 August 1808 he was commissioned by the British governor to bear dispatches from Commander-of-the-Fleet, Admiral Collingwood, to the Spanish Captain-General of the Orinoco, Don Felipe de Ynciarte. Charles Waterton now spoke Spanish fluently and was therefore considered just the man for the job. It was the most important job entrusted to a Waterton for about 250 years. Somehow, he managed to get his friend, Charles Edmonstone, included on the commission.

They sailed under a flag of truce (Spain was a French ally) and changed into a smaller, Spanish boat at the river fort of Barrancas. It was Charles Waterton's first real experience of 'that great university of natural history, the primeval forest'[13] – 'a grand feast for the eyes and ears of an ornithologist.' The inevitable accessory of any good, Watertonian journey – 'an accident . . . of a somewhat singular nature' – duly occurred, to make his happiness complete. You've heard the one about the professor stuck up a punt pole? Well. . . .

> There was a large labarri snake [fer-de-lance] coiled up in a bush, which was close to us. I fired at it, and wounded it so severely that it could not escape. Being wishful to dissect it, I reached over into the bush, with the intention to seize it by the throat and convey it aboard. The Spaniard at the tiller, on seeing this, took the alarm, and immediately put his helm aport. This forced the vessel's head to the stream, and I was left hanging to the bush with the snake close to me, not having been able to recover my balance as the vessel veered from the land. I kept firm hold of the branch to which I was clinging, and was three times over-head in the water below, presenting an easy prey to any alligator that might have been on the look-out for a meal. Luckily, a man who was standing near the pilot, on seeing what had happened, rushed to the helm, seized hold of it, and put it hard a-starboard, in time to bring the head of the vessel back again. As they were pulling me up, I saw that the snake was evidently too far gone to do mischief; and so I laid hold of it, and brought it aboard with me, to the horror and surprise of the crew. It measured eight feet in length. As soon as I had got a change of clothes, I killed it, and made a dissection of the head.[14]

The image of Charles Waterton, upside down with his head in the Orinoco, is nicely evocative of the man. It was the first of many little Munchausenisms in the South American rain forest, and the sort of thing which made *Wanderings in South America* an instant best-seller. (Though that particular episode, of course, pre-dates the *Wanderings*).

At Angostura (Bolivar), they were received by the governor with all due military pomp. Dinner consisted of 'no less than forty dishes of fish and flesh', an appalling sight to the Yorkshireman, who always hated extravagance of any kind, gluttony especially. Don Felipe went to table in full dress uniform of gold and blue, with all the gaudy accoutrements of his rank – laniard, gorget, epaulct and napkin. 'He had not got half through his soup', said the Squire, 'before he began visibly to liquefy. I looked at him, and bethought me of the old saying, 'How I sweat! said the mutton-chop to the grid-iron.'

> But at last formalities were abandoned and another firm friendship began to flower. 'Don Carlos,' said the governor, 'this is more than man can bear. Pray pull off your coat, and tell your companions to do the same; and I'll show them the example.'
>
> The next day, at dinner time, we found his excellency clad in a uniform of blue Selempore, slightly edged with gold lace.[15]

Don Carlos and Don Felipe evidently got along together like long-lost brothers.

The Spaniard claimed to have been a bit of an explorer himself, in his day, and showed the Englishman 'a superb map of his own drawing', clearly the sort of thing to impress an accidental traveller with ambitions to wander. But the best of friends must part and, mission accomplished, Lieutenant Waterton returned to base, to settle down to the more routine side of colonial life, on a cream-coloured hybrid from the region he'd just visited. 'After my return to Demerara I sent this courteous governor a fine telescope, which had just arrived from London.'

The British governor of Demerara, meanwhile – Governor Ross – was approaching retirement. He finally left the colony on 31 March, 1809, and the following day, to mark the occasion, five Sapphic stanzas in his honour, composed by Charles Waterton, appeared in the local newspaper, for the edification of All Fools and rollocking boys.[16]

The new governor – General Carmichael – a generous but quick-tempered man, was intent on rooting out all trace of corruption in the colony. The planters soon dubbed him 'Old Hercules'. In the spring of 1812, one of Charles Waterton's acquaintances was on the run from Old Hercules, 'on account of a certain bill transaction'; he was hiding up-river, waiting for a chance to escape across the border to freedom. Hercules offered £500 for his arrest but the fugitive, as it happened, was 'sorely afflicted with a liver complaint', for which illness, consistent with current advances in medical knowledge, he wished to be bled. Advised of Charles Waterton's talents in that direction, he searched for, and finally found the celebrated Yorkshire poet at the river-side house of one of the up-country planters – probably Edmonstone – where the poor man was given shelter and relieved of about 18 ounces of his life-blood, in the interests of saving his life.

Nothing in the world could have been better calculated to win Charles Waterton's sympathy than a trust in his scientific pretensions and medical expertise; so, when a boat-load of police arrived, the criminal was sent out of the back door of the house, into a field of sugar cane, whilst Charles 'disputed the passage' with the officers at the landing-stage. The outlaw made his getaway and the officers withdrew, 'muttering curses.'

Next day, a warrant arrived, ordering Charles Waterton to appear immediately at Government House, a summons made the more ominous by the fact that he'd previously omitted to call on the governor to request a passport for his projected trip into the interior – his first Wandering.

> 'And so, Sir,' boomed the general (claimed the lieutenant), 'you have dared to thwart the law, and to put my late proclamation at defiance.'
>
> 'General,' came the reply, in rich Quixotic Waterton, 'you have judged rightly; and I throw myself on your well-known generosity. I had eaten the fugitive's bread of hospitality when fortune smiled upon him; and I could not find it in my heart to refuse him help in his hour of need. Pity to the unfortunate prevailed over obedience to your edict; and had General Carmichael himself stood in the shoes of the deserted outlaw, I would have stepped forward in his defence, and have dealt many a sturdy blow around me, before foreign bloodhounds should have fixed their crooked fangs in the English uniform.'[17]

'That's brave', said the general, wrote the lieutenant, and the two men shook hands and chatted for a couple of hours about the projected expedition into the interior. The upshot was that a passport was arranged, and permission thereby granted 'to range through the whole of ci-devant Dutch Guiana for any length of time.' The date was 16 April 1812.

So Charles Waterton went back to Charles Edmonstone's house, at Mibiri Creek, to make final preparations for the trip and to enlist the necessary river-men to accompany him. But he still contrived to jeopardise his new friendship with Governor Carmichael by stepping forward in defence of another wanted man, unable to find it in his heart to refuse him help in his hour of need. Bravo!

A Dutch captain who retained an administrative post in Georgetown had failed to balance the books and fled up-river to escape the general's quick temper, just as the poor chap with liver trouble had done, only a few days before. He took refuge at Mibiri Creek, intending to travel through the interior to one of the Portuguese or Spanish settlements close to the Brazilian border, just as soon as arrangements could be made to take him there. But Charles Edmonstone suddenly received news from Government House of an impending visit by the governor himself – Warrows Place was a fashionable place to visit, despite its difficult access. The Dutchman was not at all pleased to hear of the governor's intended visit and he badly needed a means of escape as soon as possible. He was stuck up a creek with no paddle. Next afternoon, Charles Waterton took him in a small canoe to a friend's house at Waratilla Creek, nearby, and there they stayed the night, relatively safe. For some reason – probably connivance at a good story – the Yorkshire squire decided to return to Mibiri Creek next day, with his Dutch ward of honour, and on their way there, at the mouth of Waratilla Creek, what should they see approaching but an executive tent-boat – top of the range – with no less than Old Hercules himself on board, accompanied by two captains and a petty commissary – 'three of the best fellows alive.' The Squire immediately told the Dutchman to lie in the bottom of the canoe, where he'd stowed away some palm leaves in case of rain. Then, 'like robin redbreast of old, I covered this great babe of the wood with leaves.' The two boats drew alongside one another and the lieutenant and the general renewed their recent acquaintance. The former pumped the latter with questions regarding the missing Dutchman who, hiding under the palm leaves, heard for himself that, far from wanting to clap him in irons, the governor thought he was an 'excellent fellow', a real ace kid; he was looking forward to seeing him again. So the Dutchman later returned to his job in Georgetown and all ended happily for the time being.

Charles Waterton was now ready for his first long trip into the wilds of Guiana, his first of four Wanderings in South America. But before setting out, there was one more duty to perform, as one good friend to another. In late April, 1812, he attended the christening of the newly-born second child of Charles Edmonstone and his half-Indian wife, Helen, daughter of 'Princess Minda'.[18] A legend arose from somewhere – probably his own lips – that Charles Waterton held the baby in his arms, at the font of the missionary church, and vowed that one day he would marry her. 17 years and four expeditions later, he did marry the girl – Anne Mary Edmonstone – and thereby further enriched the Waterton pedigree, through a bogus Amerindian princess and the Scottish royal house.

He obviously had to find some way to prove himself worthy of her.

Chapter 5
1812-1813
Scenes of No Ordinary Variety

From an analysis of his unmethodical account of it, the outward route of Charles Waterton's first Wandering in South America can be roughly summarised as: a few days paddling up the Demerara to the rock of Seba, four days from Seba to the 'Great Fall', four days portage to the Essequibo river, six or seven days up the Essequibo to the Burro-burro, five days up the Burro-burro to the 'borders of Macoushia', three days to an Indian track into the Savannah, a trek across country for about seven days more to the banks of the river Pirara, then another four day canoe journey to the Portuguese frontier fort of Sao Joachim, where the Takatu falls into the Rio Branco, a tributary of the Amazon.

From Georgetown to Sao Joachim, as the aeroplane flies, is about 320 miles – one and a half hours. Waterton's round trip, including detours, probably covered something like 1,000 miles – perhaps more. Most of the country he travelled was almost totally unexplored by Europeans.

The main object of – or excuse for – the journey was to collect samples of curare poison from the Macusi Indians, who prepare it as a kind of lethal warhead for their blowpipe arrows.

He must have heard about the latest medical experiments with curare, during one of his trips home to England. Just 14 months earlier, Dr. B.C. Brodie had read a paper before the Royal Society, describing an experiment with a volunteer rabbit which had been inoculated with curare and then kept alive for almost an hour and a half by artificial respiration.[1] More of the poison was needed, to test Brodie's conclusions that the drug acted upon the brain and entered the circulation via 'divided veins'.

It had been known for some time that quinine could be used as an antidote to malaria; Charles Waterton badly wanted to believe that curare – an extract of bark, like quinine, and from roughly the same part of the world – would turn out to be an effective treatment against hydrophobia and tetanus. His reasons for thinking that might be the case have been described as 'cryptic'[2] but since curare in small doses was known to serve as a muscle relaxant, it was reasonable to suppose that it might be of use as a treatment for spastic diseases. He was not alone in believing it and laid no claim to invention of the theory. It was not until 1942 that curare-based drugs were first used clinically,[3] but its derivatives are now considered almost essential in open-heart surgery, organ transplant surgery and delicate brain operations, and it has also been suggested for the relief of spastic conditions such as multiple sclerosis and cerebral palsy.[4] The relief, however, is temporary, whereas the disease is permanent; and an overdose could prove fatal.

Death from curare poisoning is really a process of asphyxiation, not as pleasant as it looks. Respiratory muscles are paralysed and the victim dies from lack of oxygen. When curare is used in surgery, the patient is kept alive by artificial respiration until, at the end of the operation, the anaesthetist reverses the effect of the drug with a chemical antidote called neostigmene, derived from the West African Calabar bean. Research fuelled by Charles Waterton was partly responsible for the twentieth-century adoption of the drug itself, and of artificial ventilation to counteract it.

But the first European to give an authentic account of 'ourari' was Sir Walter Ralegh. According to Ralegh, the secret was vouchsafed him by the aboriginal Indians who clearly preferred his style of diplomacy to the brutality of the Conquistadors. 'There was never Spaniard either by gift or torment that could attain to the true knowledge', he said.[5] 117 years later, Charles Waterton was still searching for the true knowledge.

The strongest poison was manufactured by the Macusi Indians. It was therefore to 'Macoushia' that Charles Waterton bent his steps in the spring of 1812. It also happened to be the land of El Dorado, another tourist attraction for famous explorers.

He set out on 26 April, just as the main wet season was starting, and got back about four months later, as it was coming to a close.[6] The first few miles of the journey, on the lower reaches of the Demerara, were already familiar to him, but after leaving the last signs of cultivation behind, he was in a land of almost limitless rain forest inhabited only by a few freed slaves and wood-cutters. The trees, he says, 'put you in mind of an eternal spring, with summer and autumn kindly blended into it.'

> The green-heart, famous for its hardness and durability; the hackea, for its toughness; the ducalabali, surpassing mahogany; the ebony and letter-wood, vieing with the choicest woods of the old world; the locust-tree, yielding copal; and the hayawa and olou trees, furnishing a sweet-smelling resin, are all to be met with in the forest, betwixt the plantations and the rock Saba.[7]

Four-footed animals were less common than he'd been led to believe but included jaguar, ocelot, margay, tapir, labba (an edible rodent), deer, peccary, 'polecat', opossum, ant-eater, 'fox', porcupine, and various kinds of New World monkey.

'This too, is the native country of the sloth.' And thereby hangs a tale.

The sloth became one of Squire Waterton's great obsessions during his years in British Guiana; it was the one indigenous life-form he studied in any detail. Until then, it was generally believed that it was some sort of sorely afflicted freak, condemned to a life of constant pain and misery on account of its "enormous defects". Little or nothing was known about its life history, and mounted specimens and illustrations usually showed it upright. Charles Waterton turned accepted truth upside down.

Another of his specialities was snakes. In Demerara, there are rattlesnakes, whipsnakes, lanceheads, bushmasters, anacondas et al. The Spaniards made extravagant claims for the anaconda – that it could grow up to 80 feet in length and that it was capable of killing bulls. The picture of a bearded Waterton wrestling with a mad bull entwined in the coils of a giant anaconda is pure fantasy; no enraged bull or boa constrictor would dare undertake to wrestle with a bearded Charles Waterton. Real encounters between snakes and Charles Waterton – in Yorkshire as well as in Guiana

– were just as fantastic as any painting. Snakes and sloths figure large in the story of Charles Waterton.

As for birds:

> Demerara yields to no country in the world in her wonderful and beautiful productions of the feathered race. . . . Every now and then, the maam or tinamou sends forth one long and plaintive whistle from the depth of the forest, and then stops; whilst the yelping of the toucan, and the shrill voice of the bird called Pi-pi-yo, is heard during the interval. The Campanero never fails to attract the attention of the passenger; at a distance of nearly three miles, you may hear this snow-white bird tolling every four or five minutes, like the distant convent bell. From six to nine in the morning, the forests resound with the mingled cries and strains of the feathered race; after this, they gradually die away. From eleven to three all nature is hushed as in a midnight silence, and scarce a note is heard, saving that of the campanero and pi-pi-yo; it is then that, oppressed by the solar heat, the birds retire to the thickest shade and wait for the refreshing cool of evening.[8]

In Guyana as a whole, over 70 species of animal and over 700 of birds are still there to delight the eyes and ears, despite modern "developments".[9] In the course of his expeditions, Charles Waterton made the acquaintance of a representative selection of them.

Unfortunately, many of the species he mentions are not positively identifiable; despite a fondness for Latin quotations, he had an almost paranoiac aversion to Latin nomenclature. Taxonomy smacked of the closet to him. (Though taxidermy, of course, never did.) Latinised names gave him 'jaw-ache'.

The result is confusion worse confounded. What, for instance, is a Pi-pi-yo? There have been over twenty editions of *Wanderings in South America* since its first publication in 1825, and brief text notes have been added by J. G. Wood, W. A. Harding, Gilbert Phelps and L. Harrison Matthews. Wood had no idea what a Pi-pi-yo could be; Harding thought that it was probably the greenheart bird, *Lipangus cineraceus*; Phelps said it was the screaming piha, *Lipangus vociferans*; and Harrison Matthews is almost certain that the bird is *Pitangus sulphuratus* – the tyrant flycatcher or kiskadee (an English name for the bird which the early French colonists called 'Qu'est-ce-que-dit', after its call-note.) If Charles Waterton had given a scientific name, there'd be no problem.

But Latin names apart, the contemporary idea of a 'true and proper description' of birds was: first shoot your bird, then identify it. In that respect, at least, he was slightly ahead of his time:

> Having killed a pair of doves in order to enable thee to give mankind a true and proper description of them, thou must not destroy a third through wantonness, or to show what a good marksman thou art; that would only blot the picture thou art finishing, not colour it.[10]

Passenger pigeons were doves, harmless as the serpent is wise. 'But beware of men.' The picture was soon blotted out completely for the passenger pigeon.

At an Indian hut in a forest clearing, below the rock of Seba, Charles Waterton acquired his first sample of curare. The Indian told him that it had already killed some wild hogs and tapirs but the pragmatic amateur from Yorkshire wanted to see for himself. The poison was therefore tested on a 'middle-sized dog'.

To the Amerindians, hunting dogs – probably descended from animals introduced by Portuguese settlers in the sixteenth-century – are greatly prized possessions. So the Squire almost certainly had to pay for the death of this middle-sized dog with some form of merchandise. And to colour the picture for science, its death throes had to be given a true and proper description.

> He was wounded in the thigh, in order that there might be no possibility of touching a vital part. In three or four minutes he began to be affected, smelt at every little thing on the ground around him, and looked wistfully at the wounded part. Soon after this he staggered, laid himself down, and never rose more. He barked once, though not as if in pain. His voice was low and weak; and in a second attempt it quite failed him. . . . His heart fluttered much from the time he laid down, and at intervals beat very strong; then stopped for a moment or two; and then beat again, and continued faintly beating several minutes, after every other part of his body seemed dead. In a quarter of an hour after he had received the poison, he was quite motionless.[11]

The main source of supply for the poison was Macusi country and he was still about a fortnight's journey away from there. First he had to leave the Demerara and walk through the forest for a day and a half, to the banks of the Essequibo. The Indians carried the canoe by a longer route, which took four days.

Whilst waiting for the canoe, the accidental traveller, as he now began to call himself, passed the time by climbing a few trees. The view was wonderful. 'Ascend the highest mountain, climb the loftiest tree, as far as the eye can extend, whichever way it directs itself, all is luxuriant and unbroken forest.'

Without climbing irons, some of those loftiest trees – the towering emergents of the tropical forest (the giant moras, for instance, with branch-less trunks 150 feet or so from ground to crown) – would probably have defied even his climbing skills. The Royal Engineers use block and tackle nowadays, to hoist commercially-sponsored Young Explorers onto the aerial walkways of the forest roof – Operation Raleigh.

Down below, hovercrafts now ride over the South American rapids, like humming-birds hovering on their own updraughts of air. Charles Waterton might have found a hovercraft useful on the Essequibo; the rapids were almost impassable going up-river.

He was on the Essequibo, negotiating those rapids, for about a week, and covered only about 80 or 90 miles in that time. Mind and body survived on fresh meat and cassava bread, whilst heart and soul, not for the first time, nor the last, overflowed with joy at the beautiful surroundings. He describes Essequibo scenery with great enthusiasm in a later Wandering.

A few hours after passing the rapids, the expedition left the Essequibo and paddled up the Burro-burro for another four days, until they came to a small Indian settlement 'within the borders of Macoushia' – curare country.

Here, the diet was supplemented by pacu, a large, vegetarian fish reputed to have a delicious, nutty flavour. The Indians decoy it to the surface with seeds of the crab-wood tree and then shoot it with bow and arrow. They sometimes stupefy fish with the juice from narcotic roots.[12] At home, by his own lake, the 27th Lord later took to fishing in a similar fashion for pike, a large, carnivorous fish with a memorable turpentine flavour of reconstituted offal – ideal for Lent. He learned a great deal from the Indians.

But the main objective of the voyage was curare – 'wourali' as Waterton himself called it – and this was his first encounter with the wholesale supplier at point of distribution. If he managed to get any poison at all from the village, he was compelled to pay a high price for it because, as they took pains to point out, 'it was powder and shot to them, and very difficult to be procured.'

There was, of course, something else to engage his attention:

The first reliable map of Guiana was the one drawn up by Sir Robert Schomburgk in 1837, reproduced as the Great Colonial Map 50 years later. Even today, there are no decent relief maps. At the time of Waterton's Wanderings, the most up-to-date chart[13] actually showed the position of the mythical Lake Parima, or White Sea, beside which El Dorado – the 'gilded king' – was supposed to have lived it up for a space, in his fabulous city of Manoa. According to this map, Charles Waterton's party was now within 4 or 5 days walking distance of Lake Parima. He was impatient to take a look.

El Dorado used to have his body smeared with oil and would then roll from head to toe in gold dust, like an aboriginal Goldfinger, before taking a bath in the lake for the entertainment of visiting stock-jobbers from other tribes. Manoa was a source of unimagined wealth – 'for the greatness, for the richness and for the excellent seate, it farre exceedeth any of the world', said Ralegh. The price of Ralegh's failure to find it, as every schoolboy used to know, was the loss of his head on the scaffold at Westminster.

Charles Waterton was a gullible man but he did have enough critical sense to realise that El Dorado was almost certainly a fairy story. The Macusis told him that Lake Parima was still in existence but then spoiled the story by insisting that ships went there – clearly impossible. He heard of a race of Indians with long tails, roughly equivalent to the Ewaipanoma Ralegh was told about – 'the men whose heads do grow beneath their shoulders.'[14]

Nevertheless, a vestige of El Dorado's fascination remained for El Squire, if only as a lure to find some sort of logical foundation for the fable. As he reached the 'borders of Macoushia' it was clearly as much in his mind as wourali.

The journey, now, was made partly by river and partly on foot. Two or three days upstream of the first Macusi settlement, the canoe was left under the trees and the expedition struck out along an Indian path through the forest, to the northern edge of the great Rupununi savannah, now a beefburger supply depot. In 1812, the ranchers had not yet arrived:

> The finest park that England boasts falls far short of this delightful scene. There are about two thousand acres of grass, with here and there a clump of trees, scattered up and down by the hand of nature. . . . Nearly in the middle there is an eminence, which falls off gradually on every side; and on this the Indians have erected their huts. To the northward of them the forest forms a circle, as though it had been done by art; to the eastward it hangs in festoons; and to the south and west it rushes in abruptly, disclosing a new scene behind it at every step as you advance along.
>
> This beautiful park of nature is quite surrounded by lofty hills, all arrayed in superbest garb of trees.[15]

After crossing this northern end of the savannah, they entered the tree-clad hills, where – 'prettily situated' – was another Indian village. Here they spent the night.

For half of the next day's journey, they were up to their knees in standing water; a detour along the western foothills was necessary. They were delayed most of all by a river – a wide, turbulent river, swollen with rain, where 'the alligators are numerous and near twenty feet long.' The Indians carefully examined the water for alligators – prodding it with long sticks for half a mile up and down stream – before, one by one, they considered it safe to swim across. Rafts carried the baggage across.

After spending the night on the bank of the river, they set out again next morning and nine hours later reached a few more Indian huts on rising ground. The village overlooked a 'spacious plain' which, covered with rain water, took on 'somewhat the appearance of a lake.' He began to wonder whether this could be the place which had given rise to the Parima legend. Geological surveys do suggest that the plain may once have been the bed of a lake and at least two twentieth-century commentators seem to think that Charles Waterton deserves the credit for proposing a credible explanation.[16] Others cite Humboldt and believe that Parima was simply Lake Amuku, a permanent freshwater lake in the same area.[17] El Dorado himself is thought to have been the Peruvian Inca, Atahualpa.

From this village, a letter was sent on ahead of the party to the commander of the Portuguese frontier fort at Sao Joachim, requesting permission to see the fort. 'Pray excuse this letter written without ink,' says the Englishman, in Spanish, 'because an Indian dropped my ink-bottle and broke it.'

The expedition proceeded slowly behind the letter, a few days after it was sent, with the intention of intercepting its reply at some point in the journey. About three hours after leaving the village, they came to the banks of the river Pirara and the last leg of the outward journey began. A canoe freshly made for the purpose, probably from the bark, or 'woodskin', of a purple-heart tree, was provided by some Portuguese soldiers from the fort.

By now, they were close to the Amazon jungle itself, and the weather got wetter and wetter. Though just a couple of degrees from the equator, the nights were cold and stormy and the days sunless. All except the negro member of the party had already had 'a severe inflammatory complaint', for which all hands had been diligently bled in the arm, and the self-appointed surgeon himself had also been 'seized with a dysentry', after living for three days on the fruit of the coucourite tree and mouldy cassava sprinkled with pepper.[18] All these events, he delicately avoided mentioning in the *Wanderings*, 'lest they should savour too strongly of self', a scruple which, luckily, he was very seldom afflicted with at other times. What he does mention in the *Wanderings* is that the 'strength of constitution' which had obviously already let him down, 'at last failed.' As they paddled up the Pirara river, exposed day and night to the cold and rain, 'a severe fever came on', which stayed with him, off and on, for three years.

As well as his lancet, he of course had a mosquito net with him, but its purpose, in those days, was simply to ensure sound sleep, not to prevent malaria. The disease, he was sure, was brought on by the 'chilling blast and pelting shower', with no sun by day to dry out his hammock. To make matters worse, the commander's reply to his letter said that orders had been received not to admit any strangers to the frontier and that the request to visit the fort must therefore be declined.

It seemed he was at the mercy of the elements. However, the commander did arrange to meet him at a spot some distance from the fort and as soon as he learned that the Englishman had fallen ill, he made double quick time to be there. He took it upon himself to ignore the instructions of his superiors, and personally escorted the invalid to Fort Sao Joachim where he was kindly nursed back to better health with rest and good food. In six days, the 'sick English gentleman' was back on his feet.

Exactly how much longer Charles Waterton stayed at the fort, he omits to say – we've so far accounted for a little over a month of his four month safari and the remainder of his own account is given over almost entirely to his observations on curare, of which he'd now collected a 'sufficient quantity'– The return journey is covered by a few brief 'Remarks', appended like an afterthought. Details have to be pieced together from a series of subsequent afterthoughts, in various unlikely places.

But first – curare.

The preparation of curare, with all the rituals and taboos attending it, had probably been seen by only one other European in the history of American exploration – Alexander Humboldt.[19] Charles Waterton's description of the process, as contained in the *Wanderings*, was used as the basis for all subsequent nineteenth-century accounts, though hindsight suggests that he was either guilty of a hoax or the victim of one.

He'd obviously witnessed the preparation in that last Indian settlement on El Dorado's native shore. He certainly acquired some curare there which 'proved afterwards to be very strong.' His Indian instructor told him there were six basic ingredients: wourali vine (later identified as *Strychnos toxifera*); 'A root of a very bitter taste'; the glutinous juice from two kinds of bulbous plant; Triturated fangs of lancehead and bushmaster snakes; very strong pepper; and a few specimens of two species of stinging ant. Boil over a slow fire and pour into a new calabash or pot to dry. Hubble, bubble, toil and trouble. An exhibit in Wakefield Museum is labelled, in his own handwriting, 'Pot in which the Indian boils the Wourali poison. From Guiana, 1812.'

He went to some pains to point out that all ingredients except the wourali were probably mumbo-jumbo. But this failed to prevent accusations from later explorers, notably from Richard Schomburgk, that he was guilty of false teaching.

> It was to be expected that the many mysterious details of the earlier travellers in British Guiana, of Waterton for example, were too inrooted amongst the Colonists for them to believe the simple method of preparation with which my brother [Sir Robert Schomburgk] furnished them on his return.[20]

The Macusis never allowed the Schomburgks to be present during the manufacture of curare and Richard Schomburgk confidently implied that Charles Waterton himself could not possibly have been an eye-witness. No-one after Humboldt, he claimed, could have seen the manufacture personally. 'Their information is always supported only by the accounts of the Indians, who naturally take care to keep the manufacture of the poison as dark as possible.'

In the 1850s, Francis Buckland examined a solution of curare poison under the microscope and found no trace of snake fangs or venom at all; only vegetable matter.[21] But many details described for the first time by Charles Waterton – no women allowed, fasting throughout the operation, abandonment of the shed afterwards etc. – have since

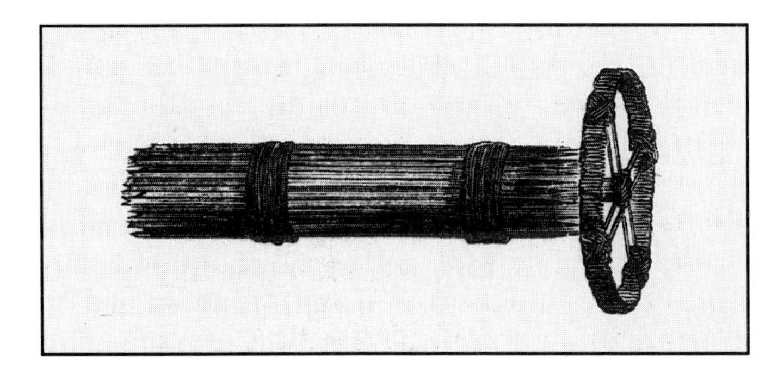

Figure 5. Blowgun and arrows rolled and tied. (Wakefield MDC. Museums and Arts.)

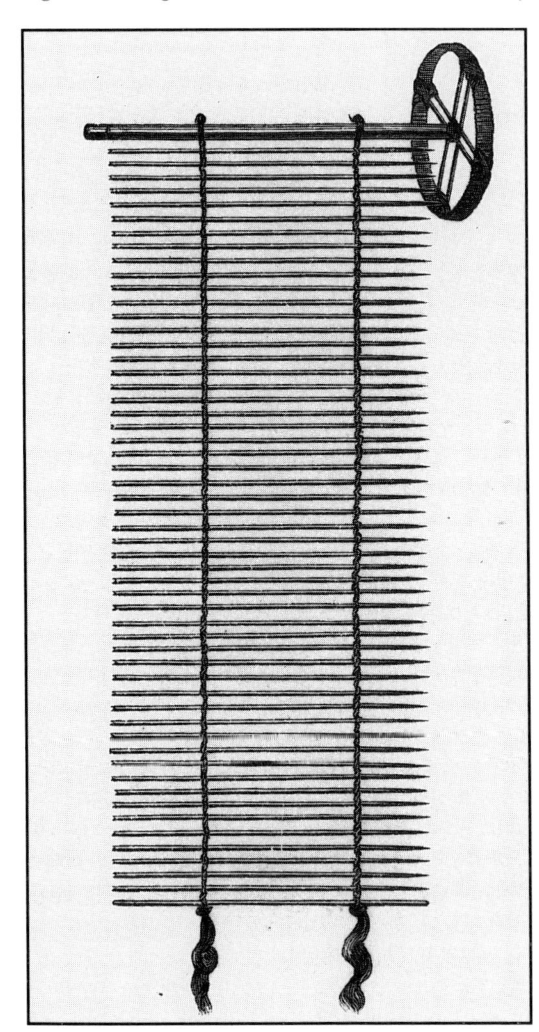

Figure 6. Blowgun and arrows strung. (Wakefield MDC. Museums and Arts.)

been confirmed many times.

Waterton's account of the Indians' use of curare is unimpeachable. The usual tool for killing smaller game was, and still is, the blowpipe – 'one of the greatest natural curiosities of Guiana.' It was made from the hollow stems of a species of reed which grows in the region. The finished article was about ten or eleven feet long. The brittle leaves of the coucourite palm were used to make the blowpipe arrows, sharpened with the jaw-bone of a piranha fish and tipped with curare. Five or six hundred of them could be carried over the shoulder, in a basket-work quiver with a lid made of tapir skin. One of the quivers – full of arrows – which Waterton brought home with him, is now in Wakefield Museum. Others were given to his friend, Dr Harley of Harley Street.[22]

The effect of curare was always lethal, even on the sloth – of all animals, the most tenacious of life. It takes a lot to slow down

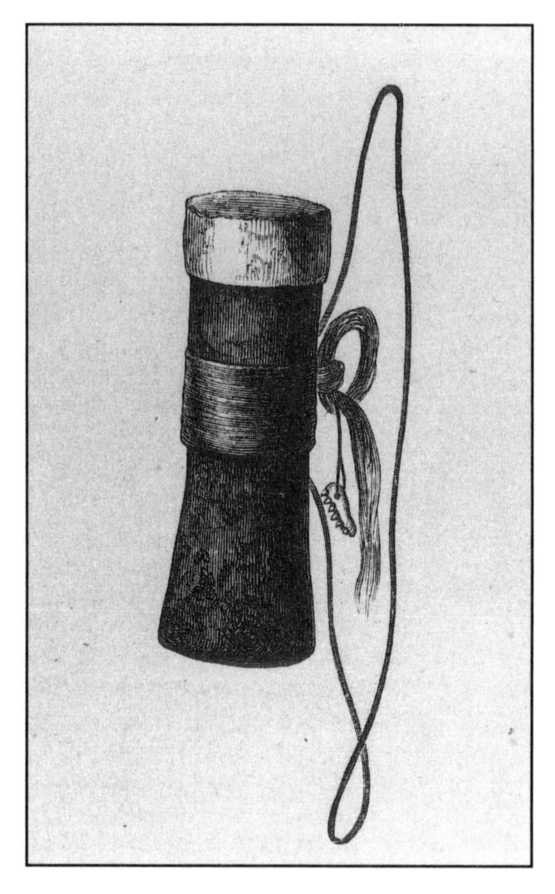

Figure 7. Quiver and blowgun. (Wakefield MDC. Museums and Arts.)

a sloth. But ten minutes after being inoculated in the leg with curare, a three-toed sloth 'stirred' and one minute later it was dead. No fuss, no cry of pain. 'From the time the poison began to operate, you would have conjectured that sleep was overpowering it, and you would have exclaimed, "Pressitque jacentem, dulcis et alta quies, placidaeque simillima morti."' Or words to that effect.

But many Europeans – Sir Joseph Banks was one of them[23] – were doubtful about the ability of curare to kill anything much larger than a sloth, a doubt which Charles Waterton dispelled completely by another experiment:

> A large well fed ox, from nine hundred to a thousand pounds weight, was tied to a stake by a rope sufficiently strong to allow him to move to and fro. Having no large coucourite spikes at hand, it was judged necessary, on account of his superior size, to put three wild hog arrows into him; one was sent into each thigh just above the hock, in order to avoid wounding a vital part, and the third was shot traversely into the extremity of the nostril.
>
> The poison seemed to take effect in four minutes. . . . In five and twenty minutes from the time of his being wounded, he was quite dead. His flesh was very sweet and savoury at dinner.[24]

The experiments were obviously carried out some time after returning to civilization – perhaps in Georgetown – and the number of species of animal since killed by curare in the interests of civilization is formidable. Apart from dogs, sloths and oxen, martyrs to progress include rabbit, donkey, chicken, lizard, rat, frog, guinea pig and pigeon. They died with varying degrees of expedition, roughly according to size, unless kept alive by artificial respiration. In surgery, the drug is now administered in minute doses – generally less than 0.5mg/kg – and its effect lasts for about an hour.[25]

Details of Charles Waterton's return journey from Sao Joachim have to be pieced together from one of his essays, from a few pages of 'Remarks', from a letter to *Loudon's Magazine of Natural History*, published in 1833, and from a letter to the Mayor of

Nottingham, written in 1839. Needless to say, they contradict one another.

Shortly after leaving the Portuguese fort, he was gripped with a fresh attack of fever. A Portuguese soldier accompanied him, and his one remaining black servant, as far as the river Pirara, where three Carib Indians were enlisted as guides for the rest of the journey. He'd lost his six original Indian guides 'through orders misunderstood.'

Very slowly, the five men trekked north, backtracking. At a small Indian settlement, they found the six mislaid Indians lying sick in their hammocks. The Squire himself was by this time 'a mere skeleton, with four bad sores on my legs and body, caused by the stings of venomous insects.'[26] He was in no condition to see or do much and nothing else is recorded of the journey until they reach the Essequibo rapids, several weeks later. After three months of rain, the river was in full spate.

> The roaring of the water was dreadful; it foamed and dashed over the rocks with a tremendous spray, like breakers on a lee shore, threatening destruction to whatever approached it. . . . Nothing could surpass the skill of the Indian who steered the canoe. He looked steadfastly at it, then cast an eye on the channel, and then looked at the canoe again. It was in vain to speak. The sound was lost in the roar of waters; but his eye showed that he had already passed it in imagination. He held up his paddle in a position as much as to say that he would keep exactly amid-channel; and then made a sign to cut down the bush-rope that held the canoe to the fallen tree. The canoe drove down the torrent with inconceivable rapidity. it did not touch the rocks once all the way.[27]

The letter to the Mayor of Nottingham, in 1839, gave an apparently contradictory account, to explain the lack of triturated snake fangs and stinging ants in his luggage:

> In coming down the rapids of the river Essequibo, our canoe was upset, and though we saved ourselves and part of the baggage, I lost all the collected materials of which the poison is composed. They had been carefully folded up in leaves, as is the Indian custom, but by the time that I had managed to get upon the nearest rock they had disappeared forever.[28]

One way or another, they managed to get through the rapids but a violent storm followed and the fever returned, then abated again, then returned. Judging by the time it seems to have taken to make the return journey – about three times as long as the outward trip – he must have been delayed by the fever for about six or seven weeks. At some point in the journey, he lay in his hammock, in an Indian hut, oblivious of the fact that some female jigger fleas, or chegoes, their wombs distended with eggs, had burrowed into his back. His faithful black companion spotted them and the Squire handed him his penknife, with instructions to 'start the intruders'.

> Sick as I was, I wished an artist were present at the operation. The Indian's hut, with its scanty furniture, and bows and arrows hanging around; the deep verdure of the adjoining forest; the river flowing rapidly by; myself wasted to a shadow; and the negro grinning with exultation, as he showed me the chegoes' nests which he had grubbed out, would have formed a scene of no ordinary variety.[29]

There must have been other scenes of no ordinary variety, abridged in his 'Remarks' to three words – 'delays and inconveniences'. But eventually, the intrepid explorer, half dead with malaria, arrived at the Demerara home of his friend, Charles Edmonstone, there to renew the acquaintance of his plighted troth, Anne Mary – now all of four months old. 'I was so wayworn, sick, and changed,' he wrote, in that letter to *Loudon's*

himself, but then she began to recover. It was Lord Percy – the first president of the London Veterinary College – who arranged for the convalescing animal to be put out to grass; and where better than at Walton Hall? There she was christened Wouralia and there she remained, fed on the finest pasture and 'sheltered from the wintry storm', for almost 25 years after she'd risen from the dead.

She couldn't possibly have found a better home. On the Squire's own property, all domestic and farm animals were treated like VIPs. Elsewhere, he could be unkind at times, even to donkeys, but at Walton Hall, horses, dogs, cats, pigs, cattle and donkeys testified to the belief – apparently Jesuitical – that charity has no excess. The stables were specially adapted so that the horses could neck with one another, the kennels ('palazzi carni') so that the dogs could see everything that was going on, the sties so that the pigs could sunbathe, and the field gates so that cattle could talk to one another over the fence without damaging the hinges.[3] For many years, there was a bull in Walton Park which was so tame that the Squire could sit on its flank and feed it with crusts of bread.[4] And cats, of course, were treated like royalty.

He liked cats more and more as he got older – chiefly as friends, also as allies. In cold weather, he'd light a fire for them in the saddle-room, where they could come and go as they pleased through a window which was kept permanently ajar for their benefit. Some of them were also welcome in the hall itself. The abiding principle at Walton Hall was simple: the best-fed cat makes the best mouser (or ratter), an idea which probably extended even to the margay he'd brought back from Guiana – a reputedly fierce species of wild cat closely related to the ocelot. It was just a young one when he acquired it – fresh out of the nest – and despite what was generally thought at the time, margay kittens are usually quite easy to tame. But as it grew, it presumably became less trustworthy. No record exists of how it died, nor of how long it lived at Walton Hall, but the 1824 portrait of Charles Waterton bears testimony to what happened to it after death – its head became a travelling sample of the Waterton Method.

In life, the margay would follow him around the park like a spaniel. Inside the Hall, it would lie in wait on the staircase, to spring like an arrow at the rats emerging from the skirting boards. The whole place was now completely over-run with rats; 32 doors had been gnawed by their teeth and many of the oak window frames were damaged beyond repair. Hercules himself, said the Squire, would have had trouble driving this harpy back to Stymphalus, but then Hercules was no Charles Waterton! Eventually, with the help of cats, owls, weasels and windhovers, the rats were defeated.

The barn owls, especially, could now look forward to better times:

> Up to the year 1813, the Barn Owl had a sad time of it at Walton Hall . . . it was a well-known fact [according to his old housekeeper, Mrs Bennett] that if any person were sick in the neighbourhood, it would be forever looking in at the window, and holding a conversation outside with somebody, they did not know whom. The gamekeeper agreed with her in everything she said on this important subject; and he always stood better in her books when he had managed to shoot a bird of this bad and mischievous family. However, in 1813, on my return from the wilds of Guiana, having suffered myself and learned mercy, I broke in pieces the code of penal laws which the knavery of the gamekeeper and the lamentable ignorance of the other servants had hitherto put into force, far too successfully, to thin the numbers of this poor, harmless, unsuspecting tribe.[5]

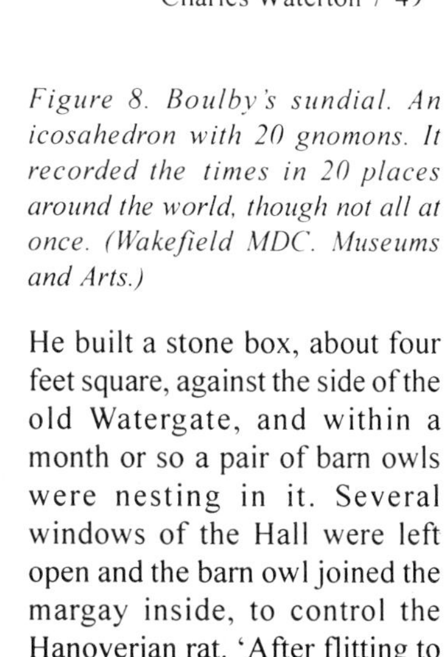

Figure 8. Boulby's sundial. An icosahedron with 20 gnomons. It recorded the times in 20 places around the world, though not all at once. (Wakefield MDC. Museums and Arts.)

He built a stone box, about four feet square, against the side of the old Watergate, and within a month or so a pair of barn owls were nesting in it. Several windows of the Hall were left open and the barn owl joined the margay inside, to control the Hanoverian rat. 'After flitting to and fro, on wing so soft and silent that he is scarcely heard, he takes his departure from the same window at which he had entered.'

But not all methods of pest control were biological; poison was also used. Arsenic was mixed with brown sugar and oatmeal – 'well triturated in a mortar' – and poured into underground traps 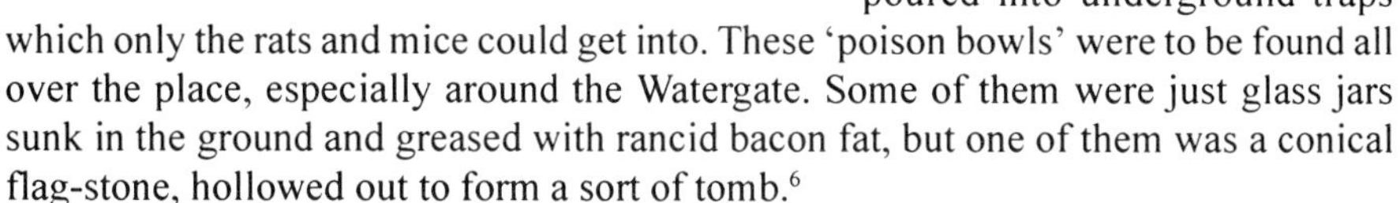which only the rats and mice could get into. These 'poison bowls' were to be found all over the place, especially around the Watergate. Some of them were just glass jars sunk in the ground and greased with rancid bacon fat, but one of them was a conical flag-stone, hollowed out to form a sort of tomb.[6]

He stopped up the entrances to rat runs, fixed cast-iron grating to the mouths of sewers, and relaid the pavement around the Hall. Even so, it was many years before the war against rats was won and for something like a quarter of a century, he could hear them scrabbling inside the walls. One of the runs crossed a hollow plinth in the drawing room. Desperate remedies were called for:

> Having caught one of them in a box trap, I dipped its hinder parts into warm tar, and then turned it loose behind the hollow plinth. The others, seeing it in this condition, and smelling the tar all along the run through which it had gone, thought it most prudent to take themselves off, and thus, for some months after this experiment, I could sit and read in peace, free from the hated noise of rats.[7]

In the summer months, at Walton, the rats took to the fields anyway, to breed – 'nine or ten young ones, twice or thrice in the year'. All hedgerows in the park were left uncut, so that the birds could nest undisturbed and the berries could grow 'for the blackbird or the poor man',[8] but this suited the rats down to the ground. They nested in the uncut hedge bottoms and plundered the birds' nests 'with a ferocity scarcely conceivable in so small an animal.' This was certainly their worst crime of all, to Charles Waterton, and the one which did most to reinforce his hatred for the 'baleful

Figure 9. 'The Poacher entrapped into the Lock-up by Mr Waterton.' From Richard Hobson, Charles Waterton, His Home, Habits and Handiwork.

presence'. But it was a quarter of a century before he could claim with any confidence that the whole estate had been cleared of 'this insatiate and mischievous little brute' – 25 years of all-out warfare, interrupted by occasional adventures abroad.

Apart from the rats, the biggest problems when he got home from Guiana were the poachers and the foxes. There was no estate wall yet and very little to stop either man or beast getting at his birds. (The height of the wall, when it was finally built, was partly governed by the jumping powers of the fox.) At this time, in fact, he may still have been a hunting man himself, though he claimed to have given it up years earlier. Hobson declared that it was on a journey home from the hunting field, in 1813, that the Squire bought the sundial which is still *in situ* on the Hall Island – the phenomenal 'Boulby's sundial' (after the local mason who made it) which records the time in ten places at once.[9] You can still make them out – some of them – by Time's fell hand defaced: Dublin, Rome, Madrid, Boston, Carthagena, Demerara, Amsterdam and Philadelphia – strange places, crammed with observation.

Whether or not he did hunt foxes at this time, he certainly had no time for them at Walton Hall. But their sins were trivial compared with the crimes of the local 'savages'. Anything was fair game to the savages – pheasants, partridges, pigeons, rooks, herons, kingfishers, even woodpeckers. The stories of the various confrontations he had with poachers begin at about this time.

Some tailors returning home from a fiddling party one Saturday night were caught raiding the rookery at Walton Park by the Squire's keeper. There were nine men altogether, one of them an apprentice. Somehow or other, the keeper managed to detain them overnight and presented them to the Squire next morning.

'If you please, Sir, I've catched eight tailors and a half stealing young rooks.'

'Well,' said the Squire to the keeper, 'after all this noise on a Sunday morning you have not managed to bring me a full man, for we all know in Yorkshire that it requires nine tailors to make a man . . . I can't prosecute eight ninths and a half of a man.' (He was referring to the passing knell – nine pulls on the bell-rope, plus one for each year of the man's life.)

Insofar as that story could be called witty, the real wit was probably the feed man, humbly setting up the punch-line for his master. Nevertheless, assuming it happened, and was not just a copper-in-the-cop-shop story, the episode does show Charles Waterton in a good light. Convicted poachers at that time could expect to be treated like rebel cricketers today – transported to Botany Bay. Many landowners protected their property, i.e. their sport, with the kind of man-traps and spring-guns which could legally maim, or possibly kill, trespassers. Squire Waterton only took to using spring-guns 19 years after they'd been made illegal.[10] He was too considerate to use them legally.

There were certainly some desperate characters around in the early 19th century, obviously made more impertinent by the callousness of the Game Laws and the low wages in nearby towns. In the summer of 1814, in a Leeds stableyard, one of them half-inched his horse's tail (useful in sofas) whilst the Squire was at mass in a nearby church.[11] In defiance of fashion and commerce, Charles Waterton's horses generally kept their tails as fly whisks.

Another desperado, a real hard case called Fur (Latin for thief), was cheeky enough to try his luck on the Hall Island itself. (No great feat – you just walk across the bridge when no-one is looking and help yourself.) But Walton Hall is full of sash windows, 25 of which directly face the footbridge. Someone was looking. The Squire went out to confront Fur and rubbed him up the wrong way as diplomatically as possible. He managed to inveigle him onto the narrow step of the watergate, then quickly shut the doors behind him and kept him there, in durance vile, until he agreed to mend his wrong ways. Fur threatened to drown himself there and then but settled instead for a gentleman's agreement. 'He is said never afterwards to have poached on the Squire's manor.'[12]

On another occasion, he had a fight with a poacher:

> One night, on going my rounds alone, in an adjacent wood I came up with two poachers: fortunately one of them fled, and I saw no more of him. I engaged the other; wrenched the knife out of his hand, after I had parried his blow, and then closed with him. We soon came to the ground together, he uppermost. In the struggle, he contrived to get his hand into my cravat, and twisted it until I was within an ace of being strangled. Just as all was apparently over with me, I made one last convulsive effort, and I sent my knees, as he lay upon me, full against his stomach, and threw him off. Away he went, carrying with him my hat, and leaving with me his own, together with his knife and twenty wire snares.

That was the case for the prosecution, in a natural history essay called *Tight Shoes, Tight Stays, and Cravats*.[13] Moral of the story: cravats 'keep our jugular veins in everlasting jeopardy.' Lesson drawn: he continued to wear one.

The report of another skirmish with a trespasser comes from an 'authentic anecdote' disclosed by the author of *Tom Brown's Schooldays* – Thomas Hughes.

> When he had succeeded in closing the bridle road through his park, which

> interfered so grievously with the comfort of the birds and beasts to whom it was devoted, there were still persons who persisted in using it, amongst whom a butcher of Wakefield was conspicuous, a sturdy Protestant tradesman who had ridden along it ever since he could remember and openly avowed his intention of continuing to do so. One evening, after sunset, he turned his horse's head as usual along the accustomed path, and jogged comfortably along, defying Squire and Pope in his mind, until he came under the dark shadow of some trees which overhung the roadway. Suddenly a whoop, which made his heart leap, sounded in his ear, and, dropping from a tree or springing from the ground – which of the two he could never rightly tell – a something alighted on his horse's quarters, just behind the saddle. The next moment his arms were pinioned to his side by an embrace which made him powerless, and his frightened steed broke into a gallop which soon brought him to the park boundary. The gate was open, and, as he passed through it, his arms were suddenly released, and he was again alone on his horse, while another whoop rang in his ears as he galloped on towards Wakefield. He reached home safely, a feeble and repentant Protestant butcher, and from that day the bridle-road through Walton Park saw him no more.[14]

The stunt is too unbelievably *Boys Own* not to be true. He seldom had much regard for probable consequence, confident in the knowledge that God was on his side as long as he kept the faith. And the risk to life and limb was just as great when no third party, apart from God, was involved.

It was in the autumn of 1814 – long before the advent of breech-loaders, and a few years before guns were banned altogether on Walton estate – that he came close to blowing his own brains out with a shotgun. He was out in the park with his brother-in-law, Robert Carr, when the gun went off as he was pushing the wadding onto the powder with the ramrod. There was no charge of shot in the barrel but the ramrod, wadding and powder went through his forefinger, between the knuckle and first joint, severing the tendons, and leaving the wound 'black as soot'.

The victim was thrilled to bits. It was a golden opportunity to test his skill as a doctor on the most faithful and regular patient on his panel – Charles Waterton. He went to a neighbouring house and poured warm water through the wound, until it was clean, then he set about replacing the loose tendons, which were hanging out of the wound, and binding up the finger into something like its original shape. Then it was time for the lancet. He opened a vein with his other hand and took away 22 ounces of blood, to complement the smaller amount he'd already lost.[15] He'd learned venesection from the 'celebrated' Dr John Marshall, in Guiana, and late in life, he claimed to have blooded himself one hundred and sixty times in all. The average extraction was about twenty ounces. 160x20/16x8 = 25 gallons. No wonder he felt the cold.

After using the lancet, he went home to apply another of his favourite remedies – 'a very large poultice'. This was removed twice a day, to keep down the swelling, and eventually he regained full use of the finger for those 'delicate manipulations' in which he 'notoriously excelled', *viz.* taxidermy. His fingers were reputed to be 'as nimble, as pliable, and as sensitive as those of a well-bred lady',[16] as one lady, at least, was eventually in a position to testify.

So life at home, with one thing and another, and a few things besides, was not

Figure 10. 'Waterton's World.' Richard Bell. (Richard Bell)

entirely devoid of incident. Apart from thrills and spills, and the occasional scourge of remedies, there was also the continual challenge of large-scale estate management, including an ambitious plan to clean out the lake and to alter the drainage so that it might be better appreciated by the birds and the fish. This pre-occupied him for much of the year 1814 and merited no less than seven pages of description in his notebook,[17] but forty-odd years later, after the mud had taken over instead of the pondweeds, he was compelled to change the drainage more radically still; finally, he got it right.

1815 brought another pre-occupation – a plague of field mice. Waterloo was trivial by comparison. A few years earlier, he'd planted two acres of the park with a few hundred oak saplings and larch trees but the mice gnawed the bark so badly that in the spring of 1816 – just before he left the country again – very few of the oaks put forth any buds.[18] His holly bushes and apple trees were also badly affected. The local kestrels did their best to keep down the mice but the local sportsmen and gamekeepers had to be persuaded not to keep down the kestrels. The margay probably also lent a paw (assuming it was still alive) but the plague itself, like all except human plagues, proved its own best regulator.

Apart from all this, there were also occasional trips to London, for the curare experiments, and at least one visit in the steps of Robert Waterton, 400 years earlier, to the home of Lord Percy – Alnwick Castle. A letter to Bryan Salvin, of Burne Hall, Co. Durham, was sent from Alnwick by 'Free Percy' on 30 August 1815. (It was written at Kielder Castle the previous day.) The Salvins were an old catholic family who could claim uninterrupted possession of Croxdale Manor since roughly the time that Robert Waterton passed through. Croxdale estate bordered on Tudhoe school, where Charles Waterton had first met the Salvins 21 years before.

The letter is chiefly interesting for a characteristic little Quixoticism near the end, which conveniently introduces the next chapter. He mentions having passed the lodges of Croxdale Manor by stage-coach about a week previously, and getting out on top of the mail to see once more the country through which he'd often rambled as a child.

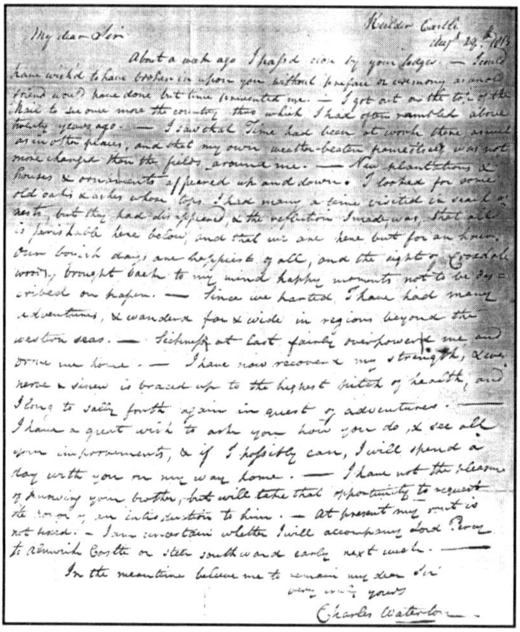

Kielder Castle
Aug[t] 29th 1815

My dear Sir

About a week ago I passed close by your lodges. — I could have wished to have broken in upon you without preface or ceremony as an old friend would have done but time prevented me. — I got out on the top of the Mail to see once more the country thro' which I had often rambled above twenty years ago. — I saw that Time had been at work there as much as in other places, and that my own weather-beaten [illegible] was not more changed than the fields around me. — New plantations & houses & ornaments appeared up and down. I looked for some old oaks & ashes whose tops I had many a time visited in search of nests, but they had disappeared, & the reflection I made, was, that all is perishable here below, and that we are here but for an hour. Our boyish days are happiest of all, and the sight of Croxdale woods brought back to my mind happy moments not to be described on paper. — Since we parted, I have had many adventures, & wandered far & wide in regions beyond the western seas. — Sickness at last fairly overpowered me and drove me home. — I have now recovered my strength, & ever nerve & sinew is braced up to the highest pitch of health, and I long to sally forth again in quest of adventures. — I have a great wish to ask you how you do, & see all your improvements, & if I possibly can, I will spend a day with you on my way home. — I have not the pleasure of knowing your brother, but will take that opportunity to request the honor of an introduction to him. — At present my [illegible] is not fixed. — I am uncertain whether I will accompany Lord Percy to Alnwick Castle or steer southward early next week. —

In the meantime believe me to remain my dear Sir
very truly yours
Charles Waterton

Figure 11.Charles Waterton's 'Stonyhurst hand,' 1815. To Bryan Salvin of Croxdale Hall. (Durham Record Office, Londonderry State Archives. Courtesy Captain G. M. Salvin.)

Boyhood days are happiest of all, he says, instructively, then: Since we parted, I have had many adventures, & wandered far and wide in regions beyond the western seas. Sickness at last fairly overpowered me and drove me home. I have now recovered my strength, & ever nerve & sinew is braced up to the highest pitch of health, and I long to sally forth again in quest of adventures.[19] A few months later, on 19 March, 1816, in the brave, unquenchable spirit of the great don himself, (he'd read Cervantes in Spanish) the intrepid knight errant of Walton Hall boarded ship at Liverpool and sallied forth once more in quest of adventures, beyond the western seas, for the second of his four sensational Wanderings in South America.

Chapter 7
1816-1817
In Quest of the Feathered Tribe

Charles Waterton's own recollections are not trustworthy. Events, names, points of the compass and dates – dates especially – are often either hopelessly confused or partially remembered. His facts are casual reminiscences, traveller's tales. He was a 'private rover'.

'Early in 1817,' for instance, he was persuaded not to join an expedition to the Congo – so he says. Sir Joseph Banks told him that the steamboat appointed for the trip up-river 'failed to meet expectations.' The only contemporary Congo expedition on record – to look for a link between the Congo and Niger rivers – came to grief late in 1816. The commander of the expedition, Captain Tucker, died of fever on October 10th, 1816, by which time most of his men were already dead. Waterton sailed for his second South American Wandering in March, 1816, and heard of the failure of the Congo expedition whilst he was 'on the other side of the Atlantic.' So for 'early in 1817', read 'early in 1816'.

Before the 'scientific gentlemen' left for the Congo, Charles Waterton was invited by Sir Joseph Banks to 'impart certain instructions' to them at Soho Square. He gave them a few helpful words on the best method of preserving birds – the Waterton Method – with a short, Jesuitical homily on hedonistic temperance and the dangers of sleeping in wet clothes,[1] all which proved of incalculable value to the doomed expedition. On March 9th, 1816, ten days before the Squire himself sailed for South America, Banks wrote to him from Soho Square and imparted certain instructions of his own:

> I trust that experience has taught you to forbear from trespassing too much into the swampy regions which lie near the roots of the hills from where the rivers descend to the plains. I am clear that no European constitution, however hardy it may be, can resist the noxious miasmata which pervade the atmosphere of these deleterious tracts & I do not know that any object of natural history that aught to tempt you to hazardous exploits in the hope of attaining it, is exclusively found in that hazardous district.[2]

The intention was to sail for Brazil, then to cross the Amazon jungle itself, to enter British Guiana by the back door and look once more for Lake Parima – a land and river journey of about 2,500 thousand miles through the densest tropical rain forest on Earth. The plan was ultimately changed but the hazardous exploits were only very slightly compromised.

Port of arrival in Brazil was Pernambuco. (Recife.) He was not favourably impressed – its Jesuits' college had been disbanded. First view, from the sea, was pleasing enough, but closer inspection revealed the unmistakeable signs of religious persecution – the streets were unclean. 'Impurities' from the houses, and 'litter' from the beasts of burden, created a cloud of 'unsavoury dust' when the wind blew, and noxious miasmata deleterious to the European constitution. Despite all which, the captain-general of Pernambuco walked through the streets 'with as apparent content and composure as an English statesman would proceed down Charing Cross.'

The Marquis de Pombal was chiefly to blame. He'd died the year Charles Waterton was born but the effects of his victimisation of the Society of Jesus survived. Under Pombal's rule, most Jesuits in Portugal had been either deported to Rome or clapped in irons; his influence inevitably extended to Brazil, then a Portuguese colony. The college at Pernambuco was closed down, its staff dispersed, and 'some time after, an elephant was kept there. . . . Virgil and Cicero made way for a wild beast from Angola.'

Robert Southey, the Poet Laureate, saw it all differently. According to Southey, these same Jesuit fathers of Pernambuco were 'missioners whose zeal the most fanatical was directed by the coolest policy'[3]. Charles Waterton blew his cool. A lengthy tirade against the 'ungenerous laureate', the 'ungrateful Englishman', is rounded off with a sturdy defence of the idolatry of which the catholic church stood accused. The attack on Southey is sustained for about 600 words in the *Wanderings* – 'incomparable nonsense' etc. – and then abruptly ends: 'The environs of Pernambuco are very pretty.'

He chose to stay in a little village called Monteiro, six or seven miles from Recife. He shot 58 of the most brilliantly coloured birds of the region, and nearly put an end to his own life in the process. He was in an abandoned orange grove, near Monteiro, and noticed a commotion in the lower branches of one of the trees. The noise was made by 'six or seven blackbirds with a white spot betwixt the shoulders'. (Some species of Icterid.)

> In the long grass underneath the tree, apparently a pale green grasshopper was fluttering, as though it had got entangled in it . . . having quietly approached it, intending to make sure of it – behold, the head of a large rattlesnake appeared in the grass close by: an instantaneous spring backwards prevented fatal consequences. What had been taken for a grasshopper was, in fact, the elevated rattle of the snake in the act of announcing that he was quite prepared, though unwilling, to make a sure and deadly spring. . . . It was he who had engaged the attention of the birds, and made them heedless of danger from another quarter; they flew away on his retiring; one alone left his little life in the air, destined to become a specimen, mute and motionless, for the inspection of the curious in a far distant clime.[4]

By the time the rainy season arrived in the province, the birds of Pernambuco had started to moult. Since he wanted an attractive display of specimens to illustrate the glory of Creation, rather than a scientific display of moult sequence to abet the closet naturalists, he decided it was time to leave. The journey to Maranham (Sao Luis) would take about forty days on horseback, and the rains would probably damage his specimens, so he abandoned his intended route and returned to the coast instead, where he looked for a ship to take him north, across the line, towards the Guianas. Most of the available vessels were slave ships, which obviously would not do, so he bided his time awhile and eventually took passage for Cayenne, in French Guiana, aboard a Portuguese brig which was assumed to be engaged on more respectable business. It was not a lot more comfortable than a slave ship, as it turned out, and the best place to sleep, he soon discovered, was on top of a hen-coop on deck. 'Even here, an unsavoury little beast called bug was neither shy nor deficient in appetite.' Unlike most English travellers of the period, however, Charles Waterton was not the man to be unduly worried by trivial inconveniences such as bugs; in due course, he arrived in Cayenne with the minimum of fuss or complaint.

He paid a few social calls in Cayenne and was taken to see the famous spice plantation at La Gabrielle. He also made a canoe journey towards the offshore rock called Le Grand Connetable – the Constable.[5]

That canoe trip took two days and nights, in 'wretched accommodations', and turned out to be just the kind of hazardous exploit which Sir Joseph Banks had been naive enough to warn him against.

He hired the canoe, with seven black men to paddle it, in Cayenne town; they set out along the Oyapoc estuary towards the open sea at six o'clock in the evening. Nights on the Cayenne coast are often uncomfortably cold in the best of circumstances, as Humboldt had discovered a few years earlier.[6] Wet nights in an open boat are much worse. It rained all night. Dawn found them on the sea-coast, to windward of the Constable, and by ten o'clock the ebbing tide had left them high and dry on a vast, alluvial mud-bank where they remained all day long, under a blazing sun, the temperature, presumably, around 100 degrees Fahrenheit. The turning tide freed them from the mud but was too fierce to allow them to continue their journey towards Le Grand Connetable. They returned, instead, slightly the worse for wear, through 'another night of hardship', to Cayenne town. The torrential rain had given the Squire a serious complication of sore throat called 'inflammation of the oesophagus', so he put himself on a temporary invalid diet of white bread soaked in weak tea, the same broth that he virtually lived on during Lent almost every year thereafter.

He apparently spent those few days of convalescence, free of hazardous exploit, slurping his weak tea and white bread in the polite company of Cayenne society. The social round included an introduction to an Englishman named Howe, an amateur taxidermist with an especially impressive box of specimens – 16 of his own teeth. '"These fine teeth," said he,' says the Squire, '"once belonged to my jaws; they all dropped out by my making use of the savon arsenetique for preserving the skins of animals."'[7]

Arsenic soap was widely used by taxidermists as an antiseptic, but Charles Waterton preferred the famous Waterton Method – a solution of corrosive sublimate in alcohol. His teeth, and his exhibits, remained in good condition all his life.

His adventures in French Guiana were not yet over. Shortly before he left Cayenne, he experienced the kind of comic-strip bump on the head which is clearly de rigueur for any soi disant chercheur d'aventures: he sat under a tree and a branch fell on him. The practical joker was a knife-grinder insect which had eaten its way through the cinnamon branch whilst the Squire's thoughts, for once, were not on higher things. The branch obviously had to be sent back to Walton Hall, for the inspection of the curious in a far-distant clime.[8]

When opportunity came to leave Cayenne, he sailed for Paramaribo, the capital of

Surinam, aboard a Yankee brig in the command of a man who had evidently served in the 1812 war against the British navy. Like many of his kind, this man still bore a grudge:

> Whenever he saw a vessel in the distance, he would take it for a British cruiser and remark, "there goes the old serpent from whose sting, thank heaven, we are now for ever free."[9]

The insult to snakes was charitably forgiven.

From Paramaribo, Charles struck inland to the Corentyn river, then travelled by public road to New Amsterdam, where he stayed a couple of days before carrying on to Georgetown – not quite the triumphant entry he'd planned.

But in Georgetown, he was back amongst old friends. He clearly regarded Demerara as his second home, and its 'liberal inhabitants' – black, brown and white – as blood brothers. No doubt he also renewed the acquaintance of his future wife, now a three-year-old toddler. But the forest primeval beckoned and Charles Waterton had other, more pressing engagements to fulfil.

> Let us now, gentle reader, retire from the busy scenes of man, and journey on towards the wilds in quest of the feathered tribe.

He'd already discovered that the best places to find the feathered tribe were not in the depths of the forest but on its edges – 'the sides of rivers, lakes, and creeks, the borders of savannahs, the old abandoned habitations of Indians and wood-cutters.' It was in areas like these that he spent the next 6 months or so, collecting more than 200 specimens of 'the finest birds'.

'First place' went to the humming-bird:

> See it darting through the air almost as quick as thought! – now it is within a yard of your face! – in an instant gone! – now it flutters from flower to flower to sip the silver dew – it is now a ruby – now a topaz – now an emerald – now all burnished gold!

Waterton was no universal man, in the Goethe/Humboldt/Davy mould; he was never in quest of anything as grand as a quintessential unity in nature, a philosophical synthesis – philosophy was drawn from the catholic faith, the universal church. In his explorations of the 'far extending wilds', he was principally employed in a search for the most glamorous manifestations of what he saw as God's omnipotence. And the humming-bird won first prize in the beauty contest.

He was curious about other aspects of them, naturally, as his essays prove, but taxidermy was obviously the prime mover. At the last count, there were 319 species of humming-bird altogether, and one of them belongs to a very distinguished family: Waterton's woodnymph, *Thalurania watertonii* – the only creature Charles Waterton ever collected which was previously unknown to science. He later used his first-hand knowledge of humming-birds, plus a bit of ignorant presumption, as a weapon for bringing down some of the closet naturalists; his aim was not always perfect but he usually did manage to hit the target.

Firing from the hip was, in truth, a speciality of his, and not reserved solely for his sport with the pedants; it was whilst studying humming-birds in the jungle, that he used the technique on 'a martin of the foumart family'.[10] (i.e. a marten.) The animal was going about its own business in the undergrowth when it suddenly spotted the

Squire of Walton Hall and took to the trees in fright, but the intrepid explorer chased it, shouting at the top of his voice, and fired from the hip whilst it was leaping from one branch to another. 'Wonderful to relate, down dropped the flying martin (sic); dead as Julius Caesar.' Many a hunted closet naturalist was later brought down in almost identical fashion.

Next to the humming-birds, perhaps the most beautiful birds in Demerara are the cotingas, of which he counted five species on the lower reaches of the Demerara river. Some of them inevitably had the honour of being martyred to Squire Waterton's gun; they now share the most brilliant of display cabinets in Wakefield Museum, with a selection of manikin, oriole, tanager and 'fig-eater' – almost as colourful today as on the last day of their lives almost 200 years ago. One of them is perched on his finger in the famous 1824 painting of Charles Waterton which belongs to the National Portrait Gallery – a Guianan red cotinga, *Phoenicircus carnifex.*

Very closely related to the cotingas, and sometimes classed with them, are the spectacular bell-birds, of which the most spectacular of all is the white bell-bird, or campanero. Waterton claimed that the campanero's voice could be heard three miles away, a power doubted by Sidney Smith[11] but probably not greatly exaggerated, if at all. The bell-bird is not recommended for suburban aviaries.

Next on the avian cat-walk came the toucans and toucanets. About eight years previously – 'while eating a boiled toucan' – the best method of preserving its bill had suddenly occurred to the Squire: hollow it out, coat it with gum arabic and chalk, and touch it up with paint. Judgement, caution, skill and practice, he says, ensure success. His own specimens prove it.

Toucans belong to the same bird group as the woodpeckers – *Piciformes*. There are something like 15 species of woodpecker in the Guianese forests, and Charles Waterton declines even to attempt any specific details. But he does make a plea for their conservation. Owners of standing timber – in Europe and in America – claimed that woodpeckers encouraged the decay of their trees by boring holes in the trunks and letting in water. Defence counsel put words in his client's beak:

> "I never wound your healthy trees. I should perish for want in the attempt. The sound bark would easily resist the force of my bill, and were I even to pierce through it, there would be nothing inside that I could fancy, or my stomach digest. I often visit them it is true, but a knock or two convinces me that I must go elsewhere for support; and were you to listen attentively to the sound which my bill causes, you would know whether I am upon a healthy, or an unhealthy tree. Wood and bark are not my food. I live entirely upon the insects which have already formed a lodgement in the distempered tree. . . . Millions of insects engendered by disease are preying upon its vitals. Ere long it will fall a log in useless ruins. Warned by this loss, cut down the rest in time, and spare, O spare, the unoffending woodpecker."[12]

Almost 30 years later, a pair of woodpeckers nested in Walton Park and were killed by a predator unknown in the South American jungle – primitive savages.[13]

A bird family which is still usually classed with the toucans and woodpeckers is the jacamars, though they're more likely to be confused with the kingfishers. In terms of beauty, Charles Waterton ranks one of them – probably the Paradise jacamar, *Galbula dea* – with 'the choicest of the humming-birds'.

Another which 'ranks high in beauty' is the 'houtou' – from Waterton's description, probably the blue-crowned motmot, Momotus momota. The motmots are a beautiful, yet cryptically coloured, family which, unlike the jacamars, probably is related to the kingfishers. Their long tails have a spadular tip, like a jam-ladle, which is created by preening the loose, sub-terminal barbs until they fall out. Charles Waterton's extrapolation from this – motmot to man – makes curious, self-deprecating reference to contemporary hair styles:

> While we consider the tail of the Houtou blemished and defective, were he to come amongst us, he would probably consider our heads, cropped and bald, in no better light.[14]

The Squire retained his own cropped head, in aggressive defiance of Victorian fashion, until his dying day.

Another gratifyingly gorgeous bird of the Demerara forest was the 'Boclara', Arawak name for trogon. Charles Waterton's description most nearly fits the white-tailed, or yellow-bellied trogon, so called on account of its yellow belly and white tail. It always breeds in burrows and invariably makes a nest of wood chippings, sometimes inside a termitary.[15]

Trogons, like the motmots, are close relatives of the kingfishers, which Europeans generally regard as amongst the most spectacular of birds. But the six kinds of kingfisher which Charles Waterton identified in Demerara were disappointing to him. 'In the scale of beauty,' he says, the 'English kingfisher' would outweigh all six of them together. They therefore get very little space in Waterton's *Wanderings in South America.*

But drab birds in the tropics are the exception rather than the rule, certainly amongst males. Other beauties to tempt the wandering eye included the cassiques, troupiales, tanagers, manakins, macaws, parrots, parakeets, sugar-birds and flamingoes.

The cassiques are chiefly notable for their nests – tubular pouches, a yard or more in length, suspended from the tips of twigs, like stockings on a washing–line. The birds breed socially – like stockings – and Waterton often found colonies of two separate species sharing the same tree. He labelled his museum specimens yellow-backed and red-backed cassiques, now more usually referred to as the yellow-rumped cassique, *Cacicus cela*, and the Brazilian red-rumped cassique, *Cacicus haemorrhous*. Rump was expurgated from Victorian and Regency bird names, and he evidently omitted it from his museum labels in deference to the more delicate sensibilities of some of his female visitors. He used the word freely enough in his published writings.

He describes four species of cassique in all, though the last two are more correctly oropendolas, sometimes known as giant cassiques. The nests of these birds can sometimes reach six feet in length – as tall as the Squire himself.

Oropendolas and cassiques are members of the Icteridae, one of the most complex of all bird groups. Other icterids common in the American tropics include American orioles and blackbirds and the various kinds of troupiale – pretty little yellow and black creatures, of which Waterton also mentions four species, though not by specific name – he's seldom as precise as that.

He'd encountered the true troupiale, *Icterus icterus* – the bugle-bird of the pet trade – whilst canoeing up the Takatu river during his first expedition. The Portuguese called it the nightingale of Guiana, he says. The slightly smaller species, which 'built in the roof of the woodcutter's house', is probably the yellow oriole, *Icterus nigrogularis*, of which there is a specimen in Waterton's collection, marked simply 'Oriole from Guiana.' His description of two other species are too brief for positive identification. Like most naturalists in central America, Waterton was quick to notice the attractions for birds of the giant fig trees. 'Wherever there is a wild fig tree ripe, a numerous species of birds called Tangara is sure to be on it.' The Tangaridae are brilliantly-coloured, finch-like birds, of which he counted, or guessed, or was told about, 18 species. In fact, about 40 species have been recorded in Guyana, a few of which are included amongst the specimens in Wakefield Museum.

Often, on the same fig tree, he found four species of manakin, a family closely allied to the cotingas. Again, he fails to name them specifically but very briefly describes them in order of size. They're probably the white-breasted manakin, *Manacus manacus*; the orange-headed manakin, *Pipra aureola*; the white-crowned black manakin, *P. leucocilla*; and the golden-headed manakin, *P. erythrocephala*. There are 15 or 16 species of manakin native to Guyana as a whole.

Nectar-feeding sugar-birds also frequent the fig trees. Waterton describes two species adequately enough for identification as the turquoise honey-eater, Dacnis cayana, which is still sometimes seen on the outskirts of Georgetown itself, and the blue honey-creeper, *Cyanerpes cyaneus*, an exquisite little bird which has the distinction of laying black eggs in an almost transparent nest.

The 'six or seven species of small birds' mentioned in the *Wanderings* belong to the group loosely known as ant-birds. In Guyana as a whole, there are, in fact, twelve species of ant-bird, plus three of ant-catcher, nine of ant-creeper, eight of ant-thrush and five of ant-wren. The one species Waterton describes is obviously the white-fronted ant-catcher, *Manikup albifrons*. There is a specimen of this species in his collection, idiosyncratically labelled 'Puffback'. He noticed that the bird is only to be found in the wild in association with one particular kind of ant, the migrations of which it follows. Schomburgk confirmed the observation without, of course, mentioning Waterton's name. He identified the insect as the wander ant.[16]

One more relatively unspectacular family gets special treatment in the *Wanderings* – the goatsuckers, or nightjars. They don't milk the goats, says Charles Waterton, nor the cow herds. 'Poor injured little bird of night. . . .'Nor are they birds of ill omen, he says, as the Indians and negroes believed. 'You will forgive the poor Indian of Guiana for this. He knows no better; he has nobody to teach him.' The European had no such excuse.

Four of the sixteen species of nightjar native to Guyana have four different call notes, transcribed by Waterton as: 'Who are you?' 'Work away.' 'Willy come go.' And the famous 'Whip poor Will.' 'It is very flattering to us', said Sidney Smith, 'that they should all speak English! – though we can-not much commend the elegance of their selections.'[17]

Game birds are plentiful in Demerara and here, says the Squire, apropos of the English Game Laws, 'no-one dogs you, and afterwards clandestinely inquires if you have a hundred a year in land to entitle you to enjoy such patrician sport.' Wood quail, tinamous, curassows, guans, horned screamers and trumpeters all get a mention. He sometimes saw flocks of two or three hundred trumpeters, but never addressed his mind to the apparently esoteric theory, since endorsed by Monty Python, that their call-notes proceed from the anus.

He also came across large flocks of vultures. 'It is a fact beyond all dispute', he says, 'that when the scent of carrion has drawn together hundreds of the common vultures, they all retire from the carcass as soon as the King of the Vultures makes his appearance.' That seemingly innocuous sentence caused a furore. A few years later, when John James Audubon had the unspeakable gall to dispute what was beyond all dispute, and claimed that vultures hunted by sight, not by scent,[18] one of the great epistolic punch-ups of all time was to ensue. The argument has only recently been resolved.[19]

These, then, are some of the birds of Demerara – just 'a handful from a well-stored granary.' Others get an honourable mention – egrets, flamingoes, boatbills, bitterns, water-hens, coots, muscovy duck, curlews, sandpipers, rails, gulls, pelicans, jabiru storks, cranes, snipe, plover, geese, ibises, darters, eagles, hawks, falcons, shrikes and owls – 'all worthy the attention of the naturalist, all worthy a place in the cabinet of the curious.' But none of them worthy of any such reliable guide to identification as a scientific name.

The five or six months he spent admiring this well-stored granary seems to have been a more healthy, less adventurous experience than the first, or any of the subsequent wanderings. But in the early spring of 1817, with the rains about to begin in Guiana, and a European summer to look forward to, he took passage for home on board the West Indiaman, *Dee*, with 'above two hundred specimens of the finest birds' and a few sundry curios from Guiana and Brazil.[20]

He still wasn't satisfied. One of his reasons for trying to reach the Constable rock, off the coast of French Guiana, had been to collect a specimen of the tropic-bird – a sea-bird closely allied to the gannets and boobies. The attempt had failed, but now, shortly after leaving port for home, he saw a tropic-bird sitting on the sea, within gun-shot range of the ship. He 'fired at him with effect' and offered '"A guinea for him who will fetch the bird to me."'[21] A Danish sailor jumped overboard and swam towards the dead bird as the ship went 'smartly through the water' and swiftly left bird and sailor behind. It was no use trying to lower the jolly-boat; it was filled with lumber and tied down securely for the passage home. So, after trying unsuccessfully to tack, the captain brought the ship's helm slowly round to windward, in an effort to search for the man overboard, by which time everyone felt certain that he'd gone down to Davy's locker for good. (All this trouble on account of Charles Waterton!) At last, however, someone spotted the great Dane, 'buffeting the waves with the dead bird in his mouth', like a spaniel retrieving a pheasant. The Squire eventually received the bird from the sailor's cold hand, and awarded a golden sovereign, much as if it were a chocolate drop for a gun-dog.

As with the cool Captain Bolin who, 14 years earlier, had broken the embargo on shipping in Malaga harbour, the 'adventurous Dane' was a man after Charles Waterton's own heart – a model of cool intrepidity. The dead bird he retrieved – a yellow-billed tropic-bird, *Phaeton lepturus* – now sits imperiously, in a case all of its own, staring at a poor, innocent goatsucker, down in the cellar of the City Museum, Wood Street, Wakefield, Yorkshire.

Chapter 8
1817-1820
Delicate Manipulations

In the great days of sail and great Danish sailormen, the homeward journey across the western sea was long and tedious. Wise passengers found something suitable to wile away the hours – deck quoits, marlin fishing, bird shooting, Dane drowning, that sort of thing. Charles Waterton retired to his cabin and wrote an essay on preserving birds for cabinets of natural history, eventually published as an appendix to the first edition of *Wanderings in South America.* One passage in the essay gives a clue to the origins of the Waterton Method:

> Some years ago I did a bird upon this plan in Demerara. It remained there two years. It was then conveyed to England, where it stayed five months, and returned to Demerara. After being four years more there it was conveyed back again through the West Indies to England, where it has now been near five years, unfaded and unchanged.[1]

If the essay was written on board ship, in 1817, as he claims, that puts the first use of corrosive sublimate at about 1804 or 1805, but the oldest dated specimen in his collection – a scarlet ibis – is labelled 1810. The colouring is still bright and clear but, unlike his other specimens, the neck is now almost completely bare of feathers, suggesting that perhaps it was preserved before he found that corrosive sublimate was a good safeguard against insect attack. One possibility is that the passage quoted was added to the essay just before publication of the *Wanderings* in 1825, which would date the first use of the Waterton Method at somewhere between 1810 and 1813. Another possibility is that he seldom knew what day it was, let alone what year.

Until that time, and well into the second half of the nineteenth century, the most usual skin preservative in taxidermy had been either arsenical soap or, less often, camphor. The body was usually stuffed with straw and kept rigid with an artificial skeleton of wire. Very early specimens were stuffed with tobacco and spices and sometimes kept rigid with their own skeletons. The oldest stuffed bird in Britain – a grey parrot in Westminster Abbey – still has its own skeleton and brain.

Charles Waterton had no use for bird skeletons, still less for bird brains; his specimens were hollow, heads empty. Straw? Tobacco? Spices? Stuff and nonsense! He soaked his skins in a solution of bichloride of mercury and alcohol, which soon made them firm enough to support themselves. It also made them anti-putrescent and to a large extent repellent to insects – carnivorous beetles, clothes moths etc. He used one teaspoonful of bichloride (corrosive sublimate) to a bottleful of alcohol, and soaked

Figure 12. 'View of Walton Hall in the distance; of the ancient ruins, and of the cast-iron bridge.' From Richard Hobson, Charles Waterton, His Home, Habits and Handiwork. *(Wakefield MDC. Museums and Arts.)*

each skin for anything from three to nine hours, according to its thickness.[2] Then it was dried and moulded into shape.

It was the moulding process that involved all the skill; plus a good deal of time and patience. Towards the end of his life, he spent more than seven weeks on one specimen – a peacock which was sent to him by his Quaker friend, T. Allis.[3] The time involved, and the cost of bichloride of mercury, ensured that his Method would never be universally adopted, though it was gradually accepted as the best method of preservation yet devised. Few collections as old as Charles Waterton's have survived in anything like such good condition.

Various methods of tanning and curing skins are employed nowadays and the most common preservative in use appears to be industrial borax. False bodies are often made from balsa wood, or even from glass fibre. But the process which Squire Waterton followed was far more crafty. He was the prophet and progenitor of modern taxidermy, and its supreme nineteenth century practitioner. On 3 June, 1982, to commemorate the 200th anniversary of his birth, the Guild of Taxidermists held a day-long symposium at Walton Hall, complete with readings from his works and a full-scale bird mount competition. All tickets sold well in advance.

He spent much of his time with his specimens after he returned home from Demerara. With over 200 of them to finish and mount, he obviously had plenty to occupy his nimble fingers, notorious in delicate manipulations. But there was also the estate to attend to, with its attendant obligations to wild birds, and since the rats had obviously had a good time whilst he'd been away, special efforts were necessary to send them back to Stymphalus.

So it was a busy but idyllic summer at Walton Hall, in 1817 – the most glorious for years. Think of him pottering about in the sunshine, shooting fish with the Indian bow and arrow he'd brought back from Guiana (the lake was quite clear in 1817), or lying in his hammock between the two old oak trees which one day would frame his last resting place. In the soft evening sunlight, he'd row home across the lake and present Mrs Bennett with a full-grown, home-grown pike, impaled on an Indian coucourite arrow, to be cooked a la forêt for Friday's dinner. 'Very superior in flavour', said a friend.

All this was very fine, but the prospect of spending the winter in Yorkshire held no attraction at all for Squire Waterton. There was eternal summer in his soul and a penny or two yet in his pocket. Yorkshire summers fade, and by the middle of October, he was setting off south again, this time on a pilgrimage to Catholic form and aristocratic fashion: the Grand Tour. In his baggage was a letter to His Holiness, Pope Pius VII, which he had every intention of delivering personally.

The account of Charles Waterton's first journey to Rome is frustratingly sparse. We do know that his travelling companion was an old friend named Alexander, a Royal Navy captain, but details of the journey can only be supposed. We know, from one of the Essays, that he was in Florence some time in 1817. (It was in the Cascine – 'a kind of Hyde Park for the inhabitants of Florence' – that he learned the value of ivy 'for the protection of the feathered race'.)[4] We can also surmise that it was possibly during this first journey that he tested the powers of curare on a donkey which he bought from a little boy 'on the outskirts of an Italian city'.[5] The donkey died, the boy was upset, and the Squire took him to market to buy another. He told the story many years later, in expiation of his own conscience, but since he made several trips to Italy in his life, that episode could have occurred during any one of them. Nothing is positively recorded of this first journey in 1817.

'The greater part of travellers tell nothing because their method of travelling supplies them with nothing to be told', said Samuel Johnson,[6] who shammed a sense of inferiority because he never went on the Grand Tour. Johnson enjoined travellers not to disturb others with accounts of their travels, and by the 19th century, most Grand Tourists, including Charles Waterton, apparently took his advice literally, more's the pity.

Apart from the letter to his Holiness, Charles Waterton took with him to Rome, as a kind of visiting card, the famous red cotinga which he'd prepared a few months earlier, 'in the wilds of Guiana'. Who better to appreciate such a triumph of fine statuary and artistic skill than the first and foremost of neo-classical sculptors, Antonio Canova? He showed the bird to Canova, in the great man's studio, where a third party named Wenceslaus Peter 'appeared as it were enraptured.'[7] Peter included the cotinga in a picture he was painting of the Creation, which now hangs in the Vatican.

Rome, at that time, was bursting at the seams with foreign tourists; fine old English gentlemen bumped into old friends from pump-room or prep-school, usually in the Piazza di Spagna, a little home from home for the British school tie network. It was probably here that Charles Waterton 'fell in with' an old school friend of his named Edwin Jones, a captain in the First Royal Lancashire Militia. The two boys had been great tree-climbing friends at Tudhoe, and the mutual recollection inspired a reversion to type; they decided to climb the dome of St Peter's. Anyone can climb the 537 steps inside the dome. Captain Jones and Squire Waterton had something else in mind – a spiffing prep-school jape. They scaled the facade of the dome itself, then ascended the cross, and then climbed 13 feet higher still, to leave their gloves as tokens of conquest on the tip of the lightning conductor. For encore, they later went to the Castello di Sant' Angelo, at the other end of the Via delle Conciliazione, and clambered up onto the head of the guardian angel, 'where we stood on one leg'[8] – one leg each, that is – safe where men fall.

That was enough for the time being, but more than enough was necessary nonetheless. An article of faith in the Catholic church-climber's manual is the recognised principle that gloves on lightning conductors are not a good idea; they decrease the gathering properties of copper wire. When his Holiness the Pope learned that two pairs of Englishmen's gloves were putting the whole basilica of St. Peter at the mercy of a thunderbolt from heaven, his trust in God's mercy deserted him. The gloves had to be removed. Trouble was, no-one had the necessary whatever it was to remove them. No-one, that is, except the two overgrown English schoolboys who put them there. Whether or not they felt like chastised schoolboys as they re-ascended the dome next day, the fact is that a large crowd gathered in St. Peter's Square to watch them, and at least one of the two steeplejacks must have enjoyed that. So must the crowd.

But it was a different kind of audience Charles Waterton was really after. He had a letter to deliver. He failed to get the chance to deliver it personally and it has been claimed that the pope's refusal to see him was on account of his refusal to wear black; he wanted to appear in his royal blue Stonyhurst uniform. But the failure almost certainly had more to do with the contents of the letter itself.

Just three years earlier, Pius VII had solemnly restored the Society of Jesus to favour, after more than 200 years of suppression; but anti-Jesuit feeling was still strong and many still wanted complete papal renunciation. Waterton's letter was a defence of the Jesuit order. His mission was serious and the intention to wear what to him was the Jesuit uniform was not, in the circumstances, entirely illogical. It was not a lot more vain or frivolous than an old campaigner's medals at a military parade. In fact, years later, with the connivance of a group of naval men, including, apparently, Captain Marryat, Charles Waterton did attend an audience from a succeeding pope (Gregory XVI), by adding epaulettes to his blue Stonyhurst coat and posing as a naval captain in the Demerara Militia. So honour was eventually satisfied and papal etiquette observed.

Pope Pius VII may have received the Squire's letter, but not from his own hand. Perhaps the Cardinal Secretary had to read it before he could deign to have it personally presented to the pope. Its contents were slightly subversive. After a bit of swank about his ancestry – 'My grandmother was the last of Sir Thomas More's family in a direct line' – it proceeds to complain about moral degradation in South America within the church itself, a direct attack on Vatican policy.

> I was told that the destruction of the Jesuits was the cause of this sad and general falling off. The mandate for their dissolution, like an earthquake, had shaken education to its very centre. . . . If the shepherds forget themselves, who can blame the flock for straying. . . . I had more than once to wait half an hour for mass till the priest had finished his game at backgammon.[9]

So the refusal of an audience, in 1817, probably had nothing whatever to do with climbing escapades, nor with old school uniforms.

In the following year, according to the Stonyhurst Centenary Record, a deputation of old Stonyhurst boys, 'including Charles Waterton,' sought an audience with the Vatican Prefect of Propoganda, 'to urge the Society's cause.'[10] It's not clear whether the hearing was granted but if it was, it must have occurred very early in the year because by 19 February 1818 the Squire was back in England, where he stayed for the next two years. Which means, of course, that he must have crossed the Alps some time in January – the best of times and the worst of times, like the rainy season in Guyana.

The chosen route across the Alps was through the pass of Mont Cenis, about 50 kilometres north of Turin. A tunnel about 40 kilometres west of the pass now makes the crossing relatively safe in all weathers, but in the days of horse-drawn travel, accidents were common and sometimes fatal.

As they crossed Mont Cenis, at 10 o'clock of a January night, the accidental traveller took it into his head to inspect the baggage. He was tired of sitting still. He climbed onto the wheel arch, to haul himself up onto the roof, in persuance of a lifelong argument with inevitable consequences. 'As bad luck would have it,' consequences inevitably occurred: his knee went through the coach window and two pieces of glass, an inch long, got embedded in his leg, just above the knee cap. He put his thumb on the wound, to stop the bleeding, whilst the horses were reined in and Captain Alexander fetched a carriage lamp, to inspect the damage.

> On seeing the blood flow in a continual stream, and not by jerks, I knew that the artery was safe. Having succeeded in getting out the two pieces of glass with my finger and thumb, I bound the wound up with my cravat. Then, cutting off my coat pocket, I gave it to the captain, and directed him to get it filled with poultice, in a house where we saw a light at a distance.[11]

The wound became septic and by the time they reached Paris, after two delays on account of high fever, it was 'in a deplorable state.' In Paris he consulted a medical friend he'd known in British Guiana. This was Dr John Marshall, the man who gave him the prescription for jalap and calomel tablets which came to be known as 'Squire Waterton's Pills'. Dr Marshall treated the wound 'with exquisite skill' but the patient was impatient to get home in the travel-worn condition expected of him; he left Paris before he was fully fit.

In London, further treatment was necessary and he therefore sought the advice of a Jesuit priest, Father Scott. Father Scott inspected the wound and called in a celebrated surgeon named Carpue. To Mr Carpue's 'consummate knowledge and incessant attention' the Squire melodramatically attributed the 'preservation of the limb, and probably of life too.' The knee remained stiff for almost two years afterwards but eventually healed completely, thanks to constant exercise and steadfast refusal to use a walking stick.[12]

Those two years were spent at home in Yorkshire, attending to affairs of the world's first nature reserve and furthering the cause of nineteenth century civilization in the approved ornithological fashion.

His plans for Walton Park had obviously been greatly advanced by his visit to the Cascine, originally a pheasant preserve of the grand dukes of Tuscany. Hunting had been forbidden in the Cascine for centuries, so that it had become a sort of unofficial wild-life sanctuary, in a wild-life purgatory called Italy, long before Charles Waterton was born. But it was the ivy which most impressed him. It covered the trees like tropical bush ropes and covered the walls and carpeted the ground and was greatly appreciated by small birds as a safe roosting place.

> As I gazed on its astonishing luxuriance, I could not help entertaining a high opinion of the person, be he alive or dead, through whose care and foresight such an effectual protection had been afforded to the wild birds of heaven, in the very midst of the 'busy haunts of men'.[13]

When he got back home to Yorkshire with a bad leg, he 'began the cultivation of ivy with an unsparing hand.' It still flourishes on the ruined watergate tower, and on some of the trees in the grotto, almost 200 years later.

He also began negotiations to increase the size of his reserve. In 1819, he paid £4,600 to the Earl of Westmoreland, for part of the neighbouring estate of Haw Park,[14] apparently untroubled by the kind of problems which Lady Bracknell advised against not long afterwards: 'Land gives one a position but prevents one from keeping it up.' Charles Waterton was not greatly interested in a social position. He used land for other purposes. Throughout his life, he sought to increase the amount of land he could manage, but never spent anything like an equivalent amount on the Hall itself. Like Wordsworth at Grasmere, his living room was the open air.

The most valuable ornaments inside the Hall were certainly his specimens. In pre-photographic days, huntin, shootin and stuffin were scientifically essential – or thought to be. Museums spread like moth balls. Almost all official natural history collections were either very poor or completely pathetic, but that didn't stop people going to see them. Incipient professors went to any length – 800 miles in one case. That's the distance from Aberdeen to London, and that's how far William MacGillivray walked, in the autumn of 1819, to see the moth-eaten corpses in the British Museum at South Kensington. He later became Professor of Natural History at Aberdeen University and befriended Charles Waterton's arch enemy, John James Audubon.

The Squire also went to London in 1819, and obviously took the trouble to resent a visit to the British Museum. But the more salient purpose of his trip was to discuss plans with Sir Joseph Banks for his forthcoming third expedition to the South American rain forest. Banks now spent most of his time at his home in Spring Grove, Isleworth. He suffered badly from gout and, for 15 years or so, he'd been confined to a wheelchair. Now he was on his death-bed.

The Squire reserved his best bedside manner for the occasion. He was shown into the bedroom at Spring Grove, where he renewed the acquaintance so auspiciously begun 15 years earlier; then he exchanged with the dying man a few well-chosen words on the stuffing of quadrupeds – ideal subject for a death-bed.

> We talked much of the present mode adopted by all museums. . . . and at last concluded that the lips and nose ought to be cut off, and replaced with wax.[15]

Sir Joseph's body was not donated to science.

A few days before Christmas, the Squire wrote a polite little letter to his dying friend and thanked him for his kindness at their recent meeting. He would soon be leaving for British Guiana and hoped to spend the next two years collecting birds, insects and bats. He also hoped to write an account of his observations there.[16] Sir Joseph clung desperately to the last threads of life, waiting for Mr Waterton's scientific observations.

In February 1820, a few weeks after the death of King George III, and a whole four months before the reluctant death of Sir Joseph Banks himself, Charles Waterton set sail from the Clyde, on board the merchant ship *Glenbervie*, for the third and most infamous of his four great Wanderings in South America. It was two years, almost to the day, after returning home from his Grand Tour in Europe, and almost three years since he returned from his second Wandering, in quest of the feathered tribe.

This time, he was destined to shock the prurient carriers of scientific utilitarianism, and captivate his gentle readers, beyond all precedent. But most of them called him a liar.

Chapter 9
1820-1821
Presence of Mind and Vigorous Exertions

Before setting sail from the Clyde, he stayed with his friend Charles Edmonstone, at Cardross Park, near Dumbarton; there, it was arranged that he should use Warrows Place – the Edmonstone's plantation house at Mibiri Creek – as 'headquarters for natural history'. After leaving Demerara for the last time, a few years earlier, Edmonstone had bought back the old family estate at Cardross, which had once been part of the dowry of one of his royal ancestors, Princess Isabella of Scotland, daughter of King Robert II, first of the Stuarts.[1] Warrows Place was left uninhabited.

Charles Edmonstone's second daughter, Anne Mary – half Arowak, half Scottish – was now 8 years old. She and her sisters must surely have been on the quayside at Port Glasgow, or at least on the opposite shore at Cardross, as the *Glenbervie* crept towards the open sea with the man she was eventually to marry.

The first two weeks of the voyage were in rough seas and wintry winds. The Scottish West-Indiaman was driven off course, to the north-west of Ireland. But the weather at last changed and the Squire of Walton Hall enjoyed a pleasant trip across the western sea, to be greeted in Georgetown by another of life's little trials – an epidemic of yellow fever. Some of the older inhabitants had already died and 'the mortal remains of many a newcomer were daily passing down the streets, in slow and mute procession to their last resting place.' The slave population was also seriously reduced.[2]

He spent a few days in Georgetown, stocking up with provisions, then left for Mibiri Creek, sure in the knowledge that iron constitution and alternative medicine would see him through, as they always had. At Hobabu Creek, about half way between Georgetown and Mibiri, he stayed for awhile at the 'hospitable house' of Archibald Edmonstone, a 25-year-old nephew of Charles Edmonstone,[3] and soon enjoyed one of the epic encounters with snakes which he later quite genuinely remembered that he'd previously forgotten to mention.

As he sat down to breakfast on the first morning at Hobabu Creek, one of Mr Edmonstone's black 'employees' appeared, with news that a large snake had just caught one of the tame muscovy ducks in the swamp. (Muscovy ducks, like English monkeys, are native to South America.) The intrepid Squire sprang into action. Clutching his lance – a home-made snake pole made with an old bayonet and a long stick – he ran down to the swamp and presently found the snake – obviously an anaconda – lurking in the mud under some felled trees. He walked over the horizontal trunks of the trees, occasionally catching a glimpse of the snake, until, at last, an opportunity arose to use his lance. He missed his aim and the snake received only a slight, superficial wound.

> I had no sooner done this, than he instantly sprang at my left buttock, seized the Russia sheeting trousers with his teeth, and coiled his tail around my right arm. All this was the work of a moment. Thus accoutred, I made my way out of the swamp, while the serpent kept his hold on my arm and trousers with the tenacity of a bulldog.[4]

Somehow the grip was loosened (not easy; the fangs point backwards) and the snake, you can bet, was skinned for science or utility once the breakfast table had been cleared. Some time later, with a new pair of trousers or a snake-skin patch on the old ones, the Squire resumed his journey to Mibiri Creek, another 20 miles or so up-river. He apparently took with him a 'fine mulatto' named James, who had belonged to the deceased Robert Edmonstone, Archibald's older brother. James and Charles were bosom pals for the next eleven months.

Warrows Place was almost completely derelict. It must have looked like a set for a Hammer film. The roof was collapsing and the room where governors and generals had formerly been entertained, and Dutch fugitives concealed, was now taken over by a colony of 'vampires'. All around, the tropical rain forest was closing in.

One of Charles Edmonstone's black slaves at Warrows Place had been a man named John, who had learned 'the proper way to do birds', i.e. the Waterton way, from Charles Waterton himself. John had gone to Scotland with his master, and later became a servant in the household of Dr Duncan, one of a succession of medical lecturers at Edinburgh University who helped deflect Charles Darwin from the study of medicine. When Darwin heard that this West Indian servant of Dr Duncan was a protégé of Charles Waterton, he employed him, at a guinea an hour, to give him lessons. He thereby learned the Waterton Method via a negro slave taught by the Squire himself at Mibiri Creek. Waterton says that 'John had poor abilities and it required much time and patience to drive anything into him'[5] but Darwin found him 'a very pleasant and intelligent man' who stuffed birds 'excellently'.[6]

The Squire was obviously much more impressed with James, the mulatto he'd adopted at Hobabu Creek. This man, he says, was capable of learning anything, even taxidermy.

He was also impressed with another coloured man – Backer – who lived with his wife and children in a little hut nearby. Backer performed many a little kindness for the Squire, notably when the Englishman suffered one of his little accidents of bare-footed travel:

He was chasing a red-headed woodpecker through the forest – all woodpeckers with red heads were red-headed woodpeckers, whether in South America or in Europe – when he trod on a tree stump which pierced the arch of his foot and made 'a deep and lacerated wound there.' Back at H.Q., he washed away the blood, probed the wound for foreign bodies (he enjoyed probing wounds) then began a three-week regime of treatment with very large poultices, renewed three times a day.

> Luckily, Backer had a cow or two upon the hill; now, as heat and moisture are the two principal virtues of a poultice, nothing could produce those two qualities better than fresh cow dung boiled: had there been no cows there, I could have made out with boiled grass and leaves.[7]

After three weeks in his hammock with the cow dung, he once more 'sallied forth sound and joyful.' No bullshit.

He also endured about three weeks of self-doctoring for another little inconvenience – 'a severe attack of fever' – apparently the yellow fever which was dramatically reducing the population in Georgetown. The treatment reads like a suicide attempt:

Successive attacks of yawning, sleeplessness, head-ache, back-ache, acute thirst and hallucination called for a booster dose of purgative – 'ten grains of calomel and a scruple of jalap.' (The calomel appears to have been the same mercurous chloride he used in taxidermy.) By the following evening, his pulse rate had risen to 130 and it was time to open a vein and make 'a large orifice'. He took away 16 ounces of his life blood and next morning swallowed five more grains of calomel and ten of jalap. 24 hours later, after a fierce bout of sweating, he took a large dose of castor oil, freshly made from the seeds of a castor oil tree which grew by the door, and by this time, naturally, all symptoms of fever had disappeared – diseases run a rapid course in tropical America, as many Europeans have discovered. But it was not until the following morning – after the fever had gone – that he began to take the only safe medicine for it which was then available – Peruvian bark. (It was at about this time that the operative alkaloid in Peruvian bark was isolated and it came to be known as quinine.)

He later told more about the Waterton cure in an unpublished essay:

> Yellow fever is easily brought on by too much exercise. . . . When you begin to yawn, and are restless, and without appetite, and have a slight pain in the loins, – then is the time to take a strong purge. Ten grains of calomel, well mixed up with twenty grains of jalap, are not too much. . . . It may be made into four pills. . . . After taking them, lie flat on your back for about an hour and this position will prevent the return of the pills upwards. If the fever increases, blood may be taken away on the first day, but this requires great judgement and caution. . . . Vomiting and hickup are bad symptoms in yellow fever. When they come on, abstain from taking anything whatever, except cold water. Lie down on the ground without even a pillow under your head; and direct your attendant to give you a tablespoonful of cold water every five minutes, you taking care not even to raise your head while receiving it. This will subdue the vomiting and hickup, and fortify the stomach.[8]

Then consult a doctor.

The fever took hold of the Squire in the month of June, 1820, just as Sir Joseph Banks was finally giving up the ghost back in England. It was lucky that he declined to accompany Sir Joseph into the next world because, as he lay in his hammock one night, 'harping on the string on which hung all my solicitude', as one does, he hit upon a solution to the problem which was apparently an essential pre-occupation at death's door – the best method of stuffing quadrupeds.

More of this later.

He spent eleven months in the Demerara forests, based at Mibiri Creek, and became intimately acquainted with its 'animated nature'. His favourite subject was the sloth.

Chief source of reference on the sloth, in those days, was Comte de Buffon's *Histoire Naturelle* – 36 volumes of fact and fable – mostly fable – widely assumed at the time to be the last word. Buffon pronounced the sloth 'a bungled composition of nature', a freakish, pitiful creature, in perpetual pain, who spent many hours climbing a tree then rolled himself up like a hedgehog, when he'd eaten all the leaves, and allowed himself to drop to the ground like a ball. 'Why should not some animals be created for misery,' said Buffon, 'since, in the human species, the lives of most individuals are devoted to pain from the first moment of their existence.'[9]

But this curious idea seemed to Charles Waterton like an insult to Omnipotence. He couldn't believe that any animal was created for a life of misery. Where the sloth was concerned, the misconception arose from not studying him in his natural element. Put him on the ground and of course he's miserable; he lives in trees.

> The sloth is as much at a loss to proceed on his journey upon a smooth and level floor as a man would be who had to walk a mile in stilts upon a line of feather beds.[10]

Contemporary illustrations always showed the sloth in an upright position and it never seemed to occur to anyone before Charles Waterton, that there might be a perfectly good use for apparently ungainly limbs with long claws and no paws.

> The sloth is doomed to spend his whole life in the trees; and, what is more extraordinary, not upon the branches, like the squirrel and the monkey, but under them. He moves suspended from the branch, he rests suspended from it, and he sleeps suspended from it.[11]

In short, he passes his whole life in suspense – 'like a young clergyman distantly related to a bishop', said Sidney Smith,[12] who was in a position to know.

All this suspense was a great zoological novelty in the 1820s. No-one had dreamt that any mammal other than bats could possibly live upside down, and many people refused to believe it when Charles Waterton told them. The first appearance of a sloth in Europe, in the 1840s, constituted a small zoological triumph for him.

On the banks of the Essequibo one day, he found a large, two-toed sloth, grounded and apparently helpless. Since he seemed disinclined to believe that sloths could swim, he could not account for the animal's presence there. (The first white man to report a sloth swimming was W.H. Bates,[13] who was not yet born.) The nearest trees were 20 yards away and the sloth had no chance of reaching them in time to escape from the sentimental Squire. In such a situation, it was vulnerable to attack, not only by friendly humans but also by predators such as jaguars and harpy eagles. Squire Waterton to the rescue, Uncle Toby fashion:

> 'Come, poor fellow,' said I to him, 'if thou hast got into a hobble today, thou shalt not suffer for it: I'll take no advantage of thee in misfortune; the forest is large enough both for thee and me to rove in. . . .' On saying this, I took up a long stick which was lying there, held it for him to hook on, and then conveyed him to a high and stately Mora. He ascended with wonderful rapidity, and in about a minute he was almost at the top of the tree...I was going to add that I never saw a Sloth take to his heels in such earnest; but the expression will not do, for the Sloth has no heels.[14]

The sloth which Bates found, in almost identical circumstances, in Brazil (in 1854) – swimming across a river 300 yards wide – was caught, cooked and eaten, but Charles Waterton claimed (in 1833) that during his own eleven months in the forests of Guiana (in 1820), he 'never once tasted flesh or fish all that time.'[15] True to form, the claim contradicted his own word in the *Wanderings* (published in 1825) in which, according to his own account of Indian hospitality during this third journey, he enjoyed the kind of cordon bleu experience which could slip only the most wilfully selective of memories

– boiled ant-eater and red monkey.

> The monkey was very good indeed, but the ant-eater had been kept beyond its time; it stunk like our venison does in England, and so, after tasting it, I preferred dining entirely on monkey.[16]

As a dainty little desert to a meal like that, you can't do better than a nice fricassee of roast wasp grubs – maribuntas, if you can get them – with just a pinch of Cayenne pepper to taste. You can feed the adult wasps to a friendly toad, as Charles Waterton did.[17] Though his own stomach was 'offended' by the grubs, he readily concedes that it was more on account of the idea than the taste, which was entirely amenable to the idea.

But he drew the line at sloths. Who could harm a sloth? Rather feed it than eat it. For several months, he kept a three-toed sloth in his room at Warrows Place, as a pet, and so learned the fundamental truth about Anglican preferment.

> His favourite abode was the back of a chair; and after getting all his legs in a line upon the topmost part of it, he would hang there for hours together, and often, with a low and inward cry, would seem to invite me to take notice of him.[18]

The invitation, naturally, was accepted.

He also took particular notice of ant-eaters and armadillos, members of the same taxonomic order as the sloths – Edentata. (Though he cared nothing for that!)

Again, he seems to have been the first to match specific features of anatomy to observed ways of life. Pictures and stuffed specimens of the ant-eater ('ant-bear', as he called it) had shown it with its long, non-retractile claws pointing forward; the Squire demonstrated that it walks on its knuckles and tucks its claws backwards, out of the way, preventing them getting blunted by contact with the ground. The claws, in good condition, are an extremely effective means of self-defence. An animal caught in an ant-bear hug is doomed. They're also useful in dismantling termitaries, or – in the smaller, arboreal species – as grappling irons for clinging, like the sloth, to branches.

The long, worm-like tongue of the ant-eater is attached to its breast-bone. Waterton discovered two glands at the root of the tongue which secrete a sticky saliva and convert it into a kind of extendable fly-paper. He found similar glands in the lower jaws of woodpeckers. Aspiring taxidermists (all aspired, few attained) were solemnly instructed by Charles Waterton that if any of this secretion got onto fur or feather, they should allow it to dry; then it could be removed without leaving a stain. There are no stains on the woodpeckers in Wakefield Museum, nor on the 'Great Ant Bear From the Wilds of Guiana, 1820.'

The ant-bear, however, does bear a few marks of ill-treatment – by dogs. Luckily for them, it was already dead.

> It was shot in the forest bordering on Camouni Creek in 1820, whilst I was in those parts; and it was sent to me by the Dutch gentleman who had killed it. . . . After he shot the Ant Bear, he allowed his dogs to mouth it for more than an hour. . . . The scars may be seen distinctly, on inspecting the animal, from its eyes to the end of its nose.[19]

The skin of the ant-eater might be tough, but the flesh beneath it is relatively tender – tender as veal, said Waterton; 'something like goose', said Bates (who stewed it).[20] But the Indians won't touch it – ant-bears are unwholesome to the Indians.[21]

The armadillo, by contrast, was considered good food by Indians and Africans alike, according to the Squire. The native Indians dug them out of their burrows, in the White Sands district of Demerara, and showed the white man how to do likewise. He sometimes spent three parts of the day digging out one armadillo, and sank half a dozen pits, seven feet deep, in the process. The poor animal must have been terrified. Many Indians prize the tail of an armadillo as a token of good hunting.[22]

Those, then, are the Edentates. 'We will now take a view of the Vampire.' An affectionate view, naturally.

One of the most hurtful disappointments of Squire Waterton's sojourn in the South American rain forests, was the cruel and heartless rejection of his pure blue blood by this obviously much-beloved little 'nocturnal surgeon'.

> Many a night have I slept with my foot out of the hammock to tempt this winged surgeon, expecting that he would be there, but it was all in vain; the Vampire never sucked me, and I could never account for his not doing so, for we were inhabitants of the same loft for months together.[23]

Vampires, in fact, do not suck blood, as any Transylvanian tourist can confirm; they puncture the skin by biting it and then lap the blood with a tongue impregnated with an anti-clotting enzyme. One writer suggests that Charles Waterton's skin had been hardened by going bare-foot in the forest and was therefore too tough for the vampires,[24] but the bats at Mibiri Creek, as described in the Wanderings, were not true vampires at all; they were one of the larger kinds of American leaf-nosed bats, almost certainly *Phyllostomus histatus*, an omnivorous species which sometimes eats small mammals and birds but does not drink blood.

But Charles Waterton thought they were vampires; he was deeply wounded by the cold shoulder they gave to his foot. The snub was all the worse because what he took to be a smaller species of vampire – this time the real thing – seemed to take a special liking to the feet of some of his friends, Indian and European.

During one of his previous visits to Guiana, he took a trip up the river Pomeroon, in the north-west of the country, with a man named Tarbet, a colonial tenderfoot from Scotland.

> One morning, I heard this gentleman muttering in his hammock. . . . 'What is the matter, Sir?' said I, softly; 'is anything amiss?' 'What's the matter?' answered he, surlily; 'Why, the Vampires have been sucking me to death.'[25]

About ten to twelve ounces had been taken from his big toe, which was still leaking. The innocent observation that a European surgeon would have charged for cupping the same amount of blood, was not appreciated. The Scotsman looked up into the Yorkshireman's face and said not a word. 'I saw he was of opinion that I had better have spared this piece of ill-timed levity.'

Mr Tarbet was, in fact, very privileged – vampires seldom take human blood. These were common vampires, *Desmodus rotundus*, which generally prefer cattle or horse blood, but do occasionally choose humans if nothing better is available. The other two species of true vampire almost always choose birds to attack. They seldom, if ever, attack Scottish gentlemen or English squires.

Some creatures, however, do. The same Scottish gentleman was attacked on the following night by an army of coushie ants which were marching over the seat of the outside lavatory.[26] Charles Waterton had no trouble with coushie ants, but he did have direct, personal experience of two other insect torments – the bête rouge and the chegoe. The bête rouge is a minute species of tick which sometimes infests birds and, in large numbers, can easily be mistaken for a bright red pigment on the plumage. It also parasitises humans and gave the Squire a good deal of trouble before he discovered his own, non-proprietary method of getting rid of them – Demerara rum. (External use only.)

The bête rouge was a nuisance but his entymological bête noir was undoubtedly the chegoe, or jigger flea, *Tunga penetrans*. As with many another species of flea, the female chegoe waits on the ground for a warm-blooded host to pass by, then jumps aboard and grows fat on the victim's blood, like a motorway hitch-hiker. The usual site on a human is between the toes. The insect's abdomen swells with eggs until it reaches the size of a pea, and, if allowed to remain, causes acute irritation and, sometimes, fever. To the bare-footed Squire, it was a source of 'perpetual disquietude', though not sufficient to persuade him to wear the orthodox tropical dress of stout boots and puttees.

Each evening, he examined his feet for chegoes, and grubbed out the egg-distended abdomens with a sharp needle or penknife. Then he'd pour turpentine into the cavities, to finish off any that remained. The negro grannies who evicted chegoes from the feet of the children each evening used pepper and lime juice.

On one occasion, to satisfy his curiosity, the Squire allowed a chegoe to bore into the back of his hand. He watched it for half an hour, until it was completely out of sight, then turned it loose on the world, with a fond farewell and a merry quip. Avast, there! my good little fellow. . . we must part company without loss of time. I cannot afford to keep you and a numerous family for nothing; you would soon eat me out of house and home.[27] Saying which, he applied the point of his penknife to the point of entry, and served notice to quit. As with fleas, so with snakes:

> Gentlemen of rainbow colours, be not alarmed at my intrusion. I am not come hither to attempt your lives, nor to offer wanton molestation.[28]

Unless, that is, the gentleman was needed for dissection, in which case, any of a number of improvisations might be employed to capture him.

With the venomous bushmaster snake, or the fer de lance, the method was basic – just slide the hand slowly up the serpent's back and grasp it by the neck. What could be simpler? First, of course, you have to be sure you know which is snake and which is leaf litter – they tend to look alike.

The real difficulty arises with the large constrictor species, such as the one at Hobabu Creek. Grasp one of those by the neck (assuming your hand is big enough) and he'll probably return the compliment. But where there's a Squire, there's a way:

One Sunday afternoon, as he sat on the steps of Warrows Place, reading his pocket edition of Horace, a black man came to tell him of a large snake which his dog had discovered some distance away. The white man was delighted. Off they went, Squire flourishing his lance, negro a cutlass. Another negro, similarly armed, joined them en route. They were promised four dollars apiece for their pains.

The snake turned out to be a boa constrictor, about 14 feet long. His den was under a fallen tree, completely overgrown with 'a species of woodbine'. Charles Waterton decided to take him alive, for dissection next day – putrefaction might set in during the night if he was killed straight away. The negroes, sensibly, were terrified. They begged to be allowed to fetch a gun, but the boss wouldn't hear of it.

He took out his knife and cautiously began to cut away the bines. At last, after a quarter of an hour's work, he could see the snake's head. The Squire now rose and retreated about 20 yards, to issue battle instructions to his men. Preservation of life was clearly unimportant; what mattered was to preserve the snake's skin. In case either of the men should be tempted to defend himself against the snake, and thereby ruin the skin, the cutlasses were confiscated. Then the commander led his troops into battle. His heart, he confesses, beat quicker than usual.

With the lance held perpendicularly in front of him, the point about a foot above the ground, he advanced, slowly, towards the den which now, of course, was exposed to view. The snake had not moved. It was coiled like a spring, its head flat on the ground. Suddenly, the Squire struck. He stabbed the snake with his lance, just behind the neck, and pinned him to the ground. One of the negroes took hold of the lance, keeping it in place, whilst the Squire dived head first into the den and seized the snake's tail, 'before he could do any mischief.'

A violent struggle ensued, much like the fight with the poacher in Walton Park. Rotten sticks flew on all sides as the two wrestlers strove for superiority. The Squire was not heavy enough to keep the snake down, so he called for the other negro to throw himself on top of him. Between them, they at last managed to overpower the snake and subdue the coils which, otherwise, would certainly have squeezed the life out of one of them at least.

All that remained was to secure the jaws. Once a constrictor gets hold of something in its jaws, there's no way out again; the backward-pointing teeth grip like a gin-trap which not even the snake itself can release. Some means of tying up its mouth was vital.

But what could they use? The woodbines obviously weren't strong enough and none of the men, it seems, had any suitable twine. There was just one solution.

As at Hobabu Creek, dispensable unmentionables proved indispensable. The Squire unloosed his braces and used them for a purpose for which braces were always clearly intended – as a muzzle for a South American boa constrictor. The snake was gagged, its teeth braced, then, partly incapacitated, it was persuaded to entwine itself around the shaft of the lance and was carried back to Warrows Place by the three men, white man, of course, at the helm. In his flannel underpants – a great preserver of health – he took charge of the snake's head (and his own braces) whilst the other two men supported belly and tail, like bridesmaids at a wedding. Ten times, on the way, they were compelled to stop for a rest. The snake fought fiercely to escape but all in vain.

At Warrows Place, it was put into a large, very strong bag and left until morning. The Squire's hammock was in the loft, just above the snake, and the intervening floor was half rotten. 'Had Medusa been my wife,' he says, 'there could not have been more continued and disagreeable hissing in the bed chamber that night.' The cruelty didn't seem to occur to him. It rarely did to 19th-century naturalists.

Next morning, with the help of a few more negroes, the snake was let out of the bag and the Squire cut its throat. 'He bled like an ox.' Several days earlier, an old negro who went by the name of Daddy Quashi, had been sent on an expedition into Georgetown, to collect provisions. He returned just in time to help with the skinning of the snake. By six o'clock that evening, the dissection was complete.

This Daddy Quashi was to achieve a kind of tenuous immortality through association with Charles Waterton; he became a sort of African Sancho Panza to the English Don Quixote, squire to the Squire. Like Sancho, he was a magnificent coward, the perfect foil and antidote to his master's abject bravery. His strongest aversion of all was snakes.

The following week, his cowardice was put to the test. The Squire was 'following a new species of parakeet' one morning (he never did find it) and when the rain stopped, he left his umbrella under a tree – you don't need one when its not raining. In the afternoon, he and Daddy went to look for the brolly, Squire dressed appropriately in top-hat and braces, just in case. You can't be too sure in the South American rain forest.

Not far from the scene of his previous encounter, he came across another boa, this one no more than ten feet long. The Daddy, meanwhile, was some distance away, still searching for the umbrella. There was not a moment to lose; one knee on the ground, Charles Waterton took hold of the snake's tail and prepared for another scrap. This time, it was Queensbury rules (or an anticipation of them) – the English aristocracy always loved a good fist fight. The snake rose to enquire of the white man what business he had with his tail, and the white man gave him a good clean uppercut to the jaw. (He'd already put his hat over his fist, as a makeshift boxing glove.)

The snake, obligingly, was stunned. As it reeled under the unexpected style of attack – unheard of in the racial memory of South American boa constrictors – the Squire grabbed it with both hands around the neck and allowed it to coil itself around his body. 'He pressed me hard, but not alarmingly so.'

The Daddy, by now, had found the umbrella. Pleased with himself, he looked around for the Squire and cautiously approached the scene of the fight when he heard the commotion it caused in the undergrowth. But as soon as he saw his massa wrapped in the coils of a ten-foot boa constrictor, artless discretion proved much the better part of reckless valour. He turned on his heels and ran for his life, the Squire in hot pursuit, both shouting at the tops of their voices.

Back at Warrows Place, the snake was unravelled and Charles Waterton delivered one of those playful admonitions of cowardice through which he could congratulate himself for his own faith in God. Daddy was clearly gifted in playful subordination and begged forgiveness for running away; the sight of the snake had turned him sick. But special pleading was not necessary. Charles Waterton was no tyrant; the snake phobias of friends flattered his sense of cool intrepidity.

Rather than dissect the snake, he clearly decided to give this one a Hindu sky burial; he could recover the skeleton when the vultures had picked it clean. Somehow or other, he managed to kill the snake and with the help of Daddy Quashi and another negro, he carried the body into the deepest shade of the forest, 'impervious to the sun's rays'. There, he left it to rot.

For the first two days, not a vulture made its appearance at the spot, though I could see here and there, as usual, a Vulta Aura [turkey vulture] gliding, on apparently immovable pinion, at a moderate height, over the tops of the forest trees. But during the afternoon of the third day, when the carcass of the serpent had got into a state of putrefaction, more than twenty of the common vultures came and perched on the neighbouring trees, and the next morning, a little after six o'clock, I saw a magnificent king of the vultures.[29]

When the king vulture had eaten enough, it flew to the top of a nearby mora tree, to digest the food, and then the smaller, turkey vultures began their own meal. The Squire eventually shot the king vulture.

All of this agrees perfectly with the most recent research into the feeding ecology of New World forest vultures. He spent many hours watching vultures and remained convinced that turkey vultures, at least, never feed on live animals; they much prefer putrid meat.

> If you dissect a Vulture that has just been feeding on carrion, you must expect that your olfactory nerves will be somewhat offended with the rank effluvia from his craw; just as they would be were you to dissect a citizen after the lord mayor's dinner. If, on the contrary, the vulture be empty at the time you commence the operation, there will be no offensive smell, but a strong scent of musk.[30]

Most birds have musk glands in their rumps, particularly active during the breeding season, and several people have confirmed the evidence of Charles Waterton's own nose. The nests of turkey vultures often have a very strong scent of musk.[31] This is probably a means of marking the nests for the birds' own identification or for territorial purposes – another indication of the turkey vulture's highly developed sense of smell. Citizens at the lord mayor's banquet have more complex methods of marking out territories, generally more power-oriented, though deterministically identical.

'Kind Providence has conferred a blessing on hot countries in giving them the Vulture', concludes Charles Waterton, and not only because of the bird's role as dustbin and public health officer. As far as he was concerned, at least one species of vulture was a very elegant fowl indeed:

> The head and neck of the King of the Vultures [Sarcorhamphus papa] are bare of feathers; but the beautiful appearance they exhibit, fades in death. The throat and the back of the neck are of a fine lemon colour; both sides of the neck, from the ears downwards, of a rich scarlet; behind the corrugated part, there is a white spot. The crown of the head is scarlet; betwixt the lower mandible and the eye, and close by the ear, there is a part which has a fine silvery blue appearance. . . . Below the bare part of the neck there is a cinereous ruff. The bag of the stomach, which only appears when distended with food, is of a most delicate white, intersected with blue veins, which appear on it just like the blue veins appear on the arm of a fair-complexioned person.[32]

Edith Sitwell was highly amused.[33]

But if the Squire's notion of physical beauty was controvertible, the ripple of controversy it caused was nothing compared with the great tidal wave of contention over his next scientific entertainment – his famous adventure with the cayman. Victorian England was not amused.

The distribution of the cayman is patchy. Mibiri Creek is a tributary of the Demerara river, which contains none at all. To find one, he would have to visit either the Berbice river, to the east, or the Essequibo, to the west. He chose the Essequibo.

He went into Georgetown for provisions, then, at six o'clock next morning, with

Figure 13. Mounted cayman on cayman mounted. (Wakefield MDC. Museums and Arts.)

Daddy Quashi and an Indian for company, he hoisted sail on his canoe and set off along the muddy coast of Guiana for the Essequibo estuary. By three o'clock in the afternoon, they were up-river, looking for a place to set up camp. (They were still many miles from cayman country.)

During the night, as he lay in his hammock, unable to sleep on account of sunburn, the sharp-eared Indian drew his attention to the footfall of a jaguar: 'Massa, massa, you no hear Tiger?' The animal approached to within 20 yards of camp but Charles Waterton forbad the Indian to shoot: 'It is not every day or night that the traveller is favoured with an undisturbed sight of the Jaguar in his own forests.'

Two days later, they reached the first of the Essequibo's cataracts and 'the last house of people of colour up this river.' Here, he hired a couple more men – one negro, one coloured – who professed to know a thing or two about capturing the cayman.

The whole of the next day was spent in ascending the rapids (which form an effective barrier to the spread of the cayman's range) and in hunting the pacu for food. Then, for two days more, they travelled up-river, through magnificent scenery, until, at last, they came to a place which the two hired men recommended as a good place to get a cayman.

They baited a shark hook with fish and put it on a wooden board which was towed into the river and anchored to the bottom with a large stone. A rope connected hook to shore. As the sun went down, the sounds of a jungle night began and they could hear the explosive grunts of the caymans, anything up to a mile away. But none of them took the bait.

For three more nights, they tried unsuccessfully to catch a cayman and the Squire lost faith in the methods they were using; the coloured man, in consequence, 'began to take airs'. If there was one thing Charles Waterton could not stand, it was people taking airs. He paid him off and sent him home. Then the expedition went in search of more reliable advice on cayman trapping.

This involved another day and a half's journey through the interior, to the nearest Indian settlement. (It was here that he was given his epicurean meal of boiled ant-eater and red monkey.) The shark hook was shown to one of the Indians, who laughed it to scorn and undertook to make them a more suitable trap next day.

Whilst the Indian set to work to make the trap, the rest of the party shot some young caymans, about two feet long, and Daddy Quashi boiled one for his dinner. It was 'very sweet and tender', he said, but the Squire was not hungry. He deduced, nevertheless, that there was no reason why boiled cayman should not taste as good as frog or veal. (William Buckland, another great nineteenth-century gastronome, who once claimed to have eaten his way through the entire animal kingdom, reported that cayman meat tasted like nothing so much as sturgeon or tuna.[34])

By early evening, the Indian had finished his device for catching a full-grown cayman and the Squire examined it enthusiastically. It was very simple – just four pieces of hardwood, a foot long, barbed at both ends and attached to a length of rope, like the feathers of a dart.

The hook was baited with aguti and suspended over the water from a stake driven hard into the sand-bank, a few yards away. The Indian then sounded the caymans' dinner-gong – the empty shell of a land tortoise – which he struck hard with the blunt end of an axe, just to arouse the reptiles' curiosity. Then they all went to bed.

> About half past five in the morning, the Indian stole off silently to take a look at the bait. On arriving at the place, he set up a tremendous shout. We all jumped out of our hammocks, and ran to him. The Indians got there before me, for they had no clothes to put on, and I lost two minutes in looking for my trousers and in slipping into them.
>
> We found a Cayman, ten feet and a half long, fast to the end of the rope.[35]

Repeated mention of unmentionables certainly offended some of Charles Waterton's readers; so did the cruelty of hooking a large reptile as though it were a small fish. But what stuck in the gullet of some of the author's most implacable rivals was the method he now avowedly employed to take the creature alive, so that the skin would not be damaged. His enemies simply didn't believe him.

> We mustered strong, he says. . . . We were, four South American savages, two negroes from Africa, a Creole from Trinidad, and myself, a white man from Yorkshire. In fact, a little tower of Babel group, in dress, no dress, address, and language.[36]

All except the white man wanted to kill the cayman. The Indians were all for shooting arrows into him but the Squire wouldn't hear of it. He'd come 300 miles for an unmarked specimen, and an unmarked specimen was what he intended to get. Daddy Quashi – Sancho to a tee – suggested guns. Another comic chase ensued, this time along the sand-bank, the Daddy firmly convinced – or firmly convinced of the policy of appearing firmly convinced – that if his master caught up with him, he would surely get fed live to the cayman. But in the event, it was Charles Waterton himself who came closest to the jaws of death.

His intention was to pull the creature 'quietly' out of the water and then to secure him, for dissection at leisure. He sent for the canoe, took out the mast and wrapped the sail around the end of it. The men lined up on the rope, like a tug-of-war team (Daddy Quashi at the back, furthest from the cayman) whilst the Squire knelt close to the water's edge, with the mast in his hand, ready to thrust it down the cayman's throat in case of emergency. The monster was brought to the surface, where he fought for freedom like a fish out of water. But the Squire was pitiless.

> I saw enough not to fall in love at first sight. I now told them we would run all risks, and have him on land immediately. . . . By the time the cayman was within two yards of me, I saw he was in a state of fear and perturbation; I instantly dropped the mast, sprung up, and jumped on his back, turning half round as I vaulted, so that I gained my seat with my face in a right position. I immediately seized his forelegs, and, by main force, twisted them on his back; thus they served me for a bridle. . . . The people now dragged us above forty yards on the sand: it was the first and last time I was ever on a Cayman's back. Should it be asked how I managed to keep my seat, I would answer, – I hunted some years with Lord Darlington's fox hounds.[37]

The poor beast was by now well-nigh exhausted. The jockey tied up its jaws and then fastened the feet firmly behind its back, in a permanent half-nelson. So, trussed up and vanquished, the cayman was 'conveyed' to the canoe and thereby back to camp, where the white man at last ended its misery by cutting its throat. After breakfast, he started the dissection.

The story of Charles Waterton and the South American "crocodile" remained a staple of conversation and controversy, in many a Victorian parlour, for many years after publication of the Wanderings. The author was widely assumed to be some sort of Baron von Munchausen, privileged to lie. The tone was set by one of the earliest notices of the book, in the Tory Quarterly Review:

> With every disposition to give full credit to these exploits of our entertaining 'Wanderer', we confess that this last circumstance – this new-fangled bridle, made from a pair of crocodile legs – does somewhat stagger our faith. Indeed we should doubt very much whether Lord Darlington himself, or the boldest squire that follows his hounds, could sit a crocodile, with all the advantage of the hardest bit in his lordship's harness-room.[38]

The view was generally shared and the Squire was generally dismissed as a shameless 'drawer of the long bow'. Others sought to verify his account and scholarly articles began to appear, citing obscure passages from scripture and the classics, and rare old books of field sports, written in Latin, to demonstrate not only the feasibility of riding on the back of a crocodile, but its frequent practice in ancient times.

Pliny, for instance:

> The crocodile is terrible to them that flee from him, but runs away from his pursuers; and these men alone can attack him. They swim after him in the river, and, mounted on his back like horsemen, as he opens his mouth to bite, with his head turned up, they thrust a club across his mouth and, holding the ends of it,

Figure 14. Cruikshank cartoon with the Squire's own caption. (Wakefield MDC. Museums and Arts.)

> one in the right hand and the other in the left, they bring him to shore captive, as with bridles.[39]

Charles Waterton would certainly have read Pliny, and possibly got the idea from him. Or from Herodotus perhaps:

> When they have fixed a piece of pork upon a hook, they throw it into the midst of the river, and on the banks have a living pig, which they beat. The crocodile, hearing him squeak, advances towards the noise, and, having seized the flesh, devours it. They then pull him, and having dragged him on shore, first of all fill his eyes with mud, and, having done this, he is easily dispatched.[40]

But no amount of classical allusion could satisfy Waterton's enemies. One of the most hostile of these was the naturalist, William Swainson – 'wholesale dealer in closet zoology'.[41] Swainson said, in effect, that Charles Waterton didn't do what he said that he did, but that anyone could do it anyway. He was influenced by 'the greatest possible love of the marvellous, and a constant propensity to dress truth in the garb of fiction.'[42]

The truth is that Charles Waterton made no claim to exceptional valour in this case and he later conceded that 'any old lady minus her crinolene' could have done as much; but the implication that he was a liar – dressing truth in the garb of fiction (or the other way round) – stung him to the quick.

The cayman debate continued to simmer, and briefly came to the boil again after the Squire's death. His friends defended him loyally. J. G. Wood claimed that the story was not only verifiable, it was great literature. 'A better word-picture does not exist in our language. . . not one single statement of his has ever been proved to be exaggerated, much less shown to be false.'[43] Not one.

Very shortly after this, an obscure man named James Simson – an authority on gypsies – declared at some length that very few of Charles Waterton's statements had ever been proved to be reasonable, much less shown to be true. The cayman episode was just one piece of evidence in the indictment.

> The only part of this description which can or may be considered true is, that the cayman was landed, and that Waterton, in a fit of hare-brained enthusiasm, got on its back. . . . The rest, perhaps with the exception of part of 'fighting the cayman', towards the end, is unquestionably a piece of Munchausenism that does not have even the appearance of probability. It is surprising that even a man like Waterton should have put it on record; and amazingly so that anyone should have attached the slightest credit to it.[44]

James Simson's hatred for the Squire was almost psychotic – apparently based on a fierce objection to Catholics in general and Jesuits in particular. Whether or not many Victorians read Simson, the fact is that he did apparently have the last word in the controversy. Unlike the vulture's nose, the cayman issue barely spilled over into the 20th century, and no scientific papers on the subject appear in the modern journals.

One tangible piece of evidence remains – the cayman itself. This specimen, said the Squire, will 'set decay at defiance for centuries to come, providing no accident befall it.' Over one and a half centuries later, it still is in perfect condition and can be seen at Wakefield Museum, alongside the hook which caught it, still bearing the reptile's tooth marks. They don't prove the story but do constitute what a lawyer might call substantive prima facie evidence. As Hobson said, the Squire performed more dangerous feats; why shouldn't he ride a cayman and use its front legs as a bridle? Or as the great baron himself said, 'All these narrow and lucky escapes, Gentlemen, were chances turned to advantage, by presence of mind and vigorous exertions.'[45]

Apart from shooting the rapids and almost drowning where four Indians had perished about a month earlier, then getting stuck in the mud on the Guianese coast, under a baking hot sun, and getting out twice, up to the waist in mud, to push the canoe – apart from that, the return journey to Georgetown was uneventful.

He stayed in Georgetown for just one day and apparently spent part of that day working on the cayman, a fact deducible from a letter received 41 years later, from an acquaintance named Buchanan. Mr Buchanan evidently had trouble believing the evidence of his own memory.

> I shall therefore feel very greatly obliged if you will recall the circumstances to your mind, and say whether I am right, or have overdrawn what I saw; namely that, to the house belonging to Mr [Robert] Edmonstone, opposite Mr Irving's store, you brought the Alligator from Essequibo and that you had the mouth or jaws held open by a piece of wood, and that you scraped his throat & inside yourself, preparatory to his preservation, which appeared to be considered by one or two almost impossible for any European to endure in such a climate. [46]

The reply is nowhere to be found, but the chance to perform in public what was almost impossible for any European to endure, would certainly have been too much to resist. The following morning, he set off for Warrows Place and there he finished the cayman. He also collected a few more birds.

By now, the main rainy season was setting in and the Squire decided that it was time to go home. He'd collected 'some rare insects, (though none new to science) two hundred and thirty birds, two land tortoises, five Armadillas, two large serpents, a sloth, an ant-bear, and a Cayman.'[47] He also acquired some freshly-laid tinamou eggs, coated them with gum arabic and packed them in powdered charcoal, as recommended in an article he'd read in the *Edinburgh Philosophical Journal*.

He returned to Robert Edmonstone's house, in Georgetown, spent a few more days there, and then left for England, with his precious cargo, on board the same West Indiaman which had taken him home from Demerara four years earlier. The specimens were packed in ten large 'hair trunks', stowed in his own cabin.

It was another pleasant voyage across the Atlantic and he arrived in Liverpool, in the spring of 1821, 'in fine trim and good spirits'. A Treasury official for His Majesty's Customs soon put paid to that.

Chapter 10
1821-1824
Tin, Tax and Taxidermy

Twice before, he'd travelled from Demerara to Liverpool and on neither occasion experienced any difficulty at all with the excisemen. On both occasions, his personal effects had included dead specimens and live animals. This time, many of the officials remembered him. They welcomed him like an old friend.

After disembarkation, a 'very civil officer' was appointed to accompany him back to the ship and was shown the contents of some of the boxes, prior to their conveyance to the goods depot. The ostensible reason for opening the boxes was to see that all was 'properly stowed', though curiosity on the part of the officer, and the vice of fools in the Squire, were probably also involved. But then another officer entered the cabin – a complete stranger, and 'wonderfully aware of his own consequence.' He'd been sent up from London, on the look-out for smugglers.

This authoritarian genotype – everyone knows him – put his head over the Squire's shoulder and proclaimed that none of the boxes should have been opened without his express permission. The Squire told him that the boxes had been opened every day during the journey across the Atlantic. The genotype left the cabin. 'I suspect I shall see that man again at Philippi', thought the Squire, eruditely, and so it turned out.

The ten boxes of specimens were taken to the depot and Charles Waterton went to the custom house, to sign the necessary documents and pay a suitable duty. After this, he went back to the depot and showed the collection to several more gentlemen, who 'expressed themselves highly gratified.'

The boxes were closed again and an 'inferior officer' began carrying them to the cart which was waiting on the cobblestones outside. Re-enter the superior officer, still aware of his own consequence. He was not satisfied with the valuation and the duty paid. The collection would have to be detained.

It was detained, in fact, for six weeks. All that Charles Waterton took home with him across the Pennines was a pair of live Malay fowls which had been given to him as a farewell present in Georgetown – the only items which, had suitable quarantine laws existed, obviously should have been detained.

> I have wandered through the wildest parts of South America's equatorial regions. I have attacked and slain a modern Python, and rode on the back of a Cayman close to the water's edge; a very different situation from that of a Hyde Park dandy on his Sunday prancer before the ladies. Alone and bare-foot I have pulled poisonous snakes out of their lurking places; climbed up trees to peep

Figure 15. Charles Waterton and his taxidermical 'sports.' Watercolour, Captain Edwin Jones. (Wakefield MDC. Museums and Arts.)

> into holes for bats and vampires, and for days together hastened through sun and rain to the thickest parts of the forest to procure specimens I had never got before.[1]

Only to be thwarted by an over-zealous tax-collector in Liverpool!

Amongst the items detained were the tinamou eggs[2] which had been so carefully packed in charcoal, and which he hoped to hatch under a broody hen at home, and so introduce a new game-bird into Britain. (Tinamous are rather drab, partridge-like birds and are now quite commonplace in zoos and butterfly farms.) 27 years later, a correspondent from Manchester – one of Charles Waterton's many communicative readers – conveyed to him a piece of well-authenticated folk-lore on the subject which might have been of some use to almost anyone else: a long sea voyage 'enlivens porter but is death to eggs.'[3] After six weeks detention in Liverpool, they must have been dead twice over.

During those six weeks, the Squire kicked his heels at Walton Hall and made unsuccessful recourse to the highest authorities he could think of, on behalf of his collection. He could handle a wild boa constrictor but the tight coils of bureaucracy defeated him.

Eventually, Commissioners of Customs received a letter from Treasury Chambers, in London, signed J. R. Lushington, and dated 18 May 1821. A summary of its contents was forwarded to Charles Waterton. Any specimens he intended to present to public institutions might pass duty free, but:

> With regard to the specimens intended for his own or any private collections, they can only be delivered on payment of the ad valorem duty of 20 per cent; and I am to desire you will give the necessary directions to your officers at Liverpool, in conformity thereto.[4]

The Squire went to Liverpool and paid up. His friends in the custom house had kept his collection at a constant, well-heated room temperature and none of the specimens had been damaged in any way. But his own feelings were very seriously damaged.

> Stung with vexation at the unexpected contents of this peremptory letter, I determined not to communicate to the public the discovery which I had made of preparing specimens upon scientific principles.[5]

So there!

The discovery was, of course, his new method of stuffing quadrupeds – the sudden flash of inspiration which had visited him as he lay in his hammock one day, harping on the string on which hung all his solicitude. But now he wasn't going to tell anybody, see! That'd show em!

The chief object of his anger was this J. R. Lushington, Esq., spokesman for the Treasury – assumed to be the man who was wonderfully aware of his own consequence. Edward Lear had a very good friend named Lushington[6], and Lear, of course, had close connections with Liverpool through the Earl of Derby. It might have been the same Lushington. Whoever he was, he took the wind out of Charles Waterton's sails and seriously affected his plans for the future. The defeat 'cast a damp' on his enthusiasms and for three years, he 'seldom or never' mounted his hobby horse. Instead, he stayed at home, mounted his specimens, then climbed aboard his other main hobby horse in life which, ultimately, was far more important anyway – the World's First Nature Reserve.

First priority was to keep out the poachers and foxes; they'd enjoyed themselves whilst he was away. It was time to build a wall around the estate. 'We compute in England, a park wall at a thousand pounds a mile', said Samuel Johnson, thirty-odd years earlier.[7] By that reckoning, the Squire's wall should have cost about £3,000; it was about the same length as the town walls of Pisa or Assisi. But by the time the wall was completed, five years later, it cost him more like nine or ten thousand pounds, a pretty sum, even by the profligate standards of the day.

The reason it took so long, and cost so much, is priceless: he refused to use paper money or bank drafts. He much preferred what he liked to call 'solid tin' – the golden sovereigns which he kept in 'a favourite deal drawer', inadequately sealed with a lock of 'ancient and simple construction'.[8] Despite his South American sabbatical, which must have cost a bit, he managed to accumulate 500 of these solid tin guineas and engaged some masons to begin work on the wall, on the strict understanding that, when they'd completed 500 guineas-worth of work, they would be laid off until he'd saved another 500. So, by fits and starts, the work proceeded.

The income, of course, came from the estate itself, some of the responsibility for which was now delegated; he employed a manager[9] – partly to oversee the building of the wall itself. But a certain proportion of income was saved through an increasingly ascetic life-style; at the time the wall was begun, he still declined to buy anything as frivolous as a carriage, for instance[10] – there can't have been many landowners with no carriage – and he always liked to point out that the wall itself was built with the money he saved by not touching spirituous liquors. But the fact that he found enough money to buy some of his neighbour's estate – in 1819 and again in 1824[11] – suggests that he had far more solid tin shot in his locker than he liked to pretend.

Needless to say, the wall around Walton Park was no ordinary wall; the end, as well as the means, was unconventional. Most park walls were no more than five or six feet high – symbols of ownership. Spring-guns and man-traps kept human intruders out. Charles Waterton had something different in mind – something more in the nature of fortifications. He assessed the scaling power of local foxes and savages and decided that the wall should be nowhere less than eight feet high, and twice that height along the three quarters of a mile or so where his estate bordered the then much-frequented Barnsley canal. (He distrusted bargees more than any other form of human vermin.) The result was a park wall like no other park wall anywhere in England – two or three times as high in places, and two or three times as expensive overall.

Determined poachers still got in from time to time, but they got away with less and less. As for the fox, the Squire claimed that only one ever got his pads over the coping and that was only with the assistance of a 'sheep bar'. More of that later. Whilst the wall was being built, he not only tolerated foxes, he played host to at least one large family of them, though one of the cubs did eventually get it in the neck for no very good reason. He was out with his air-gun, looking for rabbits, when a family of seven foxes appeared about 50 yards away.

> I remained fixed as a statue. They were cantering away, when one of the young foxes spied me. He stopped and gave mouth. This was more than I could bear; so, as he was sitting on his hind quarters, I took aim at his head, and sent the ball quite through the wind-pipe. Away went the rest and left him to his fate.[12]

Many years later, the Squire was still trying to wash the blood from his hands:

> When I reflect on the wanton and wilful murder I then committed upon so cherished a quadruped, my heart misgives me, and I fancy, somehow or other, that the sin is still upon my conscience.[13]

Nevertheless, after the wall was built, foxes were banned on the estate and gradually, under the wall's protection, an acceptable natural balance was established, with assistance from the Squire, on the most advanced ecological principles.

As befits the future proprietor of Elysium, he now decided it was time to keep a notebook – a small, padlocked volume with thick, un-lined pages, like a modern photo album. At first, it was just the zoological notes of a student – painstaking memos of the Linnaean system (including jaw-ache) written in a meticulous Stonyhurst hand and obviously synthesised from published works, with a few labelled drawings of birds' feet to show varying degrees of webbing. Then he made the first of two entries in his notebook on the dodo, which he believed was still in existence. Eventually, at the ripe old age of 40, he made what appears to have been his first, traditional, phenological nature note:

> On the fourth of July, 1822, about half past one o'clock in the afternoon, in the Park at Walton Hall, I heard the cuckoo repeatedly.[14]

1822 was clearly a momentous year at Walton Hall; apart from hearing the cuckoo repeatedly, Charles Waterton also entered the glorious annals of aviation history and grew immortal in his own despite.

He was perched 'at an elevation of several yards', on the eaves of one of the highest farm buildings on the estate, strapped to a pair of Do-It-Yourself wings and about to throw himself trustfully into space, when someone turned up and persuaded him to use a lower, less ambitious launch-pad into posterity.[15] When he finally crashed to earth, he suffered nowt more serious than a 'foul shak' (bad leg). The cuckoo, almost certainly, made a suitable comment.

Figure 16. Charles Waterton. Reproduced from Illustrated London News, August 28 1844.

Like most great bird-men of the day, Charles Waterton assumed that manned flight would be by some sort of manually-operated ornithopter, roughly on the Daedalus model. If it had occurred to him to soar like a bird, rather than flap like one, he might have beaten the Wright brothers into the history books. But in the event, all he managed to achieve, apart from a foul shak, was the satisfaction of verifying an old, classical axiom: 'Let not the cobbler go beyond his last.' The cobbler's wings remained down at heel.

In the following year – 1823 – he almost suffered another foul shak – or the same shak re-fouled – during one of his many perilous safaris across the local common – a frightening place after the comparative security of the equatorial jungle. He was minding his own business on his native heath when he was suddenly attacked by two dogs – 'an insignificant female cur' and 'a stout, ill-looking, uncouth brute, apparently of that genealogy which dog fanciers term half bull, half terrier.' The dog came at him from the front, the bitch from the rear. The Squire kicked out viciously at both and eventually got rid of them just as two passing masons were running to his assistance. He mentioned the episode simply to illustrate 'the advantage to be derived from resisting the attack of a dog to the utmost',[16] a useful little tip.

His own dogs were apparently treated well – 'pallazzi carni', one visitor called the kennels[17] – but dogs in general were never cossetted like his favourite cats. Cats, for instance, were allowed inside the Hall; dogs were not – though a good terrier might have worked wonders with the rats.

The war against the Hanoverian rat continued without dogs. Owls probably took more than their share, and in this same year – 1823 – the barn owls on the watergate tower successfully reared two broods.[18] They were allowed to fly in and out of the open windows of the Hall more or less as they pleased, and may have caught some of the rats inside the Hall itself, unbeknown to the Squire.

The rats were certainly still there, even in the innermost sanctum; it was at about this time that he caught one of them in the drawing-room and gave him a new black coat – 'tarred but not feathered', said Dr Hobson, predictably. For a few months after

Figure 17. Silhouette of Alexander Wilson. Made in Peale's Museum, 1800. (Paisley Museum – Renfrew District Council.)

this, the smell of tar kept the rats out of the drawing-room and the Squire could sit and read in peace, undisturbed by scrabbling noises in the walls. This was an important luxury to him because he'd recently acquired an un-put-downable epic which made great demands on his fireside time and attention – *American Ornithology*, by Alexander Wilson.

Wilson was an expatriate Scot who spent the last years of his life in North America. He died there in 1813, just as the eighth of nine eventual volumes was going to press. For most of his life, he was pitifully poor and ill-fed and this fact, naturally, endeared him to Charles Waterton, though the two men never met. Wilson had met John James Audubon, however – in March 1811. He'd walked into Audubon's general goods store, in Louisville, Kentucky, and tried to get a subscription to *American Ornithology*. Audubon refused and the snub caused a feud between the two men which further endeared Wilson's memory to Charles Waterton when he, himself, drew swords with Audubon a few years later.

For now, it was sufficient that Wilson's book was about birds and that it had allegedly been written in the woods.[19] Charles Waterton was inspired. He decided that he, too, must visit North America, to see the birds for himself. Dispirited by his brush with the Treasury, he'd spent almost four years at home (just one brief sortie into Leeds, to give a lecture at the Philosophical Hall[20]) but now Wilson's work rekindled the 'almost expiring flame'. Plans were made for another trip across the Atlantic.

Before he left, however, there was one more important piece of business to transact. He was still hungry for land – strictly for the birds – so, early in 1824, he paid £4,600 to the Earl of Westmoreland, to add another piece of the neighbouring Haw Park to the estate of Walton Hall[21] – more land to be enclosed by the wall. One of the workmen on the wall took exception to being out of work from time to time, on account of the Squire's curious method of payment, and probably worked out that, if his employer could afford to buy land from the aristocrat next door and then go wandering off to America for a few months to look at birds, he could probably afford to pay the masons without laying them off intermittently whilst more money was accumulated. But there were probably other considerations, of which the workman was totally unaware. Lack of consideration for the 'labouring classes' would certainly have been out of character in Charles Waterton, except insofar as inconsistency was characteristic.

In the temporarily rat-free drawing-room of Walton Hall, he completed his 'attentive perusal' of Wilson's 'pleasing and brilliant work', then made whatever arrangements were necessary with the wall-builders and set off for New York, in the early part of 1824, 'in the beautiful packet John Wells'. It was the beginning of the fourth and last of his New World journeyings, soon to be immortalised as *Waterton's Wanderings in South America, the North-West of the United States, and the Antilles, in the Years 1812, 1816, 1820 and 1824.*

Curiously enough, i.e. typically, despite the ostensible reason for going there, his account of his travels in North America concerns just about everything except birds. And the furthest north-west he reached was Niagara.

Chapter 11
1824-1825
Sex and the Single Squire

Charles Waterton approved of the U.S.A. – of New York especially.

> I could see very few dogs, still fewer cats, and but a very small proportion of fat women in the streets of New York.[1]

The Erie canal, linking New York city to the Great Lakes, though not yet officially opened, was already in operation; the city was set to become one of the great sea ports and commercial centres of the world. 'Ere long, it will be on the coast of North America what Tyre once was on that of Syria.'

But cities never held him for long and 'on a fine morning in July', after a month or so in New York, he boarded one of the Hudson River paddle steamers and set off upstate towards Albany.

Formal letters of introduction were the nineteenth-century rule, but Charles Waterton had no faith in formal letters of introduction – for the time being anyway. He preferred 'accidental acquaintance'. Within a very short time on the steamer, he'd struck up a particularly warm accidental acquaintance with a native New Yorker whose American 'openness and candour' defied all formality.

At Albany, he took tea with this 'obliging', 'affable', 'worthy' gentleman and next morning bade him 'farewell, and again farewell', and began the long journey by horse-drawn passenger boat up Clinton de-Witt's new canal to Lake Erie, 360 miles away.

> You may either go along it all the way to Buffalo . . . or by the stage; or sometimes on one and then in the other, just as you think fit. Grand, indeed, is the scenery by either route, and capital the accommodations. Cold and phlegmatic must he be who is not warmed into admiration by the surrounding scenery, and charmed with the affability of the travellers he meets on the way.[2]

Now read Fanny Trollope's account of the same journey, six years later:

> I can hardly imagine any motive of convenience powerful enough to induce me again to imprison myself in a canal boat under ordinary circumstances. The accommodations being greatly restricted, everybody from the moment of entering the boat, acts upon a system of unshrinking egotism. . . . To anyone who has been accustomed, in travelling, to be addressed with, "Do sit here, you will find it more comfortable," the "You must go there, I made for this place first," sounds very unmusical.[3]

Now compare it with the attitude of Charles Dickens, a few years later still. Much of Dickens' travel in America was also by river steamer and canal boat, mostly on the

Potomac.[4] But he craved for slippers and a familiar bed. Conditions on board the various vessels he travelled in, he inveighed against passionately – there was nowhere to hang clothes or write, floors were dirty, passengers spat, beds were like book shelves, washing facilities were primitive, there were long delays and the boats themselves looked unsafe. Dickens and Trollope were two of a kind.

Charles Waterton had more in common with the hardy, virile instincts of Americans themselves – Thoreau, Mark Twain, even Audubon. Dickens, by comparison, was almost effete. Fanny Trollope was worse.

Waterton was one of the first Englishmen to make unreservedly friendly overtures towards America, after the bitter legacy of two recent wars between the countries. More accurately, he shared the republican distaste for England. He allied himself to the Irish Democrats in America and undoubtedly won friends through his contempt for the monarchy and 'hereditary boobies' in the House of Lords.[5] The Americans also liked his frontiersman style of life. But the finicky bellyaches of people like Mrs Trollope and Charles Dickens undid all the good work of people like Charles Waterton.

English attitudes towards America had been sharply divided for many years, but the pro-American faction was a very small minority. Samuel Johnson said he could love all mankind except an American, and Cobbett remembered that when his own father took the American side in arguments with a local gardener, the dispute grew too 'warm' for a child's ears.[6] James Fennimore Cooper – an American – could see nothing on which the English were 'so uniformly bigotted and unjust, so ready to listen to misrepresentations and caricature, and so unwilling to receive truth' as on the subject of America. Whilst travelling in Europe, he found 23 entries of American names in hotel registers, with as many insulting remarks appended by Englishmen.[7] The English, generally, clearly regarded themselves as the world's elite. Most of them clearly expected deference when travelling abroad. Some of them got it.

Charles Waterton followed a different principle. 'A man generally travels into foreign countries for his own ends', he said. 'This ought never to be forgotten; and then the traveller will journey on under the persuasion that it rather becomes him to court than expect to be courted.'[8] So he went in search of someone to court.

He felt as much at home in America as the Americans themselves. Only one thing marred his enjoyment – too many trees had been felled: 'I wish I could say a word or two for the timber which is yet standing. Spare it, gentle inhabitants, for your country's sake.'

Gentle inhabitants – then, as now – confronted with travellers in quest of the 'picturesque', often assumed a kind of cussed braggadocio with regard to the more crass manifestations of 'progress' – felling forests etc. Fanny Trollope encountered such men – cockily enthusing about industrialisation. Charles Waterton evidently did not. The gentle inhabitants were much less likely to be contemptuous of the picturesque when it was championed by a friendly eccentric than when its acolytes were jingoistic Fannies who despised everything American. But they still cut down the forests.

The Squire clearly loved everything American – the magnificent scenery, the 'spacious' and 'excellent' hotels, the affable gentlemen and, most of all, the 'highly polished' females. He found all four commodities at Niagara.

Unfortunately, he was still accident-prone; he arrived on one leg. He'd stepped out of the stage-coach, near Buffalo, and turned his ankle, an eventuality which induced practical utilisation of the scenery and poetical use of the hotel register, but which subsequently hampered relations with one particularly fair specimen of the highly polished females.

The item of scenery converted to practical use was Niagara itself – the Horseshoe Fall, nearly 200 feet high. Future generations might harness its power for hydro-electricity; Charles Waterton had a more user-friendly use. On the last occasion that he'd sprained an ankle, the doctor had advised him to hold it under a water pump two or three times a day, to reduce the swelling. If a household pump could cure a sprained ankle, how much more effective, thought he, would a cascade of water be which discharged, so it was said, 670,255 tons a minute? (About 1.5 million gallons per second.) He hobbled down to the base of the Fall, held his foot in the curtain of water, and reflected on the difference between a household pump and Niagara. The subject overwhelmed him. His ankle remained sprained.

Back at the hotel, he entered his name in the visitors' book and appended the following for the edification of future guests:

He sprained his foot and hurt his toe,
On the rough road near Buffalo.
It quite distresses him to stagger a-
Long the sharp rocks of famed Niagara,
So thus he's doomed to drink the measure
Of pain, in lieu of that of pleasure.
On Hope's delusive pinions borne,
He came for wool and goes back shorn.[9]

Bad foot notwithstanding, Niagara was everything he'd hoped for, except in the number of words it rhymed with. (Lorenz Hart chose aggra / vate and gets an encore from posterity for his 'wit'.[10]) But though Charles Waterton's company was 'in high estimation' at Niagara,[11] all hopes of painless pleasure unalloyed were ultimately spoiled by his sprained ankle, 'as will be seen in the sequel'.

The Niagara river is generally accounted one of the best places in North America to look at birds. It claims to be 'the gull capital of the world.' At least 16 species of gull hang around there, including Franklin's, Bonaparte's, Sabine's, Thayer's, Glaucous, Ring-billed, Mew gull and Laughing gull. In autumn and winter, it is also a magnificent place for wildfowl and birds of passage, en route from northern Canada to join the great Atlantic coast flyway at Chesapeake Bay, or thereabouts. Its also a good place for birds of prey, or used to be. Wilson drew attention to the bald eagles of Niagara – 'a noted place of resort for these birds'[12] – so Charles kept his eyes peeled for bald eagles but found their numbers seriously reduced by human persecution. 'Were I an American, I should think I had committed a kind of sacrilege in killing the white-headed eagle.'

He saw one or two bald eagles whilst he was there, but the gulls passed him by unremarked and as for the wildfowl, the oldsquaws of Niagara, such as he noticed, were almost exclusively human, and no loons or buffleheads at that. Certainly not trollopes. The young birds of passage, in full breeding dress, were a treat for the eyes and ears of a true-born Yorkshireman: 'Words can hardly do justice to the unaffected ease and elegance of the American ladies who visit the falls of Niagara.' Ayup!

One of these ladies – a young girl from Albany – clearly impressed him more than the rest. Before leaving Niagara, he decided to join the evening social round and attended a dance at one of the local hotels, despite the 'unlucky foot' which prevented him taking the floor. (Fact was, he hadn't learned to dance. More important things to do.) He was sitting with his foot resting on a sofa when the 'fair Albanese' entered the room 'with such a becoming air and grace that it was impossible not to have been struck with her appearance.' It was love at first sight. 'Her bloom was like the springing flower / That sips the silver dew.' But he had to be content to worship from afar, his head full of romantic poetry and his foot reclined on a sofa, unable to court and in no position to expect to be courted. Many a passing gentleman rubbed salt in the wound by assuming he was suffering from gout, an insufferable insult to a man of Charles Waterton's habits. Gout is for slobs, sprained ankles are for accidental travellers.

As his ankle began to heal, he decided it was time to leave Niagara and to join a friendly New York family, whose accidental acquaintance he'd made, for the trip back to New York city. They were to go via the Lake Ontario and St Lawrence river steamers as far as Quebec city, then back up-river and through Lakes Champlain and George to Saratoga, and so back to Albany and the already well-known beauty of the Hudson river. The fair Albanese may or may not have been on the passenger list. She is never mentioned again.

Altogether, he spent no more than a few days in Canada, and penetrated no more than a mile or two across the border. He mentions no birds. At Quebec, he saw the new fortifications; nearby, he saw the Falls of Montmorency; and at Montreal, he briefly made the unaccidental acquaintance of the professors of the college. That was all. Yet, whilst the rest of the world has virtually no lasting monument to Charles Waterton, Canada has over 400 square miles of national park named after him, right in the heart of the Rocky Mountains – a rich, teeming biosphere of beavers, porcupines, snowshoe hares, chipmunks, cottontails, jack-rabbits, wolves, cougars, coyotes, marmots, martens, ground squirrels, wood ducks, red-tailed hawks, white-tailed deer, sharp-tailed grouse, broad-tailed hummingbirds, horned larks, great-horned owls, bighorn sheep, bluebirds, chickadees, great grey owls, hairy woodpeckers and grizzly bears a beautiful mountain wilderness of scrub, grassland, forest and wetland, to which any man with a soul could be glad to give his name Waterton Lakes National Park, in southern Alberta. He never went anywhere near Alberta but an English Royal Artillery officer – Thomas Blakiston – did. Blakiston surveyed the area for an expedition through the Rockies in the late 1850s, and it was he who named the three lakes after Charles Waterton.[13] The region became a national park in 1895 and 37 years later it was merged with the Glacier Park, on the U.S side of the border, to become the Waterton-Glacier International Peace Park officially dedicated by President Hoover. Canada Post issued a special commemorative stamp, in 1982, to mark the 50th anniversary of the international merger.[14] (And the 200th anniversary of Waterton's birth.)

Waterton's route in 1824 – including the brief sortie into Canada – was all part of the regular tourist beat, an American version of the Grand Tour, much of it through staunchly Federalist territory of the United Empire Loyalists, still faithful to the mother country, 40 years or so after the original refugees settled there. But this didn't trouble Charles Waterton. He had kindly feelings towards everyone – Canadians, Yankees, Irish, French, republicans, colonialists, everyone. Fondest words of all were reserved for some of his travelling companions – the young ladies who sang delightfully and looked even better. He clearly took a fancy to the young ladies of America.

Everywhere he went, he was impressed with the 'pleasing frankness and ease and becoming dignity in the American ladies, and the good humour, and absence of all haughtiness and puppyism in the gentlemen.' Saratoga Springs was already a fashionable watering place and 'afforded a very correct idea of the gentry of the United States.' But it was back in New York city, and especially on Broadway, that he found the complete apogee of charm and elegance. The approaches to New York, down the Hudson river, were 'enchanting'. The city itself was a paragon.

He knew of no street anywhere in the world with more attractions than Broadway. Everything about it was appealing, everything the public would allow. But it was not blockbuster musicals he had in mind. Pleasures were simpler then:

> There are no steam engines to annoy you by filling the atmosphere full of soot and smoke; the houses have a stately appearance; while the eye is relieved from the perpetual sameness, which is common in most streets, by lofty and luxuriant trees.[15]

From twelve to three, on Broadway, the American ladies took their 'morning' walk. Fanny Trollope, catty as ever, joined them; she hitched up her petticoats and stuck her snooty little English nose in the air to declare that, 'If it were not for the peculiar manner of walking, which distinguishes all American women, Broadway might be taken for a French street.'[16] Charles Waterton saw it differently:

> The stranger will at once see that they have rejected the extravagant superfluities which appear in the London and Parisian fashions; and have only retained as much of these costumes as is becoming to the female form. This, joined to their own just notions of dress, is what renders the New York ladies so elegant in their attire.[17]

In a political frame of absurd national rivalry, fashion was a suitably frivolous *casus belli*. Not even a thrifty, plain-living Yorkshireman could resist the call to arms. He was pro-American and convinced that American ladies could teach their European counterparts a thing or two about haute couture, about the best way to wear an Italian straw Leghorn hat for instance, or the best way to walk, or sing. And then, of course, there was the little matter of caps. 'And any cap what ere it be, Is still the sign of some degree', according to the seventeenth-century song,[18] but American women had more sense; they had what Charles Waterton regarded as a very just abhorrence of caps:

> A mob cap, a lace cap, a low cap, a high cap, a flat cap, a cap with ribands dangling loose, a cap with ribands tied under the chin, a peak cap, an angular cap, a round cap, and a pyramid cap! How would Canova's Venus look in a mob cap?[19]

British 'head milliners' should go to New York instead of to Paris, according to this great new arbiter of fashion. 'And if they could persuade a dozen or two of the farmers' servant girls to return with them, we should soon have proof positive that as good butter and cheese may be made with the hair braided up, and a daisy or primrose in it, as butter and cheese made in a cap of barbarous shape; washed, perhaps, in soap-suds last new moon.'

New York was not only a potential capital of fashion, it was also a comparative model of law and order. Federal Drug Enforcement was yet undreamed of. Crack-induced murders were unknown. Everyone 'walked at his ease' in New York – no jostling, no 'impertinent staring', no diversionary tactics by teams of pick pockets. The Squire saw 'a gentleness in these people, both to be admired and imitated.' And you don't have to take his word only; 20 years later, Timothy Dwight, president of Yale College, saw that in New York city 'Law reigns with an entire control, and resistance to it is unthought of.'[20] William Cobbett saw it too, and he was no republican – not at that time anyway:

> Blindfold an Englishman and convey him to New York, unbind his eyes, and he will think himself in an English city . . . he will miss by day only the nobility and the beggars, and by night only the street walkers and pickpockets.[21]

The Squire found young Jonathan 'a very pleasant fellow', inquisitive but certainly not impertinent. It cannot be too strongly emphasised how greatly at odds this view was with the average Englishman's conceit of himself – anti-Americanism was a national sport. The only flaw in Jonathan's character that Charles could see was smoking: 'It is a foul custom; it makes a foul mouth, and a foul place where the smoker stands: however, every nation has its whims.'

His only complaint, otherwise, was against American climate – the sharp changes in air temperature which were liable to give strangers bad colds. He developed a cold himself in New York, though 13 years of hindsight held a hot bath to blame, rather than the climate. The Waterton cure for his cold inevitably suited the virus: six weeks of white bread and sweet tea, with eight lettings of blood to augment the diet. It was not until he went south to the tropics again – at the 'urgent entreaty' of a Doctor Hossack – that the cold and its cure were cured.

Before he took his hot bath, Charles Waterton went to Philadelphia – the intellectual 'Athens of North America.' It was clearly the place to be – a clean, elegant, liberal city of painters, scientists, writers, booksellers and publishers; a patriot stronghold where everybody who was anybody knew everybody else who was anybody else. It was here that Wilson's *American Ornithology* had been published. It was here that Audubon fell out with the scientific establishment. It was here that William Cobbett opened his bookshop and defied the patriot press by selling pictures of the English nobility. It was here that the American government had its headquarters after the Declaration of Independence. It was here that the American constitution itself was devised. And it was here that Charles Waterton had his portrait painted by the man who made George Washington's false teeth.

The artistic dentist was Charles Willson Peale, paternal head of the 'Painting Peales' of Philadelphia. At that time, he was an old man of 84, exactly twice as old as the Squire himself. As a young man, Peale had been active in politics, first with the 'Sons of Liberty', then as a radical left-wing Whig in the Pennsylvanian legislature – just the sort of pedigree to appeal to Charles Waterton. Later, Peale had renounced politics to pursue his interests in painting, the 'mechanic arts', and taxidermy. Long before the turn of the century, he'd begun a collection of natural 'curiosities' and superintended the recovery of some well-preserved mammoth bones from a swamp in the Catskill Mountains. So was born the famous Peale's Museum – the first popular museum of natural science and art in America.[22] Charles Waterton was greatly impressed.

He walked into the museum on a summer day in 1824 and there encountered one of Peale's six surviving sons, Titian. (The others were Rembrandt, Raphaelle, Franklin,

Figure 18. Titian Peale, 1799-1885. Painted by C.W. Peale. (Private Collection.)

Rubens and Linnaeus.) Titian was the naturalist of the family, a meticulously accurate though uninspired artist who had been engaged to collaborate with Napoleon Bonaparte's nephew, Charles Lucien Bonaparte, to complete the *American Ornithology* which Alexander Wilson had left unfinished at his death. Titian showed the English squire around the museum – the Marine Room, the Mammoth Room, the insect display, the geology cases, the various mechanical gadgets and Electric Machine, the Quadruped Room, the Laboratory, the Preserving Room and, most impressive of all, the marvellous Long Room Gallery with stuffed specimens ranged in tiered rows of glass cases against the walls, each exhibit shown against a painted representation of its natural habitat. Above the stuffed specimens, the gallery was lined with portraits of Revolutionary heroes and contemporary worthies of public life or private means, each painted by a member of the Peale family.

As Titian performed his conducted tour, he was treated to the mildest of mild punctilios on the quality of Philadelphian tea – a sort of Watertonian praise by faint damnation. Titian therefore invited him home for a cup of the Peale's own special brew, which the Squire declared was the best cup of tea he'd tasted in America.

Titian's father, Charles Willson Peale, was at that time convalescing from old age in Maryland; it was when he returned home, a week or so later, that he painted the now famous picture of Charles Waterton. But for the time being, as old Mr Peale enjoyed the sea breezes in Maryland, the Squire went on a hunting trip into the New Jersey countryside, near Salem, with his new friend, Titian Peale.

The Englishman insisted on paying all expenses for the trip himself – he was 'very liberal' and 'supposed to possess wealth', said Charles Willson Peale to Thomas Jefferson.[23] (The Peales moved in the very highest of egalitarian circles.) Bed and breakfast made very little dent on the supposed wealth; accommodation was less than five star, as a little bed-time adventure revealed.

He was not unused to a little discomfort at night, but there was a limit even to Charles Waterton's tolerance where bed-mates was concerned. 'Upon my word, Madam,' said he to the innkeeper next morning, 'though I am not prone to make wry faces at a fair allowance of fleas or bugs, still I must own to you that I have not yet quite made up my mind to be devoured alive by rats.'[24]

The offending mattress was examined but no rat was discovered – only a nest of

Figure 19. Charles Willson Peale, 1741-1827. 'The Artist in his museum.' Self-Portrait, 1822. (The Pennsylvania Academy of Fine Arts, Philadelphia.)

seven or eight mice. It was a false alarm. The innkeeper was clearly disgruntled; all that fuss over a few mice! Some time previously, so she said, her own shoulder had been gnawed away by rats as she slept in that same house – by implication, on the same mattress. It offended her deeply that people should confuse mice with rats. Fanny Trollope, unfortunately, never passed that way.

Charles and Titian enjoyed their hunting trip – mice apart – then returned to Philadelphia for a few more nice cups of tea and a bit of friendly chit-chat with Titian's father, now back from Maryland. As at death-beds, so at tea-tables, the conversation inevitably turned on taxidermy. Charles Willson Peale usually stretched his skins over wooden sculptures of his subjects, which gave them a 'characteristic' attitude when mounted. This may or may not have impressed Charles Waterton, but the method of mounting specimens with their jaws open, to show the teeth, failed to meet his approval. He said that it gave all the animals an unnatural grin.[25] The Peales accepted the criticism but insisted that, for educational purposes, demonstration of tooth structure was vital. Charles Willson Peale had played around with dentistry for over half a century and obviously had more than a passing interest in the subject. Long before dental plates were in general use, he'd perfected the world's first porcelain teeth and subsequently made a set for the mouth that never told a lie.[26] It was no surprise that he should want to use his dentistry skills in taxidermy.

The Peale family was clearly on very friendly terms with George Washington, as with Thomas Jefferson. Despite an aversion to sitting for his portrait, the Father of the United States had allowed Peale Senior to paint eight life portraits of him – with his teeth in – and Rembrandt Peale painted at least one 'equestrian portrait' of Washington, which was apparently presented to the U.S. senate, in Washington, at the opening of the next session.[27] Many copies of the nine originals were produced by the Peales – each with a slight variation in background or posture, though essentially the same – and one of them inevitably found its way to Walton Hall, where it hung for many years over the mantel-piece and was toasted each 4th July with a glass of God's ale.[28] Rembrandt Peale later visited Walton Hall.[29]

Charles Waterton shared George Washington's supposed aversion to sitting for his

Figure 20. Rembrandt Peale, 1778-1860. Self-portrait, 1828. (Detroit Institute of Fine Arts.)

portrait but he, too, submitted to the ordeal for the privilege of being painted by Charles Willson Peale. If Carlyle's claim be true – that a good portrait is worth a dozen biographies – this book marks completion of exactly half the task of providing a literary equivalent to Charles Willson Peale's portrait of Charles Waterton. The picture hung in the Long Room of Peale's Museum until it was eventually acquired by George Ord – Alexander Wilson's biographer – who sent it to join Washington's portrait at Walton Hall. It now belongs to the National Portrait Gallery in London.

It was the Peales, in fact, who introduced Charles Waterton to Ord – the beginning of a firm friendship which lasted half a lifetime for both men. They corresponded regularly and Ord was a favoured guest at Walton Hall when he visited England in 1832 and 1839. The bond was sealed by a mutual distaste for John James Audubon.

Audubon apparently met Charles Waterton at the house of Dr William Mease, in Philadelphia, where the Audubon portfolio was often available for inspection.[30] He also met the Peales and evidently knew Titian quite well. He'd visited the Peale Museum just a few weeks before the Squire himself was there, and Titian had seen some of his paintings four years earlier. When Audubon was taken to Titian Peale's studio, the now famous backwoodsman was less than overcome with praise for the then famous Philadelphian. He obviously judged Titian's work by his own standards and made his feelings known.[31] Titian protested to be not unduly offended by the criticism and Audubon was taken to a meeting of the Philadelphia Academy of Natural Sciences, founded by Charles Willson Peale. It was there that he met, and vehemently argued with, George Ord. A few years later, he was to arouse the more formidable fury of Charles Waterton himself, and thereby helped congeal the Squire's friendship with Ord.

Audubon's criticism of Titian Peale's nature paintings was scurrilous but probably justified; Peale was not a great painter, Audubon obviously was. In 1984, a hundred and thirty three years after his death, one copy of his Birds of America was knocked down at Sotheby's, in London, for a little matter of £1.25 million.

But the great bird painter as far as contemporary Americans and Charles Waterton were concerned, was not J.J. Audubon but Alexander Wilson. Wilson had met, and made friends with, all the influential people in Philadelphia whom Audubon had at first impetuously declined to meet and then insulted when he did meet them. Everyone, including Charles Waterton, and even Audubon himself, revered Wilson's memory. But with regard to the people of Philadelphia, Audubon seemed to share William Cobbett's view: 'a cheating, sly, roguish gang.'[32] The attitude incurred Charles Waterton's lasting hostility towards both men, especially towards Audubon. Cobbett hated the Philadelphian Democrats, including the Peales; Audubon clearly despised the Philadelphian scientific and artistic elite. Charles Waterton loved almost all Americans, but especially republicans (i.e. Democrats).

But it was now time for him to leave Philadelphia and to go back across the Delaware river to New York – elegant, fashionable, law-abiding New York, happily devoid of fat women. He always maintained a soft spot in his heart for the Peale family, for Titian Peale in particular. Pity there were no daughters.

In New York, he took the hot bath which gave him a bad cold, and then went south across the equator again, at Dr Hossack's earnest entreaty. He sailed from the United States – presumably from New York – in the early autumn of 1824, with a lastingly favourable memory of new scenes and new acquaintances and springing flowers that sip the silver dew. His parting words: 'For ages yet to come, may this great commonwealth continue to be the United States of North America.'

The intention was to sail south to Antigua, then island-hop through the West Indies for one last visit to British Guiana – not a bad way to spend the winter. The Caribbean was currently in the grip of storm-force winds, but contrary winds in the Atlantic delayed his passage so that he was able to claim, yet again, that his life was charmed by providence. 'Had our passage been of ordinary length, we should inevitably have been caught in the gale.' A more pessimistic man – perhaps a more realistic one – would have blamed the gale for causing the delay; he chose to bless the delay for avoiding the gale.

When he finally got to Antigua, 30 days out from port, he clearly resolved to take up his predominant propensity again, after a few months relative abstention. In his account of the United States, natural history – the ostensible reason for going there – takes second place to sociology and politics. Now, in Antigua, he found an American redstart. 'So what?' inferred the natives, like latterday professors of sociology and politics. He enquired after its status on the island – resident or migratory? – but no-one, it seems, was sufficiently interested to know.

He spent a dull and depressing week in Antigua, a bird of passage himself, and then left on the mail boat for Guadaloupe and Dominica. Before leaving Antigua, he received one last reminder of Anglo-American relations. Foolishly, or defiantly, or innocently, or mischievously, he decided to wear a straw boater with the republican green riband around the brim, as he went through the embarkation procedure at St John's harbour. He'd bought the hat in the United States. Antigua was a British colony and its officials were as anti-American as British officials were everywhere. He was taken for a 'damned Yankee', an away team supporter at the wrong end of the ground. The harbour master kept him standing for half an hour, like a little boy in a dunce's cap, and no-one bothered to offer him a seat.

The harbour master at Dominica was hospitable, informative and attentive to strangers. But then Dominica, though nominally British, was almost exclusively inhabited by French settlers. He construed a great difference between the French and the British islands: the French were 'happy and content', the British were 'filled with murmurs and complaints'. To all intents and purposes, Dominica was French, with a large species of edible frog to prove it. The Squire went for walks through the beautiful Dominican countryside and found not only large frogs but flying beetles larger than birds. These were rhinoceros beetles, Dynastes hercules, most of them over six inches long, with fore-wings folded back in the manner of an RAF jump-jet and one up-tilted horn sticking out from the head, like an entomological unicorn. The humming-birds which he came across in the same woods would have been about two thirds the size of the beetles.

After leaving Dominica, he sailed via Martinique, St Lucia and Barbados to Demerara where, 'without loss of time', he proceeded to the forests in the interior. He did pause a little while in the cultivated parts – long enough to notice a dramatic change in the human population since last he'd been there, four years earlier: scarcely a Dutchman was left. Some of them had died, others had crossed the border into Surinam. 'Our overbearing demeanor as conquerors soon gave them to understand that it was time for them to go elsewhere.'[33] It was time for the Squire to go into the primeval rain forest, it being now mid-winter, the height of the dry season – 'which renders a residence in the woods very delightful.'

He went back to his tropical headquarters in the forest, bagged himself a paradise jacamar in fine plumage, a cardinal grosbeak and 'a large species of owl', and renewed his efforts to make the blood of his foot an acceptable meal for the vampires. But the main prize of this fourth expedition, if not the main reason for it, was a species of animal never seen alive, before nor since – an Itouli, or Nondescript.

The Nondescript was a veritable missing link of an animal, half man, half monkey, with fashionable beard and hair-style and a proud, self-righteous old face. 'The features of this animal are quite of the Grecian cast,' said the white man from Yorkshire, 'and he has a placidity of countenance which shows that things went well with him in life.' The white man conscientiously shot the creature and prepared to carry it back to base, for the benefit of the curious in a far-distant clime. But it was too heavy to carry on his back, so he cut off the head and satisfied himself with that alone, as scientific specimen and trophy. A drawing of the head appeared on the flyleaf of the first edition of the *Wanderings*, published a few months later. At least one reader mistook it for a portrait of the author.

It was, of course, a nineteenth-century Piltdown Man – the rump of a red howler monkey, delicately manipulated to represent the face of an old man. But no-one saw the joke. 'A quackish performance', said the Quarterly Review. 'What is there to boast of in a forced change of this kind, we would ask?'[34] (And they didn't realise which end of the monkey it was!)

Even Sydney Smith – the most waggish of Whigs, and definitely pro-Waterton – thought that it was 'foolish thus to trifle with science and natural history.' According to Smith, the Nondescript was clearly the head of a Master in Chancery, 'basking in the House of Commons after he has delivered his message.'[35] The face does look vaguely familiar.

Charles Waterton had relatively extensive knowledge of the monkey tribe – of red howler monkeys in particular. Their howls kept him awake at night. Also, of course, consistent with contemporary taste, he tasted them – and ventured the advice that, boiled in Cayenne pepper, the flesh was not to be sneezed at. 'A young one tastes not unlike kid', he said, 'and the old ones have somewhat the flavour of he-goat.' But the skinned animal's resemblance to a human child gave pause to the squeamish and forbad the insertion of a knife and fork. No such scruples applied to the dissecting knife, however, and the resultant Nondescript bore resemblance, not to a child, but to a hairy old government official in a British port – Antigua perhaps, or Liverpool. Back in England, it was to cause a good deal of abuse and notoriety for the foolish trifler with science and natural history, mixed with just a little admiration for his skill as a taxidermist.

Charles Waterton left British Guiana for the last time in January 1825. He sailed for home. On board ship, this time, apart from the Nondescript, he had a few bird skins, a straw boater with a green riband around the brim, and a young, live cock-of-the-rock which an Indian had given to him shortly before he left. A couple of days before he left, he also acquired a pet jigger flea, or chegoe, 'wishful to try how this puny creature and myself would agree during a sea voyage.'

> Ere long, a pleasant and agreeable kind of itching under the bend of the great toe informed me that a chegoe had bored for a settlement. In the three days after we had sailed, a change of colour took place in the skin, just at the spot where the chegoe had entered, appearing somewhat like a blue pea. By the time we were in the latitude of Antigua, my guest had become insupportable, and I saw that there was an immediate necessity for his discharge; Wherefore, I turned him and his numerous family adrift, and poured spirits of turpentine into the cavity which they had occupied, in order to prevent the remotest chance of a regeneration.[36]

Three days after leaving Antigua, (no report of trouble this time) he also had to take leave of the live young cock-of-the-rock. The weather in the Atlantic was too cold for it. It 'shivered and died.' The voyage across the Atlantic was obviously too cold for the Squire as well; 'my old foe, an affection of the lungs, made its appearance and seemed determined to have its own way.'

The ship docked in Southampton and he took his specimens through customs procedure with no recorded trouble, apparently too ill to make a fuss about customs duty anyway. He was in no condition to argue with anyone and prepared himself for a long, anti-climactic journey by land, with no highly-polished females to sing to him on the way.

Cold, thin, sick, and therefore short of blood, he took stage for Wakefield – for his 'cold and wifeless home' – through the bleak, late winter countryside of a land currently in the grip of its deepest recession for years – acute food shortages, widespread starvation, £30 million annual interest on the National Debt, and the avowed threat of greater than ever divisions between the upper and lower classes, partly on account of the newly-formed Catholic Association and the attendant 'Catholic Question'.

All in all, he'd rather have been in Philadelphia.

Chapter 12
1825-1829
Gathering Clouds: The Happiest Man Alive

It was customary for the Squire to get home from his foreign adventures more dead than alive. With no wife at home to mop his fevered brow, his dependence on the medical profession was almost total. His faith in it was touching.

This time, it was 'the late much lamented Dr Gilbey of Wakefield' who grappled with the illness for six months and received the credit for what warmer weather obviously achieved. Come spring and the first cuckoo, and the invalid was pottering about in the park again, advertising the news in the neighbourhood that he'd shot the Devil at last, and brought him home to Yorkshire. The parish clerk was 'dreadfully shocked at Mr Waterton's profane speech',[1] but the clerk of the Barnsley Canal Company was sufficiently impressed to get his son to draw a picture of the Devil, which was sent, together with the now completed manuscript of the Wanderings, to the publishers, Mawman & Co., in London. At the end of August, the book's publication was announced in The Times:

> This day is published, in 1 volume quarto, price £11. 11s. 6d boards, Wanderings in South America, the North-West of the United States, and the Antilles, in the Years 1812, 1816, 1820 and 1824, by Charles Waterton Esq., of Walton Hall, Wakefield.[2]

He needed all his health and strength, if not a touch of profanity, to cope with the reviews.

In September, the *Monthly Review* carried long critiques of two books as one article – Waterton's *Wanderings in South America*, and *A Historical and Descriptive Narrative of Twenty Years Residence in South America*, by W. B. Stevenson, formerly Private Secretary to the President and Captain-General of Quito, and Late Secretary to the Vice-Admiral of Chile, his Excellency the Rt. Hon. Lord Cochrane, &c, &c,[3] It was painfully clear which of the two books was considered worth reading.

Charles Waterton, a 'gentleman of fortune' was 'merely an adventurous naturalist' – eccentric, enthusiastic, sentimental. W. B. Stevenson, on the other hand, was 'a shrewd observer of men and manners' who had an intimate knowledge of all the Pacific coast countries of South America. The *Wanderings* contained 'much amusement and some instruction' but was clearly not to be taken seriously. Even the Waterton Method was 'susceptible to improvement'. Some of his observations were accurate – the account of the blow-pipes and poisoned arrows, in the first Journey, had to be true because W. B. Stevenson had made similar observations in Peru. The account of the sloth, in the second Journey, delivered 'with powers of mind however inferior', nevertheless rendered Buffon obsolete, and the accounts of snakes in the third Journey were 'given with so much naivete and appearance of truth, that we cannot doubt its authenticity.' But the fourth Journey was 'almost wholly devoid of matter and interest'. The *Monthly Review* turned its more considered attention to 'the far more important publication of Mr Stevenson' – long since forgotten.

Early in the New Year, the *Literary Gazette* – a weekly publication – carried three articles on *Wanderings in South America*, in three separate issues.[4] The book was nondescript, like its frontispiece, its author clearly addicted to the malady of fine writing and rhetorical flourish – 'The style is odd, the descriptions odd, the stories odd; and, in short, the whole medley is odd, not even excepting the Natural History.' But the Squire's character remained unimpeached:

> Will no gentle lady marry our author? His laments about the wifeless condition of Walton Hall are most melancholy and might melt the sternest She in England. We are sure he is a tender-hearted person, and would make the matrimonial state very pleasant.[5]

In March, the *Quarterly Review* gave an equally benign character reference and a slightly more appreciative notice of the book itself.[6] It was compared favourably with *The Compleat Angler*. What faults it had were comparatively trivial. The *Quarterly's* opinion of the Nondescript – 'a quackish performance' – has already been quoted. It was silly tricks like these which had made experts dismiss the duck-billed platypus as a hoax. The magazine's view of the cayman story has also already been quoted – it staggered their faith. And some of the snake stories were considered 'wondrous tough', particularly the one about the boa constrictor and the braces. But generally speaking, the *Quarterly Review*, like almost everyone else, was enchanted by Charles Waterton himself: 'We could extract a thousand small touches which prove how lavishly nature has bestowed on him the milk of human kindness.' It was the implied accusation of Munchausenism which hurt. Once established, the reputation was never entirely shaken off.

In the following month – April 1826 – the most favourable of all the first edition reviews appeared, in a religious publication called the British Critic.[7] Not really surprising; the book and the magazine were published by the same firm. 'He is for the most part concise and sententious,' said the Critic, 'but his pen is tipped with poetry.' Etc etc.

The *London Magazine* published a similar recommendation, with slight reservations,[8] and just about every newspaper in the land carried excerpts from the most bizarre adventures of the third journey. But the most famous review of them all – the one which repeatedly gets called the only favourable review – was the one written by the Reverend Sydney Smith, in the Edinburgh Review.[9]

Smith, who 'never read a book before reviewing it; it prejudices a man so', clearly made an exception where *Wanderings in South America* was concerned. He found 'force and vigour', 'considerable powers of style', 'entertaining zeal and real feeling.' Above all, he found character:

> There is something to be highly respected and praised in the conduct of a country gentleman who, instead of exhausting life in the chase, has dedicated a considerable portion of it to the pursuit of knowledge. There are so many temptations to complete idleness in the life of a country gentleman, so many

Figure 21. Drawing of the Nondescript, by F.H. Foljambe. Charles Waterton's notebook, Wakefield Museum. When it appeared as the Frontspiece of Wanderings in South America, *it was assumed to be a portrait of the author. (Wakefield MDC. Museums and Arts.)*

> examples of it, and so much loss to the community from it, that every exception from the practice is deserving of great praise.

The book's natural history proved a fertile source for Smith's own brand of entertaining zeal: political wit and satire – the Nondescript Master of Chancery, basking in the House of Commons; the sloth passing its whole life in suspense, like a clergyman distantly related to a bishop; the ecumenical vultures, sharing spoils regardless of religion, and so on.

As a leading liberal, he was also happy to see the United States praised for a change – 'a great lesson to the people of England. . . not to be defrauded of happiness and money by pompous names and false pretences.' Probably for the same reason, he shared another point of view with the Squire: the Tory journals had doubted Mr Waterton's justification for complaining about customs duty, but Sidney Smith was unreservedly on his side:

> That a great people should compel an individual to make them a payment before he can be permitted to land a stuffed snake on their shores, is, of all the paltry custom-house robberies we ever heard of, the most mean and contemptible.

Apart from the Nondescript – trifling with science – Sidney Smith's only reservations concerned style:

> Our good Quixote of Demerara is a little too fond of apostrophising – "Traveller! Dost thou think? Reader! Dost thou imagine?" Mr Waterton should remember

> that the whole merit of these violent deviations from common style depends upon their rarity, and that nothing does, for ten pages together, but the indicative mood.

Generally speaking, then, the reception of the Wanderings was sharply divided according to the political and religious colour of the periodical. Blackwoods was late in the field and claimed that it abandoned plans to carry a leading article on the book because it disliked sailing in the wake of any other vessel – 'brig, ship or sloop'.[10] Instead, it contented itself, in June of 1826, with a long review of Wilson's *American Ornithology*, recently completed by Charles Lucien Bonaparte. In that article, *Blackwoods* declared its belief that the periodical press had 'done itself infinite credit by its justice and benevolence to Mr Waterton and his Wanderings.' The constantly repeated notion that all reviews except Smith's were unfavourable, is not strictly true. But they certainly were patronising. The Squire was not unused to condescension but he probably took umbrage all the same. Luckily, he had other things to occupy his mind.

The reviews trickled in, month by month, and whether or not he saw them, he now devoted himself, less controversially, to more important affairs of estate. The last section of the park wall was completed – total cost £10,000 – and the masons were paid off with the latest accumulation of solid tin. The nature reserve could now be said to be ready, if not complete.

Before the last stone was laid, however, the last of the foxes had to be dug out and evicted, and two badgers, discovered in the process, were also set at large on the outside of the wall. (He seemed to think that the badgers were living in the fox earth; it was almost certainly the other way round; the Brocks had probably been living there longer than the Watertons.) The reason he gave for getting rid of the badgers was that they might tunnel under the wall and let in more rabbits, but he later realised how wrong he'd been; he should have kept the badgers, if only to control the wasps.[11] A number of rabbits escaped the spade anyway, and began to breed with some enthusiasm, behind the fox-proof protection of the wall. A field of turnips was much appreciated by the rabbits.

> At last the mandate was issued for their extirpation, and for the first time for many years guns were fired and dogs roamed at large within the sacred precincts.[12]

At least the poachers now had something to think about, though a nine-foot wall might have been as discouraging to some of them as a red rag to a bull. But towards the end of 1826, as he busied himself with the intricacies of nature's balance, a spectre loomed, more terrible than any predator yet encountered; a pest of such sinister proportions that it was eventually to take his mind away from Walton Park and slowly turned the milk of human kindness extremely sour indeed – John James Audubon.

Audubon landed in Liverpool, loaded with pictures and letters of introduction, on 20 July. His paintings were shown at the Royal Institution and greatly impressed the bigwigs of Merseyside society. Manchester was less enthusiastic but by the end of October, he'd arrived in Edinburgh and took that city by storm.

At the university, he lost no time in making two of his most vital contacts – Doctors Jameson and Duncan. Jameson was regius professor of natural history, Duncan was materia medica lecturer. Both men took Audubon under their wings and all three incurred measures of contempt and displeasure from Charles Waterton and Charles Darwin.

Darwin, at that time, was a student at Edinburgh and frequently attended meetings of the Wernerian Society, which Jameson had founded. (Abraham Gotlob Werner: German mineralogist – proponent of geological red herring called Neptunianism, which Robert Jameson swallowed, hook, line and sinker.)

Jameson's lectures were a nineenth-century anticipation of Librium. The effect they produced on Darwin was 'the determination never as long as I lived to read a book on geology or in any way to study the science.'[13] Duncan's medical lectures achieved a similar result for medicine. But at meetings of the Wernerian Society, held in an underground room at the university, various non-academic speakers were invited, including, apparently, J.J. Audubon.

> I heard Audubon deliver there some interesting discourses on the habits of North American birds, sneering somewhat unjustly at Waterton. [14]

The sneers were obviously in response to a 'scrubby letter' which Audubon had received from Waterton on 5 November, soon after arriving in Edinburgh.[15] The fireworks had begun. The Squire had got wind of Audubon's arrival in Britain, either from the reviews which were beginning to appear in the press, in praise of the American's portfolio, or from correspondence with George Ord of Philadelphia. The contents of the scrubby letter were never disclosed, but its effect seems to have been dramatic. Sneers were exchanged with compound interest. The great protestant 'detonating festival' was an appropriate time for it all to start.

Audubon continued to hobnob with the Edinburgh establishment – charlatans or otherwise – as conscientiously as he'd snubbed the American equivalent in Philadelphia and New York. The most influential member of the Edinburgh circle was unquestionably Robert Jameson, apparently the perfect paradigm of that perennial insult to the intelligence, the unintelligent professor. Audubon paid court like a stage-door Johnnie.

In December, he submitted an article to Jameson, in the guise of a letter, for inclusion in the forthcoming issue of the New Philosophical Journal, founded and co-edited by Jameson. It was written to Professor Jameson but aimed at Charles Waterton. The subject was vultures, more particularly the turkey vulture, Vultur Aura, commonly known as the turkey buzzard.[16]

As Lilliput went to war over the best end to crack an egg, Charles Waterton and John James Audubon became embroiled in the most savage internecine combat over the power of smell attributable to the turkey buzzard's nose. Big-Endians and Little-Endians became Nosarians and Anti-Nosarians. It was civil war amongst naturalists. The academics, naturally, took Audubon's side.

But it was apparently some time before the Squire's attention was drawn to Audubon's article because it was almost exactly five years before he replied – an eventful five years for both men.

In January 1827, whilst Audubon made professional social calls, Charles Waterton was involved in a delicate little business transaction of his own – with his friend, Charles Edmonstone. It was all far removed from vultures' noses and putrid meat.

> I heard from Robert [Edmonstone's son] about a couple of weeks ago and he informed me that he had commissioned a friend to arrange the business with you. . . . I shall be rejoiced to hear that everything is settled to your liking.[17]

The merchandise in question was Edmonstone's 14-year-old daughter, Anne Mary. 'I find I cannot live with any comfort at home unless I were married; and I have not courage enough to look out for a wife.' The Squire and Anne Edmonstone were to be engaged. His line of descent was to be saved, at the expense of any vocational hopes he may have entertained with regard to the priesthood. Anne's own thoughts on the subject don't appear to matter, though Mrs Pitt-Byrne insisted that the girl 'had no dearer wish than to trust her life to his tender guardianship.'[18] Since Mrs Pitt-Byrne could never have met Anne, and only met the Squire when he was 79 years old, the source for this piece of information is not exactly primary. The claim, however, is feasible.

In that same letter to Charles Edmonstone, the Squire suggested the possibility of visiting him early in February, when he would have a week or two to spare, to discuss the engagement and to finalise arrangements. Charles Edmonstone himself had married under a similar arrangement – a friend of his had taken to wife an Arowak Indian 'princess', and Edmonstone had married their daughter. So the Squire was now to become affianced to a member of Arowak royalty. By inference and preference, the royal palace would be a wooden hut in a forest clearing.

The most consequential part of the arrangement was that the girl should be brought up a Catholic – no problem. Charles Waterton travelled up to Cardross Park in February, as arranged, more than willing to comply with the terms he'd laid down himself. He was 'still' taking brimstone, he confided, lest any doubt remained as to his health, physical or moral. In his much-travelled portmanteau, he also carried two items of luggage designed 'to astonish their weak minds in the museum at Glasgow' – a stuffed dog's head and the Nondescript. They were probably also intended to tighten his grip on the heart of Anne Edmonstone – what teenage girl could possibly resist a stuffed dog's head or a howler monkey's backside?

After a brief stay in Scotland – long enough to conclude the deal – he returned home to Walton Hall, where, he said, he would be much occupied with the workmen until the summer, at the end of which, he proposed to set off on another of his roughly quadriennial adventures, this time to 'New Holland' (Australia). Events, however, conspired. He never went to Australia, though his direct lineal descendants now live there.

Whilst Charles Waterton was astonishing weak minds in Glasgow, John James Audubon was doing much the same thing in Edinburgh. On 24 February, he read a paper before the Wernerian Society which was shortly afterwards published in Jameson's *New Philosophical Journal* as 'Notes on the Rattlesnake (*Crotalus horridus*)'[19] It told of rattlesnakes chasing squirrels, of rattlesnakes climbing trees and robbing bird-nests, of rattlesnakes fishing, and of 30 or more rattlesnakes entwined together in a copulation orgy lasting several days. It also recounted 'a curious but well-authenticated series of facts' about a rattlesnake fang in a farmer's boot.

The farmer had been bitten through his boot as he went to inspect a field of ripening corn one day, but he thought it was just a small thorn in his sock. He died in pain that night. Twelve months later, his eldest son went to church in the dead man's boots, scratched his leg as he took them off, and died in bed. Another two years later still, a second son tried on the boots and died like his father and brother before him. Doctors cut open the boot, discovered the rattlesnake fang, still stuck in the leather, and scratched a dog's nose with it, to test its power. The dog died.

The whole paper, especially the farmer's boot story, provided more fuel for Audubon's enemies, first in Philadelphia, then, a few years later, in Yorkshire, England. Three more of his articles were published in Edinburgh at roughly the same time – the early part of 1827 – on the alligator, black vulture and passenger pigeon. These, also, were eventually torn apart by Charles Waterton et al.

In April 1827, Audubon had a hair-cut and went to London. On his way there, he stopped off at Belford, to visit the naturalist Prideaux John Selby; at Newcastle, to visit the wood engraver Thomas Bewick; and at Leeds, ten miles from Wakefield, to meet an obscure ornithologist named John Backhouse. He declined to call in at Walton Hall.

In London, Audubon was introduced to more of the relevant elite – president of the Royal Society, secretary of the Zoological Society etc – and the King himself became a subscriber to Birds of America. So did the Duchess of Clarence, later to become Queen Adelaide. The American backwoodsman had come a long way.

But in Philadelphia, the men he'd rubbed up the wrong way were now beginning to scent blood. One of the newspapers in Philadelphia carried a report on Audubon and called his Edinburgh articles a pack of lies. He was told about it all in a letter from one of his few remaining Philadelphian friends, Dr Richard Harlan. Troops arrayed on the field of battle, ready to put the boot in at the first opportunity – a boot spiked with venom from the fang of a mythological rattlesnake.

Charles Waterton, meanwhile, seemed blissfully unaware of all the fuss. He stayed at home, waged war on rats, in lieu of rivals, and dreamed of his forthcoming state of eternal bliss with his dear little Amerindian bird-girl from Dumbarton – his own little Rima.

She was now just 15 years old, ripe for admission to a suitable catholic seminary. In the spring of 1827, after formal betrothal to her father's best friend, she was packed off to Bruges, with her sister Eliza and a chaperon called John Gordon, to complete her education at the English convent in the Carmerstraat, there to prepare herself for a life of eternal bliss at Walton Hall. Strict presbyterian friends were none too happy. 'I many times laugh at the rueful countenance with which Miss Baine told us farewell, as much as to say "poor misguided girls, to what a sinful school you are going."'[20]

But it was a school to which Eliza and Anne at the very least felt a strong sense of maidenly duty right from the outset. 'I like the Convent very much for the short time I have been within its walls. I am quite happy and already feel much attached to some of the nuns; they are exceedingly kind to us.'[21]

Mr Gordon paid a visit and was told by the reverend mother and general mistress that the girls appeared not to be happy at all, but Anne was at pains to assure her father, in her first letter home, that if she felt the least unhappy, she would immediately say so – the purest of pure white lies. (Later, to Eliza: '. . . you know we used to say we were happy months before we were so.')[22]

One thing obviously troubled her much more than was called for but no less than was borne out by events: 'I once intended to have written to Mr Waterton by Mr Gordon, but write I cannot, my hand shakes so much.'[23] She was clearly unhappy about the match but obliged to go through with it, for the sake of honour and obedience, if not for love. That would follow in due course.

Any letter she might find the courage to write would have had to pass through the hands of the general mistress anyway – for approval – and any betrayal of true feelings must needs have been written in code, like premature protestations of happiness to convey unhappiness. This was clearly a somewhat tremulous affair.

The palpitations increased as the fateful day grew near, but for the time being, after a few months of unhappy protestations of happiness, the two sisters settled down to the cloistral way of life at Bruges, and impressed everyone with their humble manifestations of piety and submission to conventical discipline.

Their parents were gratified or reassured as the case may be. But they still had five children at home to provide for and benificent country squires were not all that easy to come by.

The Squire's proposed trip to Australia was cancelled. He stayed at home. Life pursued its leisurely course. On 1 December, he had a flock of nearly a hundred 'mountain sparrows' on the estate – presumably snow buntings – and in the first three months of the new year, a few 'dun-divers' were on the lake – possibly black-necked grebes but more likely juvenile or first winter cormorants from Flamborough Head. In the early spring, a moorhen nested in a brick-built nest chamber intended for ducks, whilst a mallard hatched eleven eggs on the watergate tower, a few feet away from the nests of one pair of barn owls, four pairs of starling, two pairs of house sparrow and one of woodpigeon. These were the most momentous events in what he later referred to as 'little common occurrences incidental to retirement from a busy world.'

Then, in May 1828, perfectly timed for the elevation of common occurrences, that new bi-monthly fanzine for country-lovers was launched – *Loudon's Magazine of Natural History*.

John Claudius Loudon was a landscape gardener and horticulturist who, in just two years, had made himself a small fortune with a freshly-germinated seedling of vigorous habit called *The Gardener's Magazine*. Its natural history equivalent was also destined to be a great success (for the first few years anyway) and was used as the field of battle for the unofficial natural history championship of the world – John James Audubon versus Charles Waterton.

Loudon approached Audubon and offered him eight guineas to write something for the first issue of the magazine, but the bird-painter was still smarting from the criticism his Edinburgh articles had received in Philadelphia; he resolved to publish nothing else until the last, life-size, copper-plate engraving of his great magnum opus – *Birds of America* – was finished. Loudon, therefore, cast around for someone to write a review of the 30 *Birds of America* plates which had already been circulated. (The complete work ultimately consisted of 435 plates, in 87 parts of five plates each – the text published separately as *Orntihological Biography*.)

The man Loudon found to review the Work in Progress was William Swainson, himself a painter and naturalist (and a member of Jameson's Wernerian Society) who had published a book of *Exotic Conchology* and was now engaged on the third volume of his *Zoological Illustrations*, entirely at his own expense. Swainson was to become chiefly known for a contorted new theory of nomenclature called the Quinary System – 'theory of circles', as Charles Waterton called it.[24] Swainson's review of the first six

Figure 22. Alexander Wilson, 1766-1813. Studio photograph of statue which now stands in the grounds of Paisley Abbey. (Paisley Museum. Renfrew District Council.)

parts of Birds of America duly appeared in the first issue of Loudon's magazine and left no paean unsung in its acclamation of genius.

In nothing is the inconsistency of mankind more striking, instructed Swainson, than in their treatment of genius. [Not to mention pronouns.] In every age, however enlightened, and in every kingdom, however great, innumerable are the melancholy examples of its coldness, ingratitude or apathy.[25]

He was determined that no such fate should befall John James Audubon. He'd read the two vulture articles and compared them favourably with the work of Alexander Wilson. Both men 'drank the pure stream of knowledge at its fountain-head' etc. *Birds of America* was a fitting sequel. The first 30 plates achieved a perfection at which Swainson himself had long aimed. 'It will depend on the powerful and wealthy, whether Britain shall have the honour of fostering such a magnificent undertaking.'

Needless to say, Charles Waterton was one member of the powerful and wealthy who omitted to foster. He reserved his patronage for the local poor. A royalty cheque for £100 dropped through the Walton Hall letterbox one morning and the Squire set off, like Santa Claus, to give it away. First victim was a local Protestant who, according to Hobson, had been in the habit of publicly abusing the Squire, on account of his being a Catholic. The unprincipled refusal of such men to be as abjectly grateful for gratuitous aid as their stations in life obliged, was sufficient proof to Hobson – himself a protestant – of the 'impropriety of indulging in promiscuous charity.' The Squire, of course, treated charity more charitably. This was not the last time that he gave his royalties away.

But John James Audubon and his *Birds of America* were not considered suitable objects for charity. He seemed to be doing quite well for himself anyway. Swainson's rave review of Audubon's work struck Audubon as very just - clearly a man of good taste. The two men met, established a warm friendship, and went off to Paris together, in September, to look for more subscriptions. They returned two months later, to respective homes in England, with 13 more names on the subscription list (including King Charles X) for a complete work which cost 200 guineas per copy.

By now, Audubon had about 150 subscribers in all. Waterton's *Wanderings* was also selling well. But the latest volume of Swainson's *Zoological Illustrations* was well-nigh still-born. In January, Audubon received a letter from Swainson, complaining about the scale of public indifference and, by implication, its treatment of genius – his book had been bought by a grand total of ten people. So much for the natural history market. But the most significant part of the letter is the final two sentences:

> I have had a most extra-ordinary letter from Waterton which will highly amuse you. The man is mad – stark, staring mad.[26]

The Squire had obviously read Swainson's eulogy of Audubon, in Loudon's Magazine of Natural History, and disapproved. Its contrast to the reception for Wanderings in South America could not have escaped him. Some years earlier, he'd been a friend of Swainson's father (a collector of customs at Liverpool) and had given William 'lessons' as a young man. An ex-pupil's friendship with an assumed enemy was certainly seen as betrayal.

Alliances, then, were beginning to take shape, and battle lines were forming, for the great trivial pursuit of the century. But on 1 April, 1829, Audubon went back to America for a few months, to collect a few more specimens to paint and to patch up strained relations with his wife, whom he hadn't seen for three years. Charles Waterton was also otherwise engaged – to Anne Mary Edmonstone.

The engagement was not announced in *The Times*, nor even in *Loudon's Magazine of Natural History*. A number of hopeful females therefore imagined they still had a chance. One of them – a Miss A. Nondescript – 30 years old and living in 'a little country town' – wrote him a 67-line proposal of marriage, in verse, with a six-line postscript, also in verse, somewhat in the style of a very large valentine card. She loved him for his ready wit, which could all times and seasons hit; she loved him for his well-stored mind, nor did his judgement lag behind; but ah! still more than these impart, she loved him for his feeling heart. Etcetera. She clearly hoped for nothing more in all the world than to sit at his feet and listen to his tales of snakes and vampires and chegoes and ticks and bell-birds and nightjars and sloths and that wondrous cayman which he bestrode just like a drayman. Oh, that he would just pop the question, but that, she knew, required digestion.

She clearly wasn't much of a catch – not according to herself anyway: 'Of beauty I have none to tempt thee, / And for my coffers, they are empty' but she hoped that if he felt inclined to marry, he would let his answer duly be in Morning Herald, 'which I see.'[27]

So he duly answered, in slightly more scholarly kind, with woeful tidings to impart, that ere the charming lines she wrote, a nymph had shot him thro' the heart.

> Did I know but where to send it,
> A small present I would make;
> This request should then attend it;
> Keep it for the Wanderer's sake.[28]

Poor Ignota! 30 years old, none too sure of herself, and destined to spend the rest of her days a penniless spinster in a little country town. Just one month later, Charles Waterton was to marry his nymph, Anne Mary Edmonstone, two years after she'd entered the convent in Bruges.

Anne's father was now dead. His health had taken a serious turn for the worse in the spring of '28, and he died in the autumn, after a long series of recriminations regarding his relationship with his third daughter, Helen – possibly incest, though probably not. Whatever it was, it did nothing to improve the health of his wife, who had come from much more respectable stock in the South American jungle. Her health was further endangered by family unpleasantness over her husband's will. The Edmonstone clan was clearly in crisis.

But Anne's future, at least, was now assured, although 'I tremble when I think of it. One happiness is, that it will be very private.'[29] You can read compensation for happiness, or infer that one implies many – just as you think fit. Whatever her true feelings, certain it is that, on Monday 18 May, 1829, in the eighteenth-century chapel of the seventeenth-century nunnery, in Carmerstraat, Bruges, the barely 17-year-old grand-daughter of 'Princess Minda' of Guiana walked down the aisle with the almost 47-year-old Squire of Walton Hall, Yorkshire, to be bound in holy estate of matrimony, till death did them part. Her wedding dress was a gown of fine book muslin over a passement slip, with white satin shoes and silk stockings, a white veil, and 'a very handsome chemisette'.[30] The groom doubtless wore a blue swallow-tail coat with gold buttons, yellow kerseymere waistcoat, frilled white shirt and black cravat. One happiness was, that the ceremony took place at 5.30 in the morning.

They took breakfast in the convent and set out for Ghent by the nine o'clock canal barge (or a canal barge at nine o'clock) for the first leg of their honeymoon. John Gordon and Eliza Edmonstone went with them. They looked around a few Catholic churches in Ghent, 'which we found beautiful in the extreme',[31] and naturally called in at the botanic gardens, the 'cabinet of natural history' and the beguinage. It being a special occasion, and the spirit of romance not yet dead, they also, of course, paid a special visit to the bones in the osteological museum.[32] Then John Gordon returned to Scotland and Eliza went to stay for a few days in Wakefield, with the Squire's sister and brother-in-law, Helen and Robert Carr. Robert Carr was a local solicitor.

The newly-weds continued on their honeymoon – Belgium, France, Switzerland, Italy – forerunner of many another European 'saunter' for the Squire, in the 1840s and 50s. In Paris, they found a Jesuit priest to act as father confessor but apart from that, the honeymoon appears to have been what all honeymoons are obviously intended to be – a very private affair.

Eliza, meanwhile, enjoyed a conducted tour of Walton Hall with Mr and Mrs Carr. She found Walton Lake 'quite beyond description' and wrote to her sister to convey the information. 'Lay aside your fears, Annie dear, Mrs Carr spoke in the kindest manner, and said she longed very much for your arrival in England.'[33] That blessed advent was not to occur until early July.

The poor girl's fears and premonitions were not entirely laid aside, and in due course they proved justified, but on one thing, at least, she now surely knew she could always depend – her husband's love was completely steadfast and unqualifyingly tender, the very soul of contractualised constancy. He clearly wished for nothing more than her well-being and happiness at Walton Hall, and as for himself, well, he was just, quite simply, and by his own testimony, 'the happiest man alive'.[34] An ominous state of affairs if ever there was one.

Chapter 13
1829-1830
Severed Cord

To some extent, obviously, Charles Waterton stood in loco parentis towards his wife – part of the deal. Its easy to assume, as some do, that she was no more than an unwilling partner in a marriage of convenience, a submissive pawn, a Kate conformable, like any household Kate. What evidence there is, suggests otherwise.

She was, after all, half Indian, or quarter Indian at least, and Arowak women were not generally accustomed to submission; they often held their husbands 'in a most delightful state of subjection', said C. Barrington-Brown.[1] It's reasonable to suppose that Charles Waterton was in a similar state.

Certainly Walton Hall was an improvement on Cardross Park, where illness and family strife was still making life difficult. Mrs Edmonstone, one generation removed from the Amerindian drug culture, was now hooked on the great European sedative of the day – laudanum. She was 'reduced to a mere skeleton', said Eliza, who was in some discomfort herself, from a badly swollen leg.[2]

Helen was no longer at home. She entered the convent in Bruges, like her sisters before her, and wrote to them both to protest that she was happy before she really was. Anne to Eliza: 'Poor Helen, I am afraid it will be some time before she gets accustomed: all the nice girls we knew there have gone.'[3] Eliza stayed on at Cardross Park, where there was apparently still friction with cousins, partly, it seems, over who should have care of the youngest sister, Bethia. Her mother was 'so severely overwhelmed by laudanum that she has no power of authority.'[4]

But at Walton Hall life was less bothersome. Anne settled down to the genteel, decorative duties of an English squiress, her mind occupied constantly with thoughts of her family and friends in Scotland and Bruges. Real work, of course, was out of the question – very unladylike. Despite her husband's Yorkshire brass, she was, irrevocably, a lady, even at 17. But marriage to Charles Waterton would not have been quite the conventional, ladylike vacuum of your average, run-of-the-mill squirely squaw. It's difficult to construct a day-in-the-life-of, mainly because it's difficult to imagine anything as humdrum as an average day at Walton Hall. But so far as can be told, she walked in the park, tended fox cubs, wrote letters, played the piano, said her prayers, embroidered patterns for bell sofas and samplers, and read books. Her husband had all the Latin classics in his library, plus a Spanish edition of *Don Quixote*, Wilson's *American Ornithology*, Dryden, Sterne, Buffon, Smollet, Goldsmith, Sir Thomas More – not the stuff to abduct the mind of a 17-year-old girl, even in 1829. *Wanderings in*

Figure 23. Remains of the owl house, c. 1920. (Wakefield MDC. Museums and Arts.)

South America might have been more to her taste, or better still, *Stories of a Bride*, hot from the press and written by a young girl – Jane Webb – who was later to marry the Squire's friend, J. C. Loudon. The new bride, Anne Waterton, would likely have acquired a copy of the newly-published Stories of a Bride, written by a young bride-to-be who was later to visit Walton Hall many times with her husband.

Boredom was no great problem during the first few months of marriage – it was summer and the estate was a great novelty. If Anne had any inclination at all to watch birds, and thereby, perhaps, to get broody, her husband made sure she could do it in comfort. She watched from the Hall as the hollow trunk of an oak tree was felled, and dragged by a team of cart-horses to the top of Ryeroyd Bank, overlooking the lake, where it was draped with ivy, roofed with slates, and divided into two storeys[5] – the upper storey for owls, the lower for love-birds. A cosier bird-hide has never been contrived.

There were also visitors from time to time, though the visitor she most longed for was Eliza, and Eliza was too ill to travel. Sir Richard and Lady Bedingfeld stayed for a few days, in early October, and their son and daughters stayed three days longer. But Anne pined for her older sister, the girl she'd grown up with and been close to all her short life. Their separation was clearly painful. Eliza had left a large music book at Walton Hall but it was full of those 'pretty little piano duets'[6] which were so popular at the time and Anne had no-one to play them with.

There was a very simple reason why she couldn't go to Scotland, though it was 'a subject rather delicate to a young lady' and therefore not fit for anyone's confidence but Eliza's. Certainly not yours or mine:

> The truth is, I am between three and four months gone in the family way. . . . I beg you will not let any person know I am in that way. I am always ashamed when it is mentioned to me.[7]

She was so ill with the morning sickness that 'the doctor and all the married ladies with whom I am intimate strongly advise me to keep as quiet as possible and to take very gentle exercise, in case of a miscarriage.' But it was not a miscarriage she had in store. She made sure the baby would arrive and made all due arrangements for the happy event. She asked Eliza to look out for some baby linen in Scotland, where it was cheaper than in Yorkshire, and sent a yard of cambric to a local seamstress (via Helen Carr)[8] with instructions to make as many baby caps as possible. If only Eliza could have been with her, to share in the quiet excitement of it all.

> My dearest you need not be afraid of coming contrary to Charles inclination. I would not ask you if he did not wish it. . . . It would be extremely mortifying for me to see any visitor suffer anything from his disposition, but particularly any of my sisters. No my dearest Eliza, I know Charles loves too sincerely ever to put my affections to so severe a trial.[9]

Whatever could she mean – 'suffer from his disposition'? He was up to his tricks again, that's what – climbing trees and such-like. Extremely mortifying.

But not long afterwards, Anne heard from 'cousin Archy', late of Hobabu Creek, that Eliza was in high fever and still couldn't travel anyway.[10] (Letters from her cousins were generally burnt as soon as read – they were 'so full of nonsense' – but she sent her love to all of them, regardless of the nonsense.)

As winter set in at Walton Hall and the wind howled in the casements, and still no Eliza to play with on the piano, the delicate condition became more and more obvious to everyone. She was more and more confined to an island exile. No radio. No television. No piano duets. She'd need all the popular novels she could get. How about *The Mummy, A Tale of the Twenty-second Century*. (Much like the early 19th century, with a few extra telegraphs, electrical rain-making machines, steam lawnmowers and a resuscitated mummy which flies away in the Queen of Ireland's balloon.) Author? Jane Webb. Or *The Castle of Otranto*, perhaps, or *Frankenstein, The Story of a Monster*.

Not that she really needed Gothic horrors. She was living in a museum full of them – her own little Gothick Castle in fact. Fact was stranger than fiction at Walton Hall.

The front doors of the Hall gave hint of things to come – two bell-metal moulds of her husband's face for door knockers, one laughing, the other in a mock grimace of pain. The grinning knocker is fixed and immovable, the other works – the rapper itself hits the Squire on the head, which accounts for the pained expression on the face. They were designed by his old schoolboy friend, Captain Jones, the character who had climbed the dome of St Peter's with him in 1817.[11] Both faces are said to be good likenesses of Charles Waterton.

Inside the doors, to the right, the flickering light from an open log fire cast pale shadows on the oak panels of the entrance hall, as a ghostly white owl flew silently through the open casement window, in search of rats. (Still to be heard, from time to time, scrabbling in the walls.) More faintly, in the fire's glow, near the foot of the great winding staircase, could just be discerned a 'singularly conceived and inimitably modelled representation of the nightmare.'[12] The girl would need to pass it each night, by candle-light, on her way to bed.

> This horrid incubus has a human face, grinning and displaying the frighfully formidable tusks of the wild boar – the hands of a man – satanic horns – elephant's ears – bat's wings – one cloven foot, the other that of an eagle widely expanding his terrific-looking talons, and the tail of a serpent.

The Latin tag translates as, 'Sitting on the region of the heart, I take away sleep by fear.'

At the top of the stairs, the girl's bed-time candle would light on the grinning jaws of the South American cayman, behind which was the organ gallery[13] – every Gothick

Figure 24. John Bull and the National Debt. (Wakefield MDC. Museums and Arts.)

castle has its organ gallery. This one was now given over to a collection of such ingenious sleep-inducers as crabs, lobsters, insects, snakes, an ant-eater, a sloth, and sundry fearsome figments of taxidermical nightmare intended to represent the 'Reformation Zoologically Illustrated'. John Knox was a frog, Titus Oates a toad. Bishop Burnet and Thomas Cranmer were 'reptiles of the lowest order'[14] whilst 'Queen Bess at Lunch' was a combination of lizards, newts, beetles, flies, and such-like emblems of the Devil and eternal damnation as came to hand.

One beast 'unknown in Britain since before the Reformation' was the Noctifer, or Spirit of the Dark Ages, created from the gorget and legs of a bittern and the head and wings of an eagle owl, squashed into a sort of pot-bellied, consumptive bird of night with outsize, gallinaceous clod-hoppers for feet – not a pretty sight.[15]

Amongst the snakes was the great boa constrictor which had given the Squire himself a sleepless night in his jungle bed-chamber, and a hooded cobra de capello – Cleopatra's asp – 'the most noxious offspring of Hell and darkness.'[16] The pregnant girl would naturally have wished them all a fond goodnight before retiring.

The Nondescript was there, somewhere, so was John Bull and the National Debt, or 'Old Mr Bull in Trouble', the most famous example of an art-form virtually invented by Charles.Waterton – the taxidermical version of political cartoon. John Bull is a hairy quadruped with humanoid features similar to the Nondescript's but said to have been moulded from the face of a hedgehog. The tortoise shell on his back is weighed down by purses marked 'National Debt' and 'Eight Million Pounds', guarded by 'a sort of grinning lizard all over abnormal spikes and horns' called *Diabolus Bellicosus*. John Bull's retinue includes a winged toad *Diabolus ambitiosus* an outsize millipede – *Diabolus illudens* – and a bloated blue lizard – *Diabolus caeruleus* – 'with open mouth and sharp teeth'.[17]

These were the companions of the young Mrs Anne Waterton during the long, cold winter of 1829-30. As labour pains started, she felt the need for congenial female company more keenly than ever. She wrote to Helen Carr and told her that, 'If it is not inconvenient for you to leave home at present, I should like very much to have you with me, but only on condition that it does not inconvenience you in any degree.'[18] But a pleasant little postscript suggests that life was not too bad at Walton Hall: 'Tell Mr Carr the little foxes can all see. We have been teaching them to lap and I think we succeed uncommonly well.'

By now, she was heavy with child and almost due. It had obviously been a difficult pregnancy for Anne but not unduly so and certainly not enough to cause any serious alarm. On 21 April, the baby arrived, perfectly healthy and eager for life and patrimony. Aside from a feeling of slight anti-climax, the mother also seemed perfectly well. But all, unfortunately, was not well. As the placenta broke free and the umbilical cord was cut, the raw surface of the womb had become infected. Gothic horror and nightmare now became real. There were no anti-bacterial drugs to prevent it.

Helen Carr wrote to Eliza, apparently to prepare her for bad news, but Eliza had no means of knowing Mrs Carr's address and couldn't reply. She wrote to Anne who sent her the address and then sat up in bed to scribble out a pencilled note to that same address herself – a well-nigh illegible letter, probably written in pain. It still hurts to read it, almost 200 years later.

> I expect to be carried to the sitting-room. Perhaps you would come over tomorrow. The little Edmund is well.[19]

On 27 April, 1830, less than twelve months after the wedding and only six days after the birth, Charles Waterton became a convert to the sad, Rasselasian notion that perfect human felicity is completely unattainable – 'it pleased Omnipotence to sever the cord on which hung all my happiness here below.'[20] She died of puerperal sepsis – childbed fever – and was buried three days later, in the Waterton family vault at St Helen's church, Sandal Magna. She died happy, they said – 'most happy' – secure in the knowledge that the little Edmund was well and that she could take with her to Heaven the deep, undying love of her husband, the unhappiest man alive.

A well-meaning friend in the congregation invited him to weep, if weep he must, 'o'er that care and sin from which to part / Is to the virtuous but escape from woe' – the age-old analgesic of religion.

> Weep not for Her, my friend, whose faithful love
> False Hope once promised to thy future years,
> Who, robed in innocence, in realms above,
> Unchanged in youth and loveliness appears.
>
> Is't not enough – that free from guilt & sin,
> She wins the prize which millions seek in vain?
> That the full joys of Heaven She enters in,
> Till God restore Her to thy arms again?[21]

So he wept for Man's fallen nature, to ease his burthened heart, but still he was not consoled. 'I am now desolate and forlorn and I can barely keep myself from sinking under the heartbreaking weight of grief and sorrow. . . . My heart is full of grief, full to overflowing.'[22]

He blamed himself for her death and, true to the dictates of his deep religious faith, he resolved to serve the rest of his life in 'self-inflicted penance for her soul.'[23]

Chapter 14
1830-1839
Foul Falls for the Devil

The death of Charles Waterton's beautiful young wife reinforced his faith in divine grace.

> It pleased Heaven to convince me that all felicity here below is no more than a mere, illusive, transitory dream; and I bow submissive to its adorable decrees. I am left with one fine little boy who 'looks up to me for light.'[1]

He clearly believed that he was responsible for the death – bacteria from dead specimens – and probably said as much in the confessional. His father confessor pointed out his obligations here below and the resolution was obvious – he would become a saint; it ran in the family. (He claimed to be descended from seven saints,[2] not including Sir Thomas More who was not raised to the altar until 1935.)

From now on, he decided, he would live out his life in penance. In particular, he would never again sleep in a bed. From that day until a few days before his own death, he slept on the bare floor, with a hollowed-out block of wood for pillow and a military cloak or a napless blanket for covering. Sometimes he made do with no pillow at all.[3]

The inspiration for self-deprivation was obvious: St Ignatius had slept on the floor, with a log or a stone for pillow, and Sir Thomas More had warned his own daughters against trying to get to Heaven on a feather bed – 'it is not the way'. Charles Waterton was determined to go to Heaven, to meet his wife and ancestors. To make doubly sure, he probably also wore a hair shirt.

Amongst English gentlemen, self-denial was the fashionable entertainment in 1830, but very few met Squire Waterton's stern example. Asked to explain his renunciation of beds, he would say that the last occupant of the bed in question might have been 'some lovely form divine', like his wife, but was just as likely to have been a 'horrid alderman . . . a rough-skinned, pimpled victim to turtle soup and Curacao.'[4] But the real reason was penitence.

His apprenticeship to the hard floor cost him a fortnight, after which he became accustomed to it. His grief took longer to accommodate. For a week after the death, he spoke to no-one at all, and for the rest of his life he rarely spoke to anyone about his wife; when he did, it was in the most solemn of terms, barely distinguishable from religious worship. She was his sweet, immaculate 'angel of a wife . . . all gentleness and elegance and virtue and dignity.'[5] Heaven was clearly a much better place for the presence of people like that.

In the back end of 1830, a few months after the death, he managed to acquire a collection of 148 oil paintings, at least one of which was not unconnected with his wife's memory. He bought them from a Herr Berwind, of Wurzburg, during a brief pilgrimage to Bavaria to witness the miraculous recovery of a local nun from an incurable disease. The collection was considered the 'pride and ornament' of Wurzburg, according to the Squire,[6] with paintings attributed to Holbein, Michaelangelo, Tiepelo, Rubens, Van Dyke, Paolo Veronese and many others. Whether or not the attributions were as genuine as the miracle, he snapped up the job-lot at an undisclosed bargain price and duly hung Holbein & co. at Walton Hall, amongst the sports and diversions of zoological illustration. One picture was of *The Holy Family and St Catherine*, by an obscure Flemish painter named Otto Van Veen, but pride of place over the mantelpiece in the drawing room was ultimately given to an individual portrait of St Catherine – painter unknown – which the Squire believed to be a very good likeness of his wife. Many years after her death, he would look long and lovingly at it, whenever he was 'in his thoughts',[7] and his chapel on the ground floor was dedicated to St Catherine when it was finished. Her death on the wheel was considered a suitable metaphor for his wife's death.

On his way home from Bavaria, he stopped off at Bruges, to visit Anne's convent again, and the church where, a little over twelve months previously, he and she had been wed. Circumstances now were very different. Bruges was currently in the grip of a struggle to become independent after 15 years of Union with Holland. On the Sunday after he arrived, he went to the main square of the town and personally witnessed the hoisting of the Belgian tricolor on the church tower.

> The commander of the troops ordered his men to fire on the multitude. Luckily, the greater part of the soldiers were Belgians, and they fired in the air. However, on this wanton discharge of musketry, four poor fellows in the crowd fell mortally wounded; I saw them lying bleeding on the ground. Seven or eight more were wounded by the same discharge. This irritated the multitude beyond all description, and had not the Dutchmen evacuated the town early on the following morning, they would have been cut to pieces to a man.[8]

One of the men injured in the shooting was a waiter at the Hotel du Commerce where the Squire was staying – a smart young man named Louis whose 'pretty little cherry-cheeked Flemish sweetheart' (obvious mnemonic for a pretty little tawny-skinned Amerindian angel) was greatly distressed. The Squire himself naturally got as deeply involved as possible in the disturbance and inevitably got caught in the line of fire. He took refuge in the nearest doorway but an old woman shut the door in his face and left him to his fate. So he threw some Latin at her. 'Felix quam faciunt aliena pericula cautam', he told her, indignantly.[9] (Prudence heeds the misfortune of others.) She heard it, but she heeded it not.

Back in emancipated England, he rushed off an article for Leigh Hunt's weekly *Examiner* and told the English-speaking radicals who took it, plus a few Protestant Tories who happened upon it, that 'Never, never more will the Flemmings submit to the Dutch king, or have anything to do with the House of Orange.'[10] This article marks the beginning of a fiercely argumentative phase in Charles Waterton's life, most charitably diagnosed as post-traumatic stress.

Just a few weeks after publication of the Examiner article, an independent Belgium was established at last. A claim to the throne by the Prince of Orange was rejected by the Belgians, as the foreign correspondent from Walton Hall predicted. In England, the Reform Bill went to the House of Lords but 'those hereditary boobies' knocked it on the head[11] – again, as predicted. At about the same time, the Game Act passed successfully through parliament, legalising the status of gamekeepers – 'merciless savages' – but Charles Waterton made no predictions.

Protection laws for birds other than game were still non-existent; nor were there any restrictions on the introduction of species from abroad. Audubon came back from America with 350 live passenger pigeons which he'd bought in a New York market at four cents each. He gave them to London Zoo, and to various British noblemen, including the Earl of Derby who had some success in breeding from them. But it was Audubon's account of the slaughter of passenger pigeons at Green River, Kentucky, which put the running fight with Charles Waterton in context – a fight which was now about to become public through the auspices of J. C. Loudon.

Audubon had seen 'immense legions' of passenger pigeons, which shut out the noon-day light 'as by an eclipse'. At the Green River shoot, the noise they made was like 'a hard gale at sea, passing through the rigging of a close-reefed vessel.' They arrived in their thousands all night long and broke the branches of trees by sheer weight of numbers. Men shot them down or simply knocked them out of the sky with poles.

> Persons unacquainted with these birds might naturally conclude that such dreadful havoc would soon put an end to the species. . . [but] I have satisfied myself by long observation that nothing but the gradual diminution of our forests can accomplish their decrease.[12]

The world's last passenger pigeon died in Cincinnati Zoo on 1 September 1914. It was hunting, not tree-felling, which drove the species to extinction. The Green River events were almost exactly contemporary with Charles Waterton's 'sentimental' injunction not to kill more than one pair of doves – 'that would only blot the picture thou art finishing, not colour it.'[13] Audubon, in fact, considered it a poor day's shooting if he killed fewer than a hundred birds. On some of his collecting expeditions, he shot so many birds that the pile resembled a haycock.[14] There are now Audubon Conservation Societies all over America. The only Waterton Society is in Waterton Lakes National Park, Alberta. But the great American bird painter had obviously blotted the picture he was finishing.

Shortly after Audubon arrived back in England, with 350 pigeons and his wife and son, he set about supplying the text for his *Birds of America* plates. His English was still imperfect, so was his taxonomy, so it was arranged that William Swainson should help with the writing and with the scientific nomenclature. But Swainson wanted his name on the cover – Plates by Audubon, Text by Swainson, that sort of thing. Audubon wanted sole credit. The two men fell out.

Audubon needed a new collaborator and quickly chose William MacGillivray, the man who, in 1819, had walked from Aberdeen to London to see the British Museum bird collection. He later became Professor of Natural History at Aberdeen University.

MacGillivray's fee was two guineas per 16 pages of manuscript corrected, but Charles Waterton fancied something more than mere correction was involved. He was sure that MacGillivray wrote Audubon's book.

Then, in 1831, another outrage was perpetrated – a new edition of Wilson's *American Ornithology*, 'edited' by Professor Jameson. Wilson's contribution to his own book – with footnotes by William Jardine – was prefatory and well-nigh subordinate to 'Additional Details in Regard to the Birds of America, and Birds in General', by J.J. Audubon, J. Richardson and W. Swainson.[15] The 'Appendix' so titled contained long extracts from the proofs of *Northern Zoology (Birds)*, by Richardson and Swainson, which at that time was in the process of going to press. The extract – in *Wilson's* book, remember – contained another long excerpt from Audubon's article in Jameson's own *Philosophical Journal*, on the supposed power of smell in the turkey buzzard, corroborating Swainson's, Richardson's, Jardine's and, of course, Jameson's false assumptions, and contradicting Waterton's, and Wilson's, correct ones. It also contained an equally long extract from Audubon's description of the Washington sea eagle, a new species discovered by Audubon[16] which turned out to be nothing more than an immature bald eagle – as illusory as any Nondescript but not intended as a practical joke; Swainson, Richardson, Jardine, Jameson and, of course, Audubon himself, all firmly believed it was a new species.

Many other birds were discussed in the Appendix to *Wilson's American Ornithology*, with reference to Audubon's plates, not Wilson's, whose pictures were totally omitted from most editions of the book anyway.

At about the same time, two more rave reviews of Audubon's work in progress appeared in Edinburgh and before the year was out, similar recommendations were published in London and Philadelphia. The Squire, by now, was furious. *Wanderings in South America* was selling well and was now being translated into French, but the critical reception was always tepid at best. He'd read a letter in Loudon's magazine which applauded not only Audubon's 'Shakespearean' notes on the Washington eagle, but also the turkey buzzard article which had supposedly exposed the 'fallacy' regarding its sense of smell.[17] Something had to be done, and the obvious place to do it was in Loudon's *Magazine of Natural History*. He soon became the magazine's most regular contributor. (Closely followed by 'Rusticus of Godalming', of eternal memory.)

It all began in earnest in November 1831, when the first of his subsequently collected essays on natural history appeared in the magazine. Its style set the tone. It was an attack on James Rennie, professor of natural history at King's College, London.[18] Charles Waterton lambasted Rennie's 'deficiency in bog education'; the chair in natural history was soon abolished for want of students.

Rennie had recently edited Montagu's *Ornithological Dictionary*, where he referred to Charles Waterton as an eccentric crow.[19] And his list of 'Philosophic Naturalists and Original Observers' pointedly omitted the naturalist whom some considered the most original philosophic observer of the lot – Charles Waterton. Audubon, of course, was included. 'Many writers have recourse to theory,' said the Squire. 'Others adduce facts.'

Rennie replied, superciliously, that Waterton had 'published nothing respecting the economy or faculties of animals of the least use to natural history.'[20] Which was true in all but its context.

Stupid professors drew his fire but most of Charles Waterton's zoological ammunition was reserved for Audubon. Five years and many accolades after publication of Audubon's 'Account of the Habits of the Turkey Buzzard', Waterton replied to the latest round of applause – from Percival Hunter – and gratuitously gave the article its first public thumbs down.[21] Several instances from his own experience proved that the turkey buzzard finds its food by scent and that it ignores living flesh, even though it might appear dead to the naked eye.

There was an element of hit and miss in it all, but Charles Waterton was definitely right with regard to the turkey buzzard's nose, and Audubon was wrong. The debate evolved slowly, however, and with increasing rancour, punctuated by other affronts to the delicate sensibilities of a Catholic nosarian with resolutely pro-American sympathies.

His main non-scientific bogies at this time were William Cobbett and Frances Trollope. During 1832, Cobbett ran for election to the new reformed parliament – as a candidate for Oldham – and Fanny Trollope's *Domestic Manners of the Americans* was published in London. Cobbett's chief crime was patriotism – during six years in Pennsylvania, he'd published about 20 pro-Monarchist pamphlets and claimed that, by 1800, every family in America had read some part of his writings.[22] With few exceptions, he hated America 'and all their mean and hypocritical rule.'[23] Not a man after Charles Waterton's heart.

When the election campaign got under way, the mean streets of Manchester were glorified by a long and scholarly lesson in moral and political philosophy, composed at Walton Hall, printed in Wakefield, and posted on walls all over Oldham and Greater Manchester, for the guidance of the newly-enfranchised man-in-the-street – a hard sell, populist piece of election literature, laced with political allusion to Ulysses, Don Quixote, Oedon, Medea, Julius Caesar, Hercules, Catilene, Cato, Aristedes, Cincinnates, Circe, Galen, Ajax and the wooden horse of Troy. All to no avail – the man-in-the-street still elected Cobbett.

Before going to America, Cobbett had been one of the severest critics of Charles Willson Peale's close friend, Thomas Paine. When he returned to England, Paine's disinterred corpse was in his luggage, for honourable, atheistic re-burial in his native soil. The Squire of Walton Hall drew Oldham's grateful attention to the fact. There was clearly 'a speculation afloat', he said. Cobbett was selling rings, at a guinea each, containing relics from the dead man's mortal remains.[24] But it was the anti-Americanism which chiefly offended the Squire.

Frances Trollope despised America even more than Cobbett. 'I do not like their principles, I do not like their manners, I do not like their opinions.'[25] Domestic Manners reads like a caricature of Victorian prudery, published five years before Victoria's accession. The Squire gave his friend, George Ord, advance warning:

> I have just read the first volume, and struggled half through the second, of the most arrant trash that ever came from the pen of woman. Mrs Trollope (a fit name) has published two volumes on the Domestic Manners of the Americans. The jaundiced old jade convinces me that she is either the most ignorant or the most duped of all our fools who visit America.[26]

Its most notable effect on Charles Waterton was to rouse the muse. In a fever of creativity, the very soul of the age, he composed a magnificent Augustan epic on the subject, redolent of extensive scholarship in the Latin classics but inexplicably omitted from Palgrave's Golden Treasury:

Figure 25. Rattlesnakes and mocking-birds. J.J. Audubon. Plate XXI of Birds of America.

O Naughty Mrs Trollope,
You deserve to have
your clothes all up
And receive a good wallop.[27]

But Fanny Trollope had made no personal insinuations against the Squire and her little urban peccadillos were therefore not really worth answering. Table manners and the insubordination of servants were not matters of sufficient import to Charles Waterton. Audubon was another matter.

Audubon himself was now in Labrador. His reputation was left in the care of family and friends. An extract from his turkey buzzard article appeared in Loudon's Magazine, to appreciative applause from friends,[28] but Charles Waterton called it 'lamentably faulty at almost every point.'[29] The same applied to just about everything else that Audubon wrote. *The Ornithological Biography*, on the other hand, was clearly the work of a finished scholar. William MacGillivray not only corrected Audubon's grammar; he obviously wrote the whole book. (A view which MacGillivray's biographer appears to share.[30])

A series of extremely acrimonious exchanges ensued, concerning not only authorship of *The Ornithological Biography*,[31] but rattlesnakes chasing squirrels,[32] an aerial tug-of-war between an eagle and a vulture,[33] hummingbirds flying at one week old,[34] and the Virginian partridge's ability to swim.[35]

As far as the turkey buzzard and black vulture were concerned, the Reverend John Bachman – a Lutheran minister whose daughters later married Audubon's sons – now composed a formal affidavit and managed to get it signed by the leading professors and college lecturers in Carolina; all to the effect that these species 'devour fresh as well as putrid food of any kind, and that they are guided to their food altogether through their sense of sight, and not that of smell.'[36]

Bachman conducted his own experiments, with a painting of a dead sheep and a barrow-load of offal at the bottom of his garden. The vultures attacked the painting but ignored the offal. He drew no scientific conclusions as to the birds' artistic appreciation. Then his reverence – witnessed by several more academic dignitaries – performed the less amusing Christian ceremony of perforating a vulture's eyes, to test a theory that its sight could be restored by holding its head in its own axillary down.

Food placed near the blinded bird was ignored; more supportive evidence for the anti-nosarians.[37]

This last experiment distressed Charles Waterton. (His affection for vultures has already been shown.) As for the confusion of painted canvas for mutton:

> Pitiable indeed is the lot of the American vulture! His nose is declared useless in procuring food, at the same time that his eyesight is proved to be lamentably defective. Unless something be done for him, 'tis ten to one but that he'll come to the parish at last, pellis et ossa, a bag of bones.[38]

Bachman took his apologia for Audubon elsewhere, and the Squire went off to Flamborough Head, to risk his neck on the sea cliffs, in quest of guillemot nests – a very brief pause in the argument.

Eleven times, Charles Waterton was let down the cliffs – on two separate expeditions.[39] The professional egg-collectors searched the cliffs for eggs every third day. The Squire clearly approved of them. He disapproved of the new chimney-sweeping act for much the same reason:

> When I was a lad in the nursery, had I been allowed, I should have liked no better fun than to have had an occasional spree with the little black fellow up the chimney.[40]

He joined the egg-collectors for a spree on the cliffs and was lowered down by rope, one eye cocked for dislodged stones. One of the men grinned to see such fun and revealed a gap in his teeth, caused by a falling stone from the same cliffs the previous year. The Squire was undeterred; the grandeur of the scene more than compensated for any danger to his teeth or to his neck.

Egg-collectors were innocent farmers of the cliffs, as far as Charles Waterton could tell (the eggs were sold as food) but the gunmen who hunted from the relative safety of the cliff-sidings were heartless murderers. As far as most of his contemporaries were concerned, this belief was just another of Squire Waterton's quaint little English eccentricities. The 'sentimental' prose did nothing to dispel the mistake:

> Did these heartless gunmen reflect, but for one moment, how many innocent birds their shot destroys, how many fall disabled on the wave, there to linger for hours, perhaps for days, in torture and in anguish; did they but consider how many helpless young ones will never see again their parents coming to the rock with food; they would, methinks, adopt some other plan to try their skill, or cheat the lingering hour.[41]

On 14 July, he was back on the North Yorkshire coast, this time to descend the Raincliff – '140 yards high' – in search of cormorant nests. The stench offended his nostrils but the gunmen offended much finer sensibilities; the cormorant, in those days, was persecuted for pleasure and displeasure – for 'mere wanton pastime' on the cliffs, and for protection of fishponds inland. Charles Waterton wanted nothing to do with it all. His pleasure was 'science' and the challenge of being where others feared to tread. And fish were expendable.

> I do not care if thou takest all the eels in the lake. Thou art welcome to them. I am well aware that thy stomach requires a frequent and large supply. So pr'ythee, help thyself.[42]

He was always happy to see a cormorant, unless it was Protestant.

Back home from his exploits on the cliffs, he re-addressed his mind to rattlesnakes, on whose behalf he was prepared to be sent to the treadmill for three years and, if needs be, to forfeit his ears.

Six years earlier, in 1828, Swainson had been overcome with admiration, in contemplation of Audubon's famous picture of mocking-birds defending their jasmine-entwined nest against attack by a rattlesnake – Plate XXI in *Birds of America.* The Squire was less impressed; the snake's fangs curved upwards at the tip. If anyone could produce one up-curving fang of any snake anywhere, he said, he would submit to be sent to the treadmill for three years.[43]

In a later essay, he attacked the rattlesnake picture again – this time on account of its eyes staring out of their sockets[44] – but first, his attention was attracted to Audubon's five-year-old rattlesnake article in Jameson's Journal. Someone defended Audubon's account of rattlesnakes climbing trees,[45] and the Squire promptly pounced, tongue quivering, fangs erect. Of course rattlesnakes climbed trees – the merest novice in zoology could perceive that they were perfectly well adapted to climb trees – but, 'I will forfeit my ears if any of old Dame Nature's snakes are ever seen to chase either man or beast.'[46] (Audubon's squirrel-chasing rattlesnake was probably a black racer, Coluber constrictor,[47] which is reputed to chase very small prey, over very short distances, but has never been recorded to travel any faster than 3.5 m.p.h. on any reliable road test.)

Heated arguments simmered slowly in nineteenth century periodicals. The Squire persisted with his attack on Audubon's hepatology, and now received a personal attack from Professor Jameson himself which prettily questioned his honesty: 'O Candour, whither hast thou fled? – certainly not to Walton Hall.'[48] The Squire replied by open letter, using the same Wakefield printer who had issued his Notice to the Electors of Manchester, three years earlier.[49] The 215-word title (here summarised) betrayed a healthy contempt for titles:

> A Letter to James [Robert] Jameson, Esquire, Regius Professor of Natural History, Lecturer on Mineralogy and Keeper of the Museum in the University of Edinburgh; Fellow of the Royal Societies of London and Edinburgh; of the Antiquarian, Wernerian, and Horticultural Societies of Edinburgh; Honorary Member of the Royal Irish Academy, and of the Royal Dublin Society. . . of the Academy of Natural Sciences of Philadelphia; of the Lyceum of Natural History of New York; of the Natural History Society of Montreal, &c. &c. By Charles Waterton, Esquire of Walton Hall.

Audubon's farmer's boot story, said the Squire, was considered a good joke 'fifty or sixty years back' – before either Audubon, Waterton, or Jameson were born. Jameson shouldn't have accepted it for publication. 'Pray, Sir, where were your brains? (Whither had they fled? – certainly not to Walton Hall.)' The professor had chosen to attack the Squire through praise of Audubon; the Squire would not hesitate to attack Audubon and the professor whenever occasion demanded. A second letter was addressed to the world via Jameson [50] – and the prospect of more was ominously real. There was the little matter of Audubon's bird of Washington, for instance. A polite gentleman named Neville Wood sent a 'very handsome' letter to Walton Hall, calling him to the charge, and received a promissory note by return:

> This Bird of Washington is nothing but the white-headed eagle in its immature

> plumage. . . . I may yet deem it necessary to touch upon this subject, should I be provoked to address a third letter to Professor Jameson.[51]

A third letter, thankfully, proved unnecessary; Professor Jameson wisely withdrew from the discussion. But this proved little respite for the closet naturalists; there were 'certain individuals' with whom Charles Waterton had not yet quite squared up accounts. The polemical pamphlet served his purpose well.

His next pamphlet was directed against the Reverend J. P. Simpson who had recently delivered a fierce diatribe against the Catholic religion, in front of a large congregation in Wakefield; the sermon was reproduced in the local press and the Squire felt compelled – 'an't please your reverence' – to remind Simpson that Protestants were no longer under the protection of the penal laws – the Catholic could now stand up in defence of his religion. Charles Waterton intended to make full use of the new freedom.

> For 300 years, Catholics had been persecuted.
> For adhering to the Catholic faith, my ancestor Sir Thomas More lost his head;– for adhering to the Catholic faith, my ancestor Sir Robert Waterton lost his fine estate at Methley;– my ancestor at Walton Hall was fined £20 a month for refusing to attend the new-fangled doctrines of the Reformation at Sandal Church;– my grandfather was not allowed to ride a horse valued at more than five pounds; and my father had to pay double taxes, for adhering to the Catholic faith. I myself, until the passing of the Catholic Emancipation Bill, was not allowed to give my vote at an election, solely for adhering to the Catholic faith.[52]

The Reformation had 'laid in ruins the noblest Monasteries and sacred buildings that ever adorned a Christian land.' The 'Church by Law Established' had thrown the maintenance of the poor on the country at large, and instituted the National Debt for its own support. While now obliged to allow dissent, it still kept possession of the loaves and fishes.

> Out upon it. Never will I join any political party; never will I sit down to a political dinner; never will I give another vote for a Member of Parliament, unless it be previously arranged that separation of church from state shall form a leading measure.[53]

As for Parson Simpson, By Hercules! – if his reverence were foolhardy enough to print a second series of complaints against the catholic church, he would have Squire Waterton to deal with in earnest.

Meanwhile, another parson had his nose tweaked. The Squire was in the mood. He refused to believe that a bird ever lubricates its feathers with oil from its own pygal gland – the parson's nose – or that dippers ever walk on the beds of streams. The two issues became linked and our man defended his corner on both, against all the evidence of 'authenticated facts' presented by Parson F. O. Morris, of York.[54]

Seven bad reasons were construed by the Squire for flying in the face of authenticated facts (four of them in an essay on snakes)[55] and after a couple more exchanges, Morris submitted eleven foolscap folios to refute him. J. C. Loudon accepted one page[56] and then called a halt to submissions on oil glands and dippers.

Birds do oil their own feathers, and dippers do walk on the beds of streams. You already knew that. Most people knew it in the 1830s. Even Charles Waterton's best

Yorkshire Squire, whom he terms an *Amateur.*'

The Squire borrowed a copy of the *Natural History and Classification of Birds* and identified himself as the amateur who went to Demerara for the sole purpose of procuring skins, 'neglecting all that can be truly beneficial to science.'[84] He was not pleased with William Swainson.

The amateur Charles Waterton, said the Squire, had contributed to Loudon's Magazine over 60 important papers, 'made with the greatest care, in Nature's lovely garden.'[85] The 'market naturalist', William Swainson, on the other hand, had thus far given to the world confusion and lockjaw. His theory of circles would puzzle Sir Isaac Newton himself, and scare nine tenths of students away from natural history. (Relationships within any zoological group, said Swainson, could be expressed by a series of interlocking circles, and the 'primary circular divisions of any group are three actually, or five apparently.')[86] His classification was just as bad – *Gampsonyx Swainsonii* etc.

Swainson's father was for many years a respected friend of Charles Waterton. The Yorkshire Squire had once spent an evening at the Swainson home, in Liverpool, and gave a few gratuitous lessons in taxidermy to young 'Mr William'. 'Methinks you have ill-requited the former kindness of your disinterested mentor', said the mentor. He advised Mr William to spend more time in the field.

He should climb trees, repair to mud-flats and swamps, pursue wild beasts over hill and dale, explore creeks and sea-shore, and get himself let down the tremendous precipice, like someone else we know. That's what he should do. 'Worst come to the worst, this will at least gain you the appellation of 'Amateur' from the pen of supercilious theorists – an honour not to be sneezed at in these our latter times.'[87] But Swainson stuck to the closet, a thoroughly modern man, a classic anticipation of Gradgrind. He was a self-styled 'systematic naturalist', full of sublime circles and 'continental nomenclature'. Some people appreciated that sort of thing; Charles Lucien Bonaparte, for instance, Prince of Canino and Musignano – another modern man.[88] Charles Waterton was fast becoming an anachronism, way ahead of his time.

In religious affairs, it was otherwise; the Squire was up to his neck in current affairs, through the accumulated grudges of centuries. He now got wind of a new planned assault on the Catholic faith, which brought to a focal vortex the whole religious history of his own family. That 'audacious slanderer' Seymour, of the despised Reformation Society, had been invited to preach in the parish church of Sandal Magna itself, right on the Squire's own doorstep. More than flesh could bear. A 16-page history lesson was rushed off from the local printer,[89] and Seymour's fire was stolen before the touch-paper was lit. The local electorate should know that no redress of political grievances was possible until until every last bishop was dismissed from the House of Lords, and the Law Church severed, permanently and unconditionally, from all connection with the state.

The Reformation Society duly met in Sandal church but the local squire did not attend. By that time, he was in Bavaria – a very long way from Sandal Magna.

At Hernsheim, in Bavaria, as guest of a local curate – amanuensis to the Prince of Hohenlohe – he was able to hear high mass in a Protestant church; he then witnessed a protestant congregation at prayer before statues of the Virgin Mary and St John the

Baptist. Back in England, nothing of the kind was then possible.
He returned home some time during the autumn or winter of 1837, via the ports of Rotterdam and Hull. In Rotterdam, he went shopping – for two pairs of barnacle geese and four wigeon. He took them to Yorkshire in a livestock hamper but two of the geese and three wigeon died shortly afterwards. The two barnacle ganders settled well; they ultimately mixed and even bred with some Canada geese which began to use Walton Lake as their home the following year.[90] The surviving wigeon also settled, though why it should be taken to Yorkshire is not explained.

England, by now, was deep in recession, the most miserable slump in its history. *Oliver Twist* asked for more but the Law Church, like the workhouse, washed its hands. It was very shortly after *Oliver Twist* was published – and presented on the London stage – that the first volume of Waterton's *Essays on Natural History, Chiefly Ornithological*, appeared on the world's wide stage of rumpless fowls and vultures' noses. It became very popular.

The magazines that deigned to notice the book gave it a mixed reception. Each reproduced a long extract from it, to demonstrate its value as entertainment – then at a fashionable premium – but failed to bestow any of the lavish praise which Charles Waterton had learned not to expect anyway. *The Literary Gazette* thought Master Waterton a comical fellow, as his self-portrait in the Preface proved.[91] The Athenaeum likewise cited a fairly long extract from the preface, to illustrate a style 'sufficiently original to amuse.'[92] And the *Edinburgh Journal of Natural History and Physical Sciences* quoted at length from a story in the main body of the book – on caged chaffinches blinded to make them sing – to demonstrate 'a pleasing, although diffuse and discursive style' and a kind heart, at odds with his unkind treatment of brother ornithologists.[93]

The *Spectator* called the book fresh, racy and vigorous – 'very odd, very egotistical, amusing and hasty.'[94] And Chambers' later called the author 'a man at once of talent and veracity though with some oddities in his composition.' The book could be compared, 'in point of interest,' with Gilbert White's Selborne.[95]

The *Gentleman's Magazine* announced publication but declined to review,[96] and other periodicals did the same, including, surprisingly, the *Edinburgh Review*, and *Blackwood's*.

But the Essays did well enough without critical blessing, and the Squire's constitution was sturdy enough to take his medicine like a man. He was now 55 years old and felt no more than 30. In the last 30 years or so, his blood had been cupped over a hundred times, calomel and jalap had been used liberally as a purgative, quinine or laudanam as a tonic, and rhubarb as a sure nostrum against dysentry. Yet he now clearly considered himself a paragon of health and vitality, if not a picture of it, as that famous self-portrait illustrates:

> I am quite free from all rheumatic pains, and am so supple in the joints that I can climb a tree with the utmost facility. I stand six feet high, all but half an inch. On looking at myself in the glass, I can see at once that my face is anything but comely: continued exposure to the sun, and to the rains of the tropics, has furrowed it in places, and given it a tint which neither Rowland's Kalydor nor all the cosmetics on Belinda's toilette, would ever be able to remove.[97]

According to the adverts, Rowland's Kalydor was an infallible specific against smarting of the face, and rendered shaving a pleasurable operation. It extirminated eruption, tan, pimples, freckles, redness, and all cutaneous imperfections whatsoever. But the Squire's imperfections were terminal, his tan beyond the pale. He had more important things to worry about than physical beauty.

On 4 June 1838, in Wakefield's Music Saloon, the Reformation Society showed its ugly face again, verifying the old adage, according to Charles Waterton, that the higher an ape mounts into a tree, the more it exposes its ignoble parts to view. His pet Wakefield printer had more work to do.[98]

The foreman of the meeting had pronounced the Catholic faith 'ruinous to the soul, and subversive of real godliness and Christian morality.'

The Law Church was the scourge of England, came the reply, and its founding fathers little better than a band of villains who should have been hanged. The reformed cathedrals – once adorned with 'everything that piety could suggest or magnificence conceive' – were now damp and gloomy vaults. The Reverend Sydney Smith, who 'would not pen down a falsehood, for the richest mitre that Mother Church could place on his distinguished brow,' (as witness his appreciation of Charles Waterton) had recently visited the greatest of all protestant monuments – St Paul's cathedral – and reported that it was 'constantly and shamelessly polluted with ordure.' The pews were turned into *cabinets d'aisance*, the prayer books torn up, and the monuments defaced with indecent graffiti. So much for the 'pure' doctrine of the reformed church, said the unreformed Squire. He suggested that the mess in St Paul's should be cleaned up by members of the Reformation Society, though they seemed to be more offended by the scent of incense in Catholic churches. Huge sums had been mulcted out of the English nation, to build churches 'no more like the old ones, than is a farthing candle to the noon-day sun.'

Charles Waterton begged permission to worship in the Protestant church in Wakefield, just as he'd worshipped in a Protestant church in Hernsheim, Germany. Even natural history would benefit, in his informed opinion: 'poor as a church mouse' was a Protestant axiom – Catholics fed the mice. But English Catholics were denied access to the churches their own ancestors had built, unless they vowed allegiance to the 39 Articles.

The Reformation Society could rest assured that whenever it chose to canvass for that oath, and to abuse the Catholics of Wakefield and district in the process, this particular Catholic would be ready, club in hand, to defend his faith.

But the club proved unnecessary. If the Reformation Society met again in the Wakefield area, the Squire of Walton Hall never got to know about it. The church mice remained Protestant, and no further occasion arose for them, or their descendants, to hear the Catholic mass.

And that was the end of an argumentative phase of life which lasted almost throughout the 1830s. He now applied himself to another of his preoccupations in life – a medical application for his supply of curare. Coarse controversy was temporarily forgotten; his opponents were more dumb than a fish.

There was no opportunity for self-medication with curare, not even for the lockjaw contracted from Swainson's nomenclature. But he naturally shared the medical profession's high hopes for its use in cases of 'decided' hydrophobia; like Professor Sewell of the London Veterinary College, he had such complete faith in it that he would not have hesitated to use it on himself if ever he'd caught the disease.[99]

In February 1839, twenty-five years after rising from the dead at the Veterinary College, the ass Wouralia died of old age at Walton Hall, totally beyond all hope of recall. An anonymous obituary by Charles Waterton duly appeared in the *St James's Chronicle*, and served as a timely opportunity to offer his services in case of need. He was ready, not only to supply the drug but to administer it, with a steady hand, on any creature suffering from hydrophobia, or on 'anyone of our own species labouring under a permanently locked jaw, or mad from the bite of a rabid animal.' If the treatment failed, he was ready to face trial at York assizes, for 'a cool and deliberate act of manslaughter.'[100]

Very soon afterwards, the chance to commit manslaughter arose. A Nottingham policeman went to the assistance of a dog trapped in a hole, and the dog bit him on the lip and nose. Six or seven weeks later, the police officer began to show symptoms of hydrophobia. Francis Sibson, surgeon at Nottingham General Hospital, sent an express off to Walton Hall and asked for the Squire's assistance. Trouble-shooter Waterton, the Red Adair of rabies, rushed off to Nottingham, with his wax balls of curare, pausing to collect a friend, Sir Arthur Knight, in Sheffield. By the time they arrived in Nottingham, the policeman was dead.[101]

But the trip was not completely fruitless; word got around, and a couple of days later the Squire was back in Nottingham, to demonstrate the effects of curare, and of artificial respiration as its antidote, in front of a large audience at the Medical School. Two asses were inoculated in the shoulder and five operators then proceeded to administer artificial respiration, using a modification of bellows used in the 1814 experiments in London. Inflation and deflation was repeated 16-18 times a minute, until life returned. For one of the donkeys, this took over 7 hours; it died 3 days later. The second donkey needed much less resuscitation and like Wouralia before her, she recovered fully.

Even as late as 1878, the *Lancet* was still suggesting that hydrophobia could be treated with curare but that appears to have been the last time that such a claim was made. It is now used as a muscle relaxant, or as a 'shock absorber' in convulsive therapy.[102] The Intensive Care Unit at Nottingham General Hospital is called, naturally, Waterton Ward.

Mr Waterton left a supply of the drug with Mr Sibson and returned to Walton Hall at least partially satisfied. A letter of thanks from the Mayor of Nottingham was received within a few days of his return, and answered in a tone best described as self-aggrandised deference, complete with gratuitous anecdote of his first meeting with Sir Joseph Banks and a precis of his first trip into the Guianese interior, in 1812, when the curare was acquired.[103]

Almost 170 years after it was acquired, a sample of curare resin, with scrapings from the tips of three Indian arrows in the Waterton collection, were taken for analysis at St James's University Hospital, in Leeds. It was still powerful.[104]

The apothecary now became rat-catcher again. His attitude to rats was consistent: they were filthy vermin – the Hanoverian type anyway. His attitude to mice, like his affection for donkeys, was more erratic. His concern for the church mice in Wakefield did not extend to their Catholic relatives at Walton Hall. He fed them alright, being a good Catholic – with rancid bacon fat in sunken bottle traps. Rats and mice were about to be wiped out at Walton Hall, once and for all for the third time. Repairs to the farm outbuildings, begun in 1834 and now complete, incorporated features designed specifically for the purpose – walkways paved with cement, sewer pipes fitted with iron grates (removable, in the Squire's way with words, 'at pleasure') and so on.[105]

> When I am gone to dust, if my ghost should hover over the mansion, it will rejoice to hear the remark that Charles Waterton, in the year of grace 1839, effectually cleared the premises at Walton Hall of every Hanoverian rat, young and old.[106]

In the spring of that same year of grace, he'd planned to embark on a two-year pilgrimage to Italy with his son, Edmund – now nine years old – and his two sisters-in-law, Eliza and Helen, who were now also living at Walton Hall. The thought of returning to a completely rat-free Walton Hall was the best possible going-away present.

But shortly before they were due to leave for Italy, he heard that his great friend, George Ord, was on his way from America, for a long-promised visit to Yorkshire. Ord spent the early months of summer on the Waterton estate and the friendship between the two men – borne of a common hatred for Audubon – clearly prospered. The trip to Italy was postponed.

George Ord eventually returned to Philadelphia and the Waterton family left home for the continent in early July 1839. Just before they left, they lost a couple of close friends – a pair of mute swans. One of them had died of disease, the other 'by the horns of an unruly cow'. The Jesuit fathers at Stonyhurst were told of the deaths and they offered a pair of swans from the college lake by way of condolence. Gamekeeper John Ogden went to collect the swans, and the Walton Hall inmates awaited his return with some impatience. One of the servants heard the carriage approach from the lodge and ran to the footbridge to help. But dazzling dark had fallen and so did the servant; he missed the inner gate of the footbridge and ended up in the lake, a suitably adventurous example for the Squire himself to emulate, a few years later, in Dover harbour.

He now heard his man swanning around in the lake, like a Virginian partridge, and sprang out of the window (it was nearer than the door), to administer aid. He ran to the shore, told the servant to make for the sound of his voice, and pulled him out. The swans took his place on the water with appropriate Victorian hauteur.

Domestic affairs settled, swans and servants at their proper stations, it was now time for the extended family Waterton to pack up their bags and leave for Italy's 'lovely sky'. Just one duty remained:

> I called up the gamekeeper, and made him promise, as he valued his place, that he would protect all hawks, crows, herons, jays and magpies, within the precinct of the park during my absence. He promised me faithfully that he would do so; and then, wishing him a good time of it, I handed my two sisters-in-law, the Misses Edmonstone, into the carriage; I placed myself and my little boy by them; the two servants mounted aloft, and in this order we proceeded to Hull, there to catch the steamer for Rotterdam.[107]

Chapter 15
1839-1841
Heterogeneous Aliment

At Hull, the Squire met an old friend – the captain of the brig, *Industry*, on which, as a 19-year-old squire apparent, he'd sailed for his first journey abroad, nearly 40 years previously. He found him, 'safely moored', in a house for retired sea captains, overlooking the harbour.[1] The two men reminisced, as such do, and swapped a few travellers' tales, spanning an aggregate 78 years of gales weathered and storms indulged since last they'd met. The fly on the wall was illiterate.

The great days of sail were now numbered. The Squire and his family crossed the North Sea to Rotterdam on board the regular steamship, *Seahorse*.

He always loved the Dutch – despite Calvin – and found much the same Dutch qualities in Rotterdam as American ones in New York. You could walk the streets at any time of day or night, 'without encountering anything in the shape of mockery or of rudeness.'[2] He saw no-one hurrying up the street as though the town were on fire, nor a single soul whose haughty looks would give you to understand that you should keep at a respectful distance.

From Rotterdam, the party travelled north, then south, then west, then east, via bird market, Catholic church and natural history museum – The Hague, Leyden, Haarlem, Amsterdam, Antwerp, Bruges and Ghent – to his 'beloved Aix' (Aachen). Much the same alternative saunter as for his honeymoon, eleven years earlier.

At The Hague, specimens in the museum were clumsy imitations of death alive, whilst Leyden Museum was as bad as London's. 'We might fancy that Swainson had been there, with his own taxidermy.'[3] At Antwerp, he enjoyed the museum of an old friend – M. Kats – and at Ghent, the Museum of Osteology was still 'splendid'. Helen, Eliza and Edmund enjoyed the birds and bones as much, or as little, as Anne had done.

The churches in Holland were bare and empty – a manifestation of the 'change of religion' – but the Belgian churches had kept their religious ornaments and stayed open seven days a week. They therefore drew English tourists like bees to honey. His favourite church in Bruges was naturally the eighteenth-century chapel of the English nunnery where he and Anne had married in 1829. All three of the Edmonstone girls had been 'finished' in the nunnery – 'an elegant retreat of plenty, peace and piety'[4] – as worthy as Stonyhurst itself.

Ghent, too, had its convents, two of them survivals from the 15th century called baguinages – partial retreats for baguines (nuns without vows). The Waterton equipage stopped off at a beguinage. The Squire was greatly impressed. The sisters divided their time between good works outside the convent (hence the verb to beg) and holy prayer within it, 'far removed from the caustic gossip of the tea-table, or from the dissipations of nocturnal gadding.'[5] His own 'sisters' (in-law) were, of course, far removed from dissipations, by temperament and by education.

From Ghent, the carriage rattled south-east, towards Brussels, then along the Meuse valley, through Liège, to Aix-la-Chapelle. It was now late July and the Meuse valley was 'as rich and beautiful and romantic as any valley can well be on this side of ancient paradise',[6] a circumstance enhanced, from a twentieth-century perspective, by thought of the carriage itself: Liveried servants? Brass harness? Heraldic crest on the panels? 70-seater 'Touramas' and 'Euroliners' use the same road now.

The attractions of Aix were manifold, an almost perennial magnet to Charles Waterton from now on. On this occasion, they spent two or three months in the town, at an inn called the Dragon d'Or, under the ministrations of an old widow who went by the name of Madame Van Gulpen – 'an assurance that her cheer was not administered by driblets.'[7] It was from Madame Van Gulpen that he acquired the freshwater crayfish now preserved in the cellars of Wakefield Museum. He spent three hours dissecting another crayfish, in his hotel room, but lost the fruits of his labour when the maid threw the shell onto the kitchen fire and dressed the carcass for his tea. 'Tantum ne noceas, dum vis prodesse videto', said the Squire to the maid, chattily. (See that you do no harm in your desire to help.)[8]

He and his two sisters-in-law were thoroughly charmed by Aix – by the surroundings of the town, by its 'unrivalled medicinal waters' and, almost certainly, by its religious relics. They liked to climb the Lousberg, on the outskirts of town, to enjoy the view of the Ardennes from a favourite bench at the top. In the town itself, at five o'clock in the morning, the Squire could always be found enjoying a glass of hot spring water direct from the lion's mouth at the Pump Room fountain. Later, he'd join the rest of the rank and fashion in the Pump Room gardens, or the promenades, or, more often, in the Marian Shrine of the cathedral, where the famous relics are housed – the Virgin's robe, the loincloth from the cross, the blood-stained shroud of St John the Baptist, and the infant Jesus's swaddling clothes. Charles Waterton believed unquestioningly in the authenticity of the Heiligenblut at Bruges, no less devoutly in the 'stupendous miracle' of the liquefaction of San Genaro's blood in Naples. He was certain the Santa Casa in Loretto was genuine, and never for a moment doubted that supernatural occurences took place in Britain before the Reformation. Why shouldn't he believe in Jesus's nappy?

But not all visitors to Aix were as pious. Few of them were as self-disciplined. The attractions of the town were marred, for the Squire, by the daily spectacle of shameless gluttony in hotel dining-rooms and the dissipations of nocturnal gadding in the casinos. He was not amused.

For most of the invalids who went to Aix-la-Chapelle, the efficacy of its waters was, in his belief, neutralised by resort to the 'heterogeneous display of aliment'[9] in the sal à mangér. He was never well disposed towards heterogeneous aliment, in Aix or anywhere else. Animals stuck to plain food, so should humans. During Lent, he lived for 40 days on one meal per day – dry bread and weak tea, without milk – and thrived to such an extent that the fast was usually extended for another week or so each year. He was sure the source of most human illness was to be found in the kitchen

– 'gout under the dresser, and fever in the cupboard.' If human beings followed the examples of wild beasts, and stuck to one general kind of food, 'we should be giants of strength at the age of one hundred.'[10]

In 1981, Richard Ellis, professor of anaesthesia at St James's hospital, Leeds, tried the Waterton diet – for six days. Result: irritability, giddiness, low concentration, light-headedness, and the first stages of pernicious anaemia.[11] (The Squire also experienced light-headedness but expressed it less pejoratively – 'my brain is so clear and free from all confusion that I scarcely know it exists on my shoulders.')[12]

At Aix-la-Chapelle, no-one tried the Waterton diet. Instead, their jaws belaboured their stomachs with heterogeneous aliment and all advantage of the waters was lost – 'enfeebled warriors, once all powerful on the field of Mars, now armed with knife and fork in lieu of battle-axe'.[13]

The other prurient sickness of the Cure was gambling. The casino operated three times a day. Even ladies tried their luck at the tables, 'the rose and the lily taking possession of their countenances by turns, as the dice or ball shows in their favour or against them.'[14] We know of two young ladies, Providence their guide, who were trained to prefer the less profligate thrills of the zoological museums – bet your life on it.

As winter approached, it was time to leave the delights and dissipations of Aix-la-Chapelle, to travel south, alongside the Rhine, to Freiburg. A Scottish friend – Alexander Fletcher – joined the party at Cologne, and Old Boreas, the winter wind, followed as far as Strasbourg and then gave up the chase. The autumn was mild and sunny.

At Freiburg, they stayed for a few days to enjoy the vintage – abstemiously – and the Squire met a German waiter who had written a long ode in English to Freiburg cathedral. He bought a copy of the poem and found 'matter much superior to anything that I could have expected from the pen of a German waiter at an inn'.[15] But he was subsequently denied the pleasure of immortalising the man's name, when the poem was lost, with much else besides, in a dramatic shipwreck, the account of which, when we come to it, you'll find slightly more entertaining than an ode to Freiburg cathedral.

They now drove through Switzerland and crossed the Alps by the Splugenpass,[16] towards Milan. In crossing the Alps, he experienced something of the same anti-climax that Wordsworth had felt a few decades earlier. He got out of the carriage, to walk, hoping for birds, but saw none. The nutcrackers and eagle owls, snow finches and crested tits, alpine choughs and rock partridges, declined to appear. The earth was 'one huge barren waste'.

But the descent into Italy was 'charming. . . delightful. . . perfectly enchanting' – he was clearly under the influence of Victorian girl-talk.

'I am sure we are in Italy now', said Helen archly, and she pointed out a woman searching for vermin in her daughter's hair. The Squire agreed – combs were uncommon in that part of Italy. He seemed to suffer a sudden accession of English sniffiness at this stage of the journey. Perhaps it was concern for the ladies. 'The Italians would confer a vast benefit on society', he said, 'if they would depose more fertilising matter on their fields, and less in their streets.'[17] It's difficult to imagine him with lavender water on his hankie.

They stayed in Milan for a few days then crossed the Lombard Plain, via Parma and Moderna, to Bologna, and so on to Florence. At Florence, he visited the studio of Lorenzo Bartolini, the Tuscan sculptor and self-styled non-neo-Classicist, who was currently at work on a representation of Andromache, Astyanax, and the conquerors of Troy.[18] He also met an old friend from previous journeys in Europe – Professor Nesti – who conducted the family around the 'well-stored apartments of the public museum.' There are over twenty public museums in Florence but this one was obviously 'La Specola' – the Zoological Museum. The more conventional art galleries were left to more conventional tourists.

From Florence to Rome, he was again disappointed by the bird-life – a few coots, a heron or two, the occasional blackbird at the roadside, and a 'scanty show' of hooded crows passing from tree to tree. The serins and crested larks, black-eared wheatears and great reed warblers, collared flycatchers and rock thrushes – delicious on toast – undoubtedly played hide and seek with Italian bird-catchers, indistinguishable in the field from English squires.

The party stopped overnight at Bracciano, 39 kilometres from Rome. It was arranged that the two men should leave the inn at four o'clock in the morning, to walk into Rome, and that the ladies should follow with the servants, in the carriage, after a nine o'clock breakfast.

There are Muslims who measure out the last miles to Mecca with their own prostrate bodies, like inch-worms. Charles Waterton elected to enter the Catholic capital in similarly reverend fashion, with a shoe and a sock in each pocket and nothing between his bare feet and the sacred pavements of His Holiness Pope Gregory XVI. He'd gone without shoes in the South American rain forest and assumed he could now do the same on the said pontifical pavements. (The real precedent – the principle embalmed – was clearly St Ignatius's barefooted pilgrimage to Jerusalem.)

All went 'merrily' for the first few miles, until, pausing to admire the 'transcendental splendour' of the morning planet, he noticed some blood on the road. The sole of his right foot was injured and a piece of jagged flesh hung loose 'by a string'. A similar thing had happened in the rain forest, if you remember, but Charles Waterton was slow to learn by experience – once bit, twice try. Alexander Fletcher suggested that they should hitch a ride into Rome from a passing vehicle but the wounded man wouldn't hear of it. He twisted off the loose skin, then, at last, he pulled on his shoes.

As the day got steadily warmer, he limped into Rome and took rented rooms for the family, in the Pallazo di Gregorio; this was directly opposite the Collegio di Propaganda Fide (which gave the word 'propaganda' to the English language) and very close to the house where Keats had died, on the edge of the Piazza di Spagna. A short hobble away, was the Jesuit church of the Gesù, where St Ignatius is buried, and that, in turn, was conveniently close to the bird market in the grounds of the Pantheon. Perfect.

Apart from early morning mass and bird shopping, and the odd limp up and down the Spanish steps, where most of the British tourists would congregate, the Squire's injured foot spent two undignified months on a sofa before he was fully recovered. He sat indoors and converted his rented bedroom into a taxidermical studio, a little home from home. He also found time to write to George Ord, to complain about the manners of Napoleon Bonaparte's nephew.

Charles Lucien Jules Laurent Bonaparte, Prince of Canino and Musignano – a Napoleon clone with a goatee beard – would have stood second in line to the imperial throne if his uncle had recognised the legitimacy of his birth. He was by now a distinguished naturalist – the first scientist in the family – and as such, almost certainly incurred Charles Waterton's displeasure by befriending Audubon and collaborating with Swainson. The prince's natural history museum, in Rome, ranked in deformity with museums in London and Edinburgh, Ord learned.[19] It contained 'nothing worthy of the least attention.' But the Squire was later obliged to modify his opinion of Charles Bonaparte when the prince paid him the casual little courtesy, 18 months later, of saving his life – a trivial matter (noblesse oblige) of rescue from a sinking ship.

Bonaparte's museum seems to have been the only one in Rome which he bothered to visit. The art galleries and antiquities, the 'modern schools of sculpture and of painting', held neither charms to strike the sight nor merit to win the soul. They were all too hot and stuffy anyway. He did visit Vallati's studio (Who? Vallati – 'the renowned painter of wild boars'[20]) and shipped one of his paintings back home to Yorkshire, where it vied for pride of place with the Holbeins and Tiepelos of Walton Hall. Inspired by Vallati, he also inspected another great treasure of Rome's cultural heritage, the slaughterhouse – 'probably inferior to none in Italy.'

Friday was the best day to see the slaughterhouse. In the winter, seven or eight hundred pigs were killed each day of abstinence from eating them – a spectacle not even the Colisseum could match any more.

> About 30 of these large and fat black pigs are driven into a commodious pen, followed by three or four men, each with a sharp skewer in his hand, bent at one end in order that it may be used with advantage. On entering the pen, these performers, who put you vastly in mind of assassins, make a rush at the hogs, each seizing one by the leg, amid a general yell of horror on the part of the victims. Whilst the hog and the man are struggling on the ground, the latter, with the rapidity of thought, pushes his skewer betwixt the foreleg and the body, quite into the heart, and there gives it a turn or two. The pig can rise no more, but screams for a minute or so, and then expires. This process is continued until they are all despatched, the brutes sometimes rolling over the butchers, and sometimes the butchers over the brutes, with a yelling enough to stun one's ears... Rome cannot be surpassed in the flavour of her bacon, or in the soundness of her hams.[21]

The post-Lenten diet of Charles Waterton allowed of good bacon and sound hams, and all 'well regulated' families such as his naturally conferred a blessing on the flesh when it appeared on the breakfast table. He saw no inconsistency in the blessing of dead meat from the pork shop and of the live beasts of burden who carried it there. Slaughter and sentimentality were perfectly reconcilable. In the Squire's brand of chop logic, the slaughterhouse was a fine, impressive spectacle, but the curses vented by 'village urchins' on the Barnsley canal barge-horses (even on Sundays!) were utterly despicable cruelty, a savage contrast to the public benediction of beasts of draught and burden, each January, at the door of the church of Sant' Antonio Abate, in Rome.[22] He attended this public blessing of donkeys and horses, in January 1840, and found it as joyful and exhilarating as the slaughterhouse, an act 'replete with Christian prudence'. Some of the slaughtermen probably also attended the ceremony. (One of those pig-killing skewers, incidentally, like a six-inch nail with a loop in the blunt end, can still be seen in the Waterton collection at Wakefield Museum; he took such things home as mementos.)

His foot, by now, was fully recovered, and more regular trips to the bird-market in the Rotunda were possible, for the purchase of specimens to preserve in his bedroom at the Palazzo di Gregorio. He soon became a familiar face in the bird-market, and a close friend of some of the proud Papagenos who traded there – 'good men, notwithstanding their uncouth looks.' He saw them regularly at early morning mass.

'Every species known in Italy' was sold in the bird-market – from the wren to the raven. Some were alive, others trussed up and ready for the spit. Italians, it seemed, had stomachs as tough as Daddy Quashi's, which 'could fatten on the grubs of hornets, and on stinking fish.' Anything and everything was caught, sold and eaten. The Squire's list of birds seen in the market included 'yellow-breasted chats' (whinchats), 'olive-throated buntings' (ortolans), 'gobbo ducks' (white-headed ducks), and 'mountain sparrows with yellow speck on throat' (rock sparrows). 17,000 quails passed through the Roman custom house in one day, and owls, hawks, crows, jackdaws, jays, magpies, hedgehogs, frogs, snails and buzzards were also sold in very large quantities. 'Indeed, it would appear from what I have seen, that scarcely anything which has had life in it comes amiss to the Italians in the way of food, except the Hanoverian rat.'[23]

Eating habits have changed in Italy but killing habits die hard. Migrating birds are still slaughtered and sold, and trees and flowers vandalised in the process, on a scale to gladden the heart of the most psychotic sociology professor or Jewish novelist. (In the 1970s and 80s, each autumn, 1.5 million Italian gunmen spent 18 billion lire – £70 million – on half a billion shot-gun cartridges, to kill something like 150-200 million birds.[24] Almost as many more were netted and limed.)

One of the commonest birds in Rome was the blue rock thrush, *Monticola solitarius*; Charles Waterton was sure it was the bird mentioned as a 'sparrow' in the Bible – 'the same bird which King David saw on the house-top before him, and to which he listened as it poured forth its sweet and plaintive song.'[25] A pair of blue rock thrushes bred in the walls of the Propaganda Palace, just a few yards from his own bedroom window. He badly wanted to get to the nest but figured, soberly, that 'the Romans would not understand my scaling the walls'[26] – an unusual instance of the triumph of experience over hope.

Another beautiful bird bred in the walls of Rome – the only European member of a family called roller, which Charles Waterton insisted should belong to 'the family pie'. (Four and twenty blackbirds belong to the family thrush.) The best place in Rome to see a roller, and perhaps to witness the tumbling courtship display which gave it its name, was in a grove of stone pines in the grounds of the famous Villa Pamphili Doria[27] – a Roman equivalent of the Florentine Cascine. The nightingale also bred in the Pamphili, its song, pre-petrol engine, mingling with the bells from the illuminated dome of St Peter's, overlooking the grounds. None of the birds in the Pamphili ended up in the bird-market because, as in the Cascine, they were all fully protected.

Just about everything else did end up in the bird-market. Even the air itself was treacherous. 'Idling boys' caught swifts and house martins over the streets by a process snide enough to be called 'simple and efficacious'.[28] Floating feathers were attached to silk threads and used as bait to catch birds in search of nest material; the birds snared themselves, in full flight, in little running nooses attached to the bait. 'This ornithological amusement', said the Squire, 'is often carried on in the streets of the Propoganda during the months of May and June.'

Few of the Grand Tourists witnessed the ornithological amusement. Heat and "bad air" sent them trundling out of town soon after Easter. The same mosquitos which attracted the birds emptied the streets of tourists. The boys killed the birds which ate the insects which carried the disease which drove out the tourists on whom the boys' families depended. It was called the 'fever'.

Charles Waterton and his family set bad air at defiance. With trust in God, they chose to stay on in Rome for the Month of Mary. By divine recompense, 'we had the golden oriole, the roller, the bee-eater, the spotted gallinule [spotted crake], the least of the water rails [Baillon's crake], the African redstart [black redstart?], the hoopoe, the egrette, the shrikes, and several varieties of the quail, and I procured an adult pair of the partridge of the Appenines [rock partridge] in superb plumage.'[29] Then the pilgrims from Walton Hall joined the rest of the aristocratic herd in Naples.

On the road from Rome to Naples, Charles Waterton introduced himself to a herd of buffalos. They were grazing near a little roadside inn where the Waterton horses were being watered. Some Italians warned him not to approach the buffalos – he'd get trampled to death – so he approached the buffalos.

> Having singled out a tree or two of easy ascent where the herd was grazing, I advanced close up to it, calculating that one or other of the trees would be a protection to me, in case the brutes should prove unruly. They all ceased eating, and stared at me as though they had never seen a man before. Upon this, I immediately threw my body, arms, and legs, into all kinds of antic movements, grumbling loudly at the same time; and the whole herd, bulls, cows, and calves, took off as fast as ever they could pelt, leaving me to return sound and whole to the inn, with a hearty laugh against the Italians.[30]

Instead of which, as the most laughably unamused of his posthumous critics was keen to point out, 'they might have 'treed' him, causing much inconvenience to himself and his companions.'[31] None of his companions left a diary.

Naples was currently bursting at the seams with noisy Neapolitans and fussy foreigners, with a few thousand sheep, goats and oxen for good measure.[32] The patron saint of the city is the martyred bishop of Benevento – San Genaro – the man who gave January to the calender. He was killed during the persecution of Christians, under the edict of Diocletian's adopted son, Gallerius, in AD 305. His skull rests in his own chapel in Naples cathedral, in a silver bust, with two glass phials containing congealed blood from the fatal wound. The blood liquefies – usually – during the saint's twice-yearly festivals, on the first Saturday in May and from 19 to 25 September. Charles Waterton had managed to get himself 'the very best letters of introduction', and thereby the very best ring-side seats for the miracle.

> On the 19th of September, then, in the year of 1840, accompanied by my two sisters-in-law, Miss Edmonstone and Miss Helen Edmonstone, and my little boy, we arrived at the Cathedral, and entered it just as the great clock was striking a quarter past eight of the morning.[33]

They kissed the glass phials and knelt down to wait with the rest of the faithful. At a quarter to two in the afternoon, after six hours of prayer, patience and piety had their reward. 'The blood suddenly and entirely liquefied.' The Squire kissed the phial again and did so at hourly intervals until, in the early evening, he finally left the cathedral, to watch the procession of the blood through the streets of Naples.

It was all particularly poignant for him because he'd just received some bad news from England – the death of his sister, Helen, after a long and painful illness. On 21 September, he wrote to George Ord to notify him of his 'incalculable' loss (and to complain about the condition of birds in Naples Market)[34] and two days later, he was back in the chapel of San Genaro, with a few saintly invocations on behalf of his sister's immortal soul and a few more prayers to precipitate the liquefaction of San Genaro's blood. This time it liquefied at 'a few minutes before ten o'clock', an event of such import to Charles Waterton as put snakes and alligators, and every other wild flirtation with death, completely in the shade:

> Nothing in the whole course of my life has struck me so forcibly as this occurrence. Everything else in the shape of adventures now appears to me to be trivial and of no amount. I here state, in the most unqualified manner, my firm conviction that the liquefaction of the blood of St Januarius is miraculous beyond the shadow of a doubt.[35]

He was far from uniquely struck. Hundreds of thousands of Neapolitans still confidently expect bad luck for the city if the blood fails to liquefy. In 1527, it was plague; in 1569, famine; in 1941, Vesuvius erupted and the allies bombed the city. Even the earthquake of 1980 was accompanied by holy thrombosis. (Napoli won the Italian league championship in 1987, largely through the intercession of a secular saint from Argentina – the same hand of God which had put England out of the World Cup.)

The Squire also had a letter of introduction to the Carthusian monastery of San Martino, on the Vomero hill overlooking Naples. Funiculars now take passengers up to the Vomero, but the Squire's party drove up, on an afternoon in early autumn, apparently in a hired baroccio driven by a Roman coachman with the universal cab-driver's appetite for confrontation.

The Squire delivered his letter to the monastery (now the National Archaeological Museum) and was shown the magnificent view from the cloisters, whilst the ladies were driven on to the nearby castle of St Elmo to wait for him. By the time he rejoined them, after suitable oblations in the chapels of S Martino, he found them under arrest.

> It seems that the son of Bellano who commands at the fort of St Elmo, has his organs of sight and smell so particularly refined that he cannot tolerate the least impurity on the road that leads to his domain; and thus every coachman is obliged to remove, without delay, what may drop accidentally from the caudal extremity of his horses, or pay a fine to a soldier for doing the important work for him.[36]

The hidebound hack-driver in the family's employ refused to pay any fine. The Squire now discovered him in Italian altercation with a soldier from the fort, over a gross defamation of character on one of his horses. None of the horses had passed a single motion all afternoon, said the coachman – 'non sono andati di corpo' – as the good ladies would testify. Yes they had, said the soldier, and a heated argument ensued.

A 'handsome young officer' now emerged from the fort, to enquire what all the noise was about and to offer his services as judge and jury. Counsel for the defence – Charles Waterton – proposed to prove his client's case by forensic science – by recovery of material evidence and detailed measurement of its 'component parts'. In this way, he hoped to demonstrate that the real culprit was a jackass he'd seen on the road a few minutes earlier. The handsome officer had no appetite for measuring material evidence and declared the donkey's ass guilty in absentia. He gave the ladies a look 'full of good humour', smarmy swine, and pronounced the horses Not Guilty. The coachman moved on, partially appeased, but still 'growling like a bear with a scalded head.'[37]

After the festival of St Januarius, and the battle of St Elmo, the Waterton party continued south to Reggio, on 24 September, and took a steamer for Sicily, there to enjoy the obligatory disagreement with officials in the Messina passport office, and the equally obligatory comparison of mummified corpses in the catacombs of Palermo with the stuffed zoological exhibits in British museums. Swainson had made the same comparison.[38]

Charles Waterton marvelled at the length of the horns on Sicilian cattle, remarked the probability of an interesting passage of migrant birds during April and September, then left Sicily, 'under the full impression that we ought to have remained there for three or four months.'

They returned to Naples, paid a farewell visit to Virgil's tomb, and continued north for another winter season at the Palazzo di Gregorio, in the comparative peace and quiet of Rome, pre-Lambretta. 'Rome is certainly the most quiet city I ever visited.'[39] A large part of each day in Rome was spent in prayer and study. Edmund would pop across the street into the Propoganda Fide from time to time, where Cardinal Fransoni and his secretary took him under their wings with all the fond indulgence of success assured. So did two of the holy fathers in the English college in Rome, Dr Braggs and Dr Wiseman (who later became archbishop of Westminster). The 'English angelino' became a great favourite in Rome. He spent most of his time in the convent of the Gesù, catching up on lost schooling. His father and aunts also frequented the Gesù. The earlier the mass, the better, as far as pater was concerned (a 3 a.m. service was held for the local 'sportsmen') but the ladies preferred more lady-like hours.

At high mass one Sunday, one of the ladies witnessed a fatal accident. A man who was lighting candles in a 'lofty cornice' of the church, fell from his perch and mortally wounded both himself and an old lady who was kneeling in prayer below. The prayers remained earthbound but the two souls ascended to heaven within a week.[40]

Some of the gilt hangings from the same cornice had been saved for posterity by an ingenious use of caudal deposits: The silver statue of St Ignatius, which once stood in the Gesù, close to his tomb, had been taken by Napoleon's troops and melted down for war reparations. One of the lay brothers at the convent, anticipating a return of the troops, threw the gold fringe into an 'unfrequented room', covered it with bed clothes ready for the wash-tub, then went into the street to gather a couple of bucketfuls of 'what in London would never be found exposed to the open air.' He took it into the aforementioned room, transferred it into suitable 'nocturnal crockery', and closed the doors and windows. The troops arrived, nosed around, inspired a subsequent Virgilian quote from Charles Waterton – rank winds rushing through open doors[41] – and left in a military pet of refined disgust, like English Protestants.

But the matter exposed to the open air offended Charles Waterton's sensibilities also. A friend asked for an omelette and discovered that the cook had prepared it in a pot which the Squire 'dare not exactly describe', though he'd just described it exactly enough as 'nocturnal'. His own English maid asked for a pot to pour broth into and was given the brass pan used for washing feet. And so on.

But these were mere trifles. The time was spent profitably in Rome, on holy prayer and taxidermy. He bought the market's only specimen of gobbo duck and also got 'a very handsome red-crested duck with a red back' (red-crested pochard) which was 'equally as scarce'. He prepared a green and yellow 'Roman lizard' (perhaps a green lizard) which promptly turned blue and grey, and the freshwater tortoise was so much worth the trouble of dissection that he dissected six of them. He was enjoying himself greatly.

But as midsummer approached, the heat and bad air of Rome overcame its peace and piety and the Waterton entourage considered it time to follow the lead of less preoccupied Grand Tourists, who had left the city for more healthy regions several weeks earlier. On 7 June, he wrote to George Ord and revealed plans to travel by sea to Genoa, to 'tarry awhile in the most interesting parts of the north of Italy', and to spend two or three weeks more in Aix la Chapelle, before returning to Old Mr Bull, 'who will, no doubt, be up to his ears in trouble.'[42] Before leaving Rome, one last parting kick was aimed at the latest in a long line of abominable closet naturalists. Charles Bonaparte's *Fauna Italiana* was almost complete but Charles Waterton refused to look at it. As he told George Ord, its author had spent insufficient time in woods and swamps. The Frenchman was soon to redeem himself, however – on the high seas.

On 16 June 1841, the Waterton pilgrims took diligence to Civita Vecchia, from where they were to take ship next day to Livorno.

The paddle-steamer *Pollux* – a magnificent, titanic vessel, with the power of 200 horses – left port at a little after four o'clock in the afternoon, with Charles, Helen, Eliza, Edmund and the two anonymous servants on board. The weather was serene, with 'scarcely a ripple on the sea.' The steamer chugged pleasantly towards the coast of Elba and as night began to fall, the Squire lay down to sleep on deck, as usual, under a brilliant, star-lit sky, with his travelling cloak for blanket, 'Mr Macintosh's life preserver' in his pocket, and the comforting presence of his own boots – his 'velvet cushion' – under his head. The air was pure and soft and, apart from a perceptible shortage of 'nautical discipline' on board, and a surfeit of 'unseamanlike conduct', auguries were good for a peaceful voyage to Livorno. Helen, Eliza and Edmund also slept on deck.

> Suddenly, our sleep was broken by a tremendous crash, which at first I took to be the bursting of the boiler. But I was soon undeceived; for, on looking around, I saw a huge steamer aboard of us, nearly amid-ships. It proved to be the Monjibello, of 240 horse power, from Leghorn to Civita Vecchia. She had come into us a little abaft of the paddle-wheels, with such force that her cutwater had actually penetrated into our after-cabin. . . . The *Pollux* instantly became a wreck, with her parts amid-ships stove in; and it was evident that she had but a very little time to float.[43]

As luck would have it, Charles Bonaparte, ill-mannered closet naturalist and Prince of Canino, was aboard the *Monjibello* when the accident happened, just five miles from the island where his Uncle Napoleon had spent his last days in exile. With Bonaparte judgement and cool intrepidity (and just a hint of aristocratic insolence) Prince Canino knocked the steersman aside and took the helm himself, to keep the *Monjibello* alongside the *Pollux* for the rescue of its stricken passengers. All, of course, was 'unutterable confusion'.

Edmund went down on his knees to pray to the Virgin Mary, and one of his aunts cried out, in a tone of deep anxiety, 'Oh save the poor boy and never mind me.' Charles Waterton inflated his life-jacket and prepared to do his duty.

The captain and mate of the *Pollux*, one of whom should have been on watch, were both roused from sleep in their berths. The captain was scared out of his wits; he forgot to put on his trousers – subversively expressible trousers – and arrived on deck dressed in nothing but his night-shirt and sailor's modesty. A Spanish duchess refused to move from her bed and had to be man-handled on deck, likewise dressed in her night clothes, and an Italian priest fell into the sea and broke an arm. A ship captain from Naples also fell into the sea and was drowned by the weight of his own money, sewed in a belt around his waist, whilst a man called Armstrong was struck by the *Monjibello* as it entered the *Pollux*, and lay 'sadly injured', a travesty of his own name. Several members of the crew left the sinking ship before all the passengers had been evacuated, and three people manned the ship's lifeboat and rowed away in it, just when it was needed most.

In lieu of a captain with trousers, Charles Waterton took charge in his Macintosh. Some of the *Pollux* passengers managed to clamber on board the *Monjibello*, others ran to and fro, 'bereft of all self-command'. With trust in Mr Macintosh's 'inestimable little belt', the Squire remained on board the sinking ship 'till nearly all had left her.' Panic-stricken ladies looked to this strange English bean-pole in a blown-up rubber jacket, for salvation. The Squire of Walton Hall was a model of cool intrepidity.

Eventually, he too abandoned ship and lost his travelling cloak in the process – a terror-stricken German woman clung to him for dear life as he climbed from one ship to the other. A child was held under the woman's other arm.

> I begged of her, in French, for the love of God to let go her hold, as we should both of us inevitably perish. But she was unconscious of what I said; and with her mouth half open, and with her eyes fixed steadfastly on me, she continued to grasp me close under the ribs, with fearful desperation.[44]

So the Squire cast off his travelling cloak, to free himself from her clutches, and got on board 'by means of a rope'. The woman and child were also saved.

The *Pollux* went down, 'stern foremost', almost immediately. Within a couple of minutes, her last light was extinguished in the water and she was gone for ever. 'Not a spar, not a plank, not a remnant of anything was left behind her.' Clothes, money, books, objets d'art, letters of credit, 'Palmerston passports' and an ode to Freiburg cathedral, all went to the bottom. There was no time to save anything. From the time the two ships collided to final submersion of the *Pollux* took less than 20 minutes.

The *Monjibello* was also badly damaged, and now the inevitable rumour spread that she, too, was sinking. Again, Charles Bonaparte was hero of the hour, obligations assumed by noble birth, if not imposed by it. He took one of the *Monjibello* boats to Portolongoni, on Elba, and asked permission for the crippled steamer to dock. Petty officialdom – a recurrent bane of Charles Waterton's life, even at second-hand – refused to grant permission. The *Monjibello* hove to for the night, ready to limp into Livorno at dawn, if it hadn't by then sunk.

Meanwhile, the Squire went down below, to see how things were going with the castaways. He attended to the priest who had a dislocated shoulder, replacing the bone in its socket, and was clearly taken for a qualified surgeon by all present. Someone asked him to treat the captain of the sunken ship, *Pollux*, the 'dastardly sansculotte' (trouserless bastard) who now lay writhing on the floor of the cabin, as though about to die, 'sighing, sobbing, and heaving like a broken-winded horse.' But the surgeon was not disposed to operate. He was convinced that the captain and mates of the two vessels were entirely to blame for the shipwreck. They were all asleep at the time of the collision. He declined to draw any of the man's precious blood and instead prescribed a drenching with sea water on deck, and a mouthful of salt water to gather scattered senses.

At last, the shipwrecked passengers reached Livorno, only to learn from the 'collected wisdom' of the port that they were expected to spend 20 days in quarantine because they had no bill of health. Of course they had no bill of health – it was at the bottom of the sea. Charles Bonaparte to the rescue again: for two hours, he pleaded their case with the customs men, until common sense at last prevailed and the passengers were allowed ashore, 'in appearance, something like Falstaff's regiment.' The Misses Edmonstone had lost their bonnets, the Squire had lost his top-hat, sundry others were minus stockings, shoes and coats, and two worthy priests had one shoe each. The Spanish duchess was still in her night-shift, and the captain was breechless.

They were all given temporary hotel accommodation but the Spanish duchess refused to cross the threshold until her attendants had assured her that the hotel was not about to sink. She no longer trusted Italians.

Charles Waterton's party had no such phobia. After a quick wash and brush up at the hotel, they fitted themselves out with a new set of clothes in Livorno, and hastily made new plans. The *Monjibello* was not as badly damaged as feared, so they agreed to return in her to Civita Vecchia and to take post-chaise from there back to Rome.

In Rome, they passed duty-free through the customs, acquired a new letter of credit from the bank of Prince Torlonia, and booked in for another month at the Palazzo di Gregorio, in the Via Due Macelli. Between times, the Squire was back in the bird market, to recover what virtue he might from necessity. He bought some live little owls and introduced the species into Britain.

> Thinking that the civetta would be peculiarly useful to the British horticulturalist, not by the way in his kitchen, but in his kitchen-garden, I determined to import a dozen of these birds into our own country. And still, said I to myself, the world will say it was a strange whim in me to have brought owls all the way from Italy to England.[45]

Be that as it may – and it was – on 20 July, they left Rome again, with 12 young little owls, alive and well, in a 'commodious cage' provided for the journey. Again, the first leg of the journey was by sea – this time as far as Genoa. Genoa customs let the owls through with little fuss, possibly in the belief that they were intended as a dainty little picnic lunch during the long journey to England. Then, who should they bump into again, in Genoa, but Charles Bonaparte, Prince of Canino. Greetings and gratitude exchanged, Bonaparte gave them another letter of credit, to complement the one from Prince Torlonia's bank, which in turn had been intended to replace the one from London, lost in the shipwreck. With renewed expressions of gratitude, the Waterton menagerie rolled out of town, bound for home. By this time, the Squire was suffering from a slight bout of 'dysentery', perhaps brought on by the 'fever', or by gusunders doubling as cooking pots.

At the small town of Novi Ligure, between Genoa and Alessandria, they stopped for the night at the local hotel, where the waiter presented them with the visitors' book to sign, totally innocent of Charles Waterton's reputation for embroidering facts. As in Niagara, when he sprained his ankle, he now saw fit to append a few lines of poetic autobiography, for the edification of future guests.

The *Pollux*, once so fine,
No longer cleaves the wave,
For now she lies supine,
Deep in her wat'ry grave.

When she received her blow,
The Captain and his mate
Were both asleep below,
Snoring in breechless state.

If I the power possess'd,
I'd hang them by the neck,
As warning to the rest,
How they desert the deck.

Our treasures and our clothes,
With all we had, were lost.
The shock that caused our woes
Took place on Elba's coast.[46]

From Novi, they passed pleasurably through the Piedmontese sunshine, pausing here and there to deposit dysenteric fertilising matter on Italian fields, then, via Milan and Lake Lugano (no mention) they crossed into Switzerland through the Gotthard Pass, and over the incredible Goschingen Gorge (no mention) by the road which had recently been widened sufficiently to admit horse-drawn vehicles (no mention). All 12 little owls were still alive, frowning prettily in their commodious pen at each jolt and jostle of the alpine journey. If elephants could do it, so could they.

Basle was reached 'without any loss', and all was set fair for the first Roman colonisation of Britain since AD 43. 'I should have stuck fast for money in Basle, had not Lord Brougham's brother (William Brougham Esq.) luckily arrived in the town that very day. He immediately advanced me an ample supply.'[47] It was lucky because 'a wormwood-looking money-monger' had just refused to lend him £12 on the security of a 40 guinea watch.

All now went tolerably well until Aix-la-Chapelle, where Charles Waterton decided on a literal interpretation of 'watering place'.

> A long journey, and wet weather, had tended to soil the plumage of the little owls; and I deemed it necessary that they, as well as their master, should have the benefit of a warm bath.[48]

The predictable result, naturally, was unpredicted. Five of the owls died the same night and a sixth got its thigh broke, its master knew not how. His own dysentery got steadily worse, complicated by self-medication, and it harassed him 'cruelly' for the rest of the journey to Ostend and Dover.

In London, he called in at Northumberland House, in the Strand, the town house of Hugh Percy, third Duke of Northumberland, and left a letter there for a mutual friend, Father Singleton, archdeacon of Northumberland and rector of Howick. The Squire was now in a very bad way – 'scarcely able to drag one foot in front of the other'.[49] By the time they rolled through the lodge gates at Walton Hall, he was on the brink of death, as usual. The staff expected no less.

What news? he enquired of the keeper, after two years absence from home. One of the barnacle ganders from Rotterdam had paired with a Canada goose near the boat-house, said the retainer, aware of what was expected. But the Squire was no more inclined to believe it than he was disposed to believe that income tax was a blessing and the National Debt an honour to the country[50] – nothing could be more unbelievable than that. (Except, perhaps, some of the things he did believe.)

The story was true, however; the two birds had not only mated on the Hall Island, the Canada goose had also laid eggs – all, as it happened, infertile. The following year, the little barnacle tried again – obviously put up to it – and the year after that, two eggs hatched from a clutch of five, to produce hybrids of the kind which was to fuel the contemporary obsession with freaks and varieties, filed as evidence for the theory of evolution by natural selection. The Squire stood 'convinced by a hybrid, reprimanded by a gander, and instructed by a goose.'[51]

Freaks apart, the return home was an anti-climax. Squire and owls were put into quarantine for the winter, dysentery turned into 'decided dysentery', and eventually, he was obliged to consult 'the justly celebrated Doctor Hobson of Leeds' – his future Boswell. Hobson was already an 'invaluable friend'[52] at this time, though this is his first known mention in the Waterton literature

The doctor made 'gigantic efforts' to conquer the disease[53] (with help from the family surgeon, Mr Bennett) and the bond between Hobson and Waterton strengthened into a union as strange as any of that day or this. Like the barnacle gander and Canada goose – Mopsus and Nisa – they were two different species but two of a kind – giddy as a gander and daft as a duck. Between them, somehow, they managed to avoid the Squire's otherwise quick recovery, and at last, nature triumphed. The Squire of Walton Hall became fit enough to endanger his health again, as will be seen in the sequel.

Chapter 16
1841-1844
Faith, Hope and Charity

Edmund was now eleven years old, as yet unhandicapped by any of the utilitarian dead-weights of Victorian discipline. Instead, he'd done the Grand Tour – more usually reserved for early manhood. He was therefore, perhaps, better prepared for Stonyhurst College – 'the place selected for his education' – than any of the boys in his own age group, raised on the tyranny of 'correct disciplinarians' at Tudhoe and Dotheboys Hall.

Before setting off for Lancashire, Papa pressed into the boy's hand a couple of edifying volumes to supplement his studies in rhetorics and philosophy – two important works of fine prose and erudition to reinforce his classical studies and point the road to lasting virtue: Plato? Aristotle? Dante? Shakespeare? Rousseau? Goethe? No, no, no. *Wanderings in South America* and the first series of *Essays on Natural History, Chiefly Ornithological*, by Charles Waterton.

By way of bonus, and preceptory guidance, a few instructions in the author's own hand were inscribed on some blank pages in the *Wanderings*, to help ensure the boy's superiority on this side of the grave, and his safe passage through the paths of innocence and knowledge, towards salvation on the other. Very briefly: never scoff your superiors, avoid 'particular friendships', take part in all school games, observe the college rules, and 'should any boy offer you a forbidden book to read, oh! request him not to approach you with a viper whose sting is mortal.' Needless to say, said the Squire, 'Early to bed and early to rise, makes a man healthy, wealthy and wise.'[1]

Whether or not Edmund, at eleven years old, had sufficient native wit to be embarrassed by it all, he seems to have taken the advice to heart. He took to Stonyhurst with all the unshakeable enthusiasm of his father, and apparently followed his advice rather than his example where college rules were concerned. Bounds were kept and heights left unscaled. He was always in the thick of Shrovetide football matches, and his friendships were unparticular to the extent of cultivating a predisposition to twit on other boys, or to fall out with them apparently for the pleasure of making it up again. He seems to have been a fairly receptive pupil – certainly not clever enough to be thought a dunce – and craved the good opinion of his teachers and family to an extent which suggested exaggerated respect rather than ridicule. He would certainly never have dreamt of reading any book which might be thought to subvert the Catholic faith, and clearly had no choice where getting up early in the morning was concerned, on school days or 'in retreat'. For all which information, the facts and intimations of

Figure 27. Charles Waterton. Sketch by Percy Fitzgerald. (National Portrait Gallery.)

letters home, and the reminiscences of a school friend – Percy Fitzgerald[2] – with a few details from the Stonyhurst Centenary Record, are acknowledged as primary.

The Squire had good reason to expect satisfaction from his son's education at Stonyhurst and left him there, in September 1841, with complete confidence. He then returned home to Walton Hall to await developments.

Not long afterwards, he received a letter from George Ord in re Charles Bonaparte. Ord had not corresponded with Bonaparte for 16 years – 'in consequence of some strictures of his on some of my natural history papers.' News of his noble conduct at sea changed all that; Ord wrote to Monsieur le Prince, to express his admiration and gratitude for his 'sang-froid admirable' and 'conduite heroique'.[3] Then he wrote to Monsieur le Squire, to express the reason for his admiration.

> In the course of my intercourse with you, my dear friend, I may have uttered sentiments of disrespect to Charles Bonaparte. I now recall everything that might appear harsh or unkind. . . . He now appears to me in the light of benefactor to the family of my friend. I shall love him as long as I live.[4]

Isn't that nice.

But though his friend, Charles Waterton, was alive, thanks to Charles Bonaparte, he was certainly not well – the dysentery continued its ravages upon his iron frame 'with unabated fury'.[5] He was 'hovering between life and death.'[6]

After a long struggle, he at last re-emerged into the outside world, like Rip Van Winkle, to take in the 'strife of politics and prostration of national virtue' to which John Bull had sunk during the last two years.[7] The little owls remained in quarantine.

In all the political gloom of the 1840s, there was one clear ray of sunshine – one burning light of hope which did offer genuine prospect of disestablishment at least: Newman's Tract Ninety was in circulation, price one shilling, and the Church by Law Established would never be the same again. 'In giving the Articles a Catholic interpretation,' said Newman (still a Protestant) 'we bring them into harmony with the Book of Common Prayer.'[8]

The news from Stonyhurst was also good. Edmund was second in his class and gave 'universal satisfaction'. He was such a good boy – a proper little angeletto in fact – that he even exchanged letters with Cardinal Fransoni, in Rome – 'pretty' letters from the boy, benignly exhortatory ones from the cardinal, with a few holy reliques, surplus to stock, under separate cover.[9] Not many young boys could claim to be in correspondence with a cardinal.

So the Squire's health improved day by day, dosed with the prospect of a disestablished church and his son's eternal salvation, until, on a cold and frosty morning early in the New Year, he decided to try the mettle of his iron constitution again; he stood up to his waist in cold water.

He slipped into the water when he was trimming brambles, on the bank of one of the brooks in Walton Park, and found the new position so 'commodious' that he stayed there for an hour or more. The sickness returned and 'again Doctor Hobson triumphed.'[10]

By the spring, he was back in the pink again, fast approaching his 60th birthday and full of the joys of eternal youth. The five remaining little owls had also survived the winter and 'on the 10th of May, in the year of Our Lord, 1842, there being an abundance of snails, slugs and beetles on the ground,' he released them into the grotto at the Walton Hall bird sanctuary,[11] where they became the first tentative members of a now well-established British race. The little flat-capped character you may have seen perched in broad daylight on a telegraph pole or fence post, a permanent frown of displeasure on its face, is possibly a scion of the same ennobled stock, though more probably a direct descendant of birds released later in the century, in Berkshire, Hertfordshire, Northamptonshire and Kent. For many years after its introduction, the little owl was mercilessly victimised by gamekeepers on English country estates, and one in seven ringing recoveries in England are still of birds killed by guns, despite its inclusion on Schedule A of the protected list.[12]

Shortly after release of the little owls, an interesting letter arrived from Mgr Glover, in Rome, with an update on recent improvements to the altar of St Francis Xavier, in the Gesù, and with fond regards from some of the early morning friends the Squire had made at the door of the church, including, probably, the bird-seller from whom he'd acquired the owls. News of the Squire's recent illness had reached Rome, and his funny little ways were well-known: 'You must really persuade yourself that you are turned of twenty-five', said Father Glover,[13] naively, and the advice was obviously ignored. A visit to York Minster, during the summer, provoked a violent shivering attack which prefaced a third and final bout of dysentery, counteracted yet again by Dr Hobson's 'consummate knowledge' and Mrs Hobson's 'kind attentions'. His strength returned 'apace' and soon he was coming to terms with his chronological age, as advised, by climbing trees with his usual sound and pristine vigour, 'a stranger to all sexagenarian disabilities'.[14]

By now, he was a regular visitor to Hobson's house, in Leeds, as a friend as well as a patient. Hobson was as Protestant as God Save the Queen and clearly high Tory enough to claim, ridiculously, mischievously, posthumously, that Charles Waterton was high Tory also. Readers believed him.

But differences notwithstanding, the two men clearly got along together like pen and ink – for the time being at least. Their respective homes were home from home, household servants at one another's beck and call. During one of his visits to Hobson, in August, the sky began to look ominous and the doctor suggested that his friend should stay the night. But Charles insisted on going home, driven by divine providence. He arrived at Walton Hall in the early evening, just as the rain began to fall. Seven fishermen took refuge from the rain under the Lombardy poplar tree which his father had planted in 1756, about 100 yards from the mainland side of the footbridge. (The

Figure 28. Charles Waterton and Hobson Beneath Poplar. From Richard Hobson, Charles Waterton, His Home, Habits and Handiwork. *(Wakefield MDC, Museums and Arts.)*

tree features in many old pictures of Walton Hall.)

When he spotted the fishermen, he ran out of the house and told them to go to the saddle-room for shelter; their present position, he said, was dangerous. Sure enough, as soon as the men got to the saddle-room, the poplar was struck by lightning and Charles Waterton went down on bended knee to thank God for bringing him home to save the fishermen.[15] Rather than fell it, he chose to repair the tree with bricks and mortar, a living testament to divine providence – monument, ornament and ornithological tenement.

The whole estate was now sprouting similar sports of nature and sundry 'eccentric' appurtenances of what was later called conservation – nest-boxes, decoys, yew and holly 'snuggeries', artificial duck islands, ivy tower, starling towers, kingfisher stumps, weasel retreats, stone 'sentry boxes', fish traps, pheasantry, undisturbed heronry and cultivated swamp. In 1959, Phyllis Barclay-Smith claimed that the Walton Hall bird sanctuary was 'first established' at about this time – in 1843.[16] But truth to tell, by 1843 it was already well-established.

The winter had been long and severe and the Squire's resolve to spend all his declining years in the sequestered valley of Walton Hall, 'where nature smiled and all was gay around', was undermined. The Wanderer's lawless lust returned and by the spring of 1843, he was laying plans for his next trip to Italy with his little family. With all of them, that is, except Edmund – 'poor Edmund', who would have to remain at college, acquiring the knowledge and virtue which would afterwards be his 'ornament and delight'.[17]

Before leaving for the continent, another season or two of common occurrences were spent in the happy valley. The two sisters – surrogate mothers – pined for Edmund's company and apparently asked permission of the college to have him home for awhile, but the college declined, out of 'an earnest and serious solicitude for your dear boy's advancement.'[18] How many mothers must have suffered in the interests of earnest and serious solicitude? Eliza suffered badly. Her health seemed especially vulnerable when she wanted something, and it now became 'precarious'. But the boy remained at Stonyhurst, enjoying himself far too much to worry about his aunt's physical condition. (Which turned out to be consumption.)

One or two common occurrences of this period pass into the literature. On Easter Sunday, the Squire and his brother-in-law, Robert Carr, watched a fight to the death in Walton Park, between two hares. The defeated animal was donated to Mr Carr's groom, whose wife put it into a family pie, with a few rashers of Yorkshire bacon.[19]

As the summer advanced, common occurrences continued to occur and nature wore its customary grin. But the ladies needed a break. Deprived of Edmund's company, they were obviously unhappy. Eliza's lung trouble failed to improve and in the late autumn, she and Helen were packed off to Hornsea, on the Yorkshire coast, for a good fresh restorative of cold sea air. The Squire stayed at home to look after the cat.

Tommy Pussy was used to the very best service at Walton Hall, and clearly missed his ladies-in-waiting as much as they missed Edmund. The Squire sent them news. The cat preferred meat to meal, he said, whilst he, Charles Waterton, preferred meal to meat; Tommy 'shied' Charles at meal times.

Service would not return to normal until Wednesday next, the 22 November, when the two 'Miss Cautions' were due to return home from the Marine Hotel, Hornsea. (1.25 train from Hull, change at Normanton for Oakenshaw.)[20]

In later life, Tommy Pussy became a male nurse, a volunteer auxiliary in the Walton Hall hospital for orphaned chicks. But, like many a saint immortal – like Ignatius himself, for instance – in his youth, he had the devil in him. No dunnock was safe in the district and another cat was hanged for an offence against a couple of fine cock pheasants which was committed, in fact, by the pampered T.P. He was saved from the gallows himself by his own 'engaging manners'.[21]

It seems Tommy Pussy was the only cat on the premises after the public hanging, but the place was in imminent danger of being over-run by a much more destructive animal called children. Another un-doctored tom – the servant Tom Blackamore, who lived in the lodge – already had a large family of little Blackamores and his wife was expecting another. 'Has she pigged yet?' asked the Squire, and Tom smiled and said that 'Shoo' had gotten another fine lad. The Squire said that he'd better droon some of them but Tom smiled again and said he thought the day would come when they would all be very useful,[22] an important consideration in a member of the serving class. But no child could possibly be as useful as a really good cat.

At Stonyhurst, Charles Waterton's own child, Edmund, also went by the name of Tom – Long Tom. 'A most whimsical creature,' said Percy Fitzgerald, 'full of strange antics, piety, and good nature' – a chip off the old block. He was growing like a beanstalk and apparently had an obstinate stammer which he could never quite manage to get rid of.[23] He was enjoying himself mightily at Stonyhurst College and distinguished himself as honourably as Jesuit teaching, paternal exhortation and college fees demanded. 'He is devout and attentive at his prayers,' said his school report, 'as well as diligent in his studies. His general behaviour and conduct amongst his companions is that of a good and amiable child & he is affectionately respectful to his superiors.'[24]

The ladies returned home from Hornsea and, as Christmas approached, Edmund wrote the required seasonal letter home – addressed to his aunt Eliza – with traditional instructions for the composition of his box of Christmas presents, and a petulant little twit on a couple of school friends he suspected of competing for his aunt's affections.

> I shall want in my Christmas box some strong gloves, like the pair I bought at Lockwood's & some paper, both letter & note, also some of Wardle's thistle steel pens & some quill pens, also two blank books, one a large & the other a small one & also some lip salve & envelopes. I do not go so much with C. Locke as I used; he is not near such a good boy as he was last year. So says Mr Johnson who ought to know pretty well. Christmas is only 21 days away and then Papa will be here.[25]

Edmund was second in his class this term but Nathan – 'your friend Nathan' – who came first, had been seen by Mr Clifford to cheat. 'I do not mean to say that he did copy but I do not think that Mr Clifford would tell a fib.' Usually, Edmund Waterton finished top of his class – clearly a very ordinary child.

The Christmas box duly arrived at Stonyhurst – a 'gigantic' thing, packed with all the requisite durables, plus a few delicacies such as brawn, and a 'Gargantuan' goose pie (the legendary Waterton 'grand pie')[26] on which he and his friends gorged themselves until they were all ill. Papa also duly arrived, to watch his son's long tomfoolery in the annual Christmas 'Entertainment', and to provide his own little pendant to the proceedings with a few true stories of accidental travel and cool intrepidity, and a performance of some of his most embarrassing party tricks. Other papas might produce an egg from behind the headmaster's ear; Edmund Waterton's papa did better than that – he could scratch his own ear with his big toe. He was now 62 years old.

Life and soul he undoubtedly was, but festivities this year were muted for the Squire by some sad news he'd received shortly before setting out from Walton Hall. His friend, John Claudius Loudon, who had published almost all of his essays in the Magazine of Natural History, had recently travelled to Southampton, to supervise the laying out of a cemetery, and ended up being laid out himself, aged 61.[27] The funeral took place in London, four days before Christmas, and the dead man's widow, Jane (née Webb – authoress of *The Mummy* etc, etc) was invited over to Walton Hall, where she stayed for a month or so,[28] recovering from the loss and coming to terms with severely straitened circumstances. She probably spent Christmas Day itself at the Hall, with Helen and Eliza. (They were clearly not allowed at Stonyhurst, just as the Squire himself was not allowed in the English nunnery at Bruges, except to get married.)

Jane Loudon's financial plight was largely a result of the failure of her husband's great magnum opus, *Arboretum Brittanicum*, published between 1835 and 1838. At the time of his death, he still owed about £10,000 on the book[29] – printing, stationery, wood-engraving etc – and his publishers, Longmans, were appointed as trustees to defray the debt, using the proceeds from all his other books and paying the amount in equal shares to his creditors.[30] But best means were inadequate to fit the best ends; his

Figure 29. John Claudius Loudon, 1783-1843. Engraving by unknown artist c.1812. (Westminster City Archives.)

debt to nature was all that got paid.

Jane's own books were not selling well enough to help very much. What she needed was a few kind benefactors. The Squire returned home from Stonyhurst and consoled her as best he could withkind words, fervent prayers and warm hospitality. Secretly, he had a plan.

Jane eventually left Walton Hall and went home to Bayswater, faced with the prospect of spending the rest of her life on Grub Street. Longmans did all they could to ease the debt but the circumstances, in her own words, were 'painful'.[31]

Charles Waterton had his own little problems – mites by comparison, but enough to keep him occupied whilst he hatched out his scheme to help his friend's poor widow. A violent storm had damaged the 20-foot-high window which spanned two floors on the eastern side of Walton Hall, and it was becoming obvious that it would soon have to be replaced.[32] It was also now clear that the last of the local commons was about to be enclosed – a catastrophe as far as he was concerned, though the more general view at the time was that the 'unemployed soil of Great Britain' was a potential gold-mine, 'ready to be dug'.[33]

The common was a public common and the public, in Charles Waterton's view, should be made aware of what they were losing. So he 'caused to be printed', on 24 January, a public poster, distributed gratis, in an effort to stop the rot:

> Wakefield – once Merry Wakefield . . . what is to become of thy fifteen thousand people, who will not have a yard of public land remaining, whereon to recover that health of frame, and vigour of the mind, so apt to be enfeebled when debarred from the advantage of rural air and pastime.[34]

The second volume of Essays was now almost ready to go to press, and the poster was reproduced and thrown in at the last moment, 'in the faint hope that it may operate in some degree to retard the enclosure of the few commons which still remain to us.' He figured it might also help to sell the book.

His financial position was modest but sufficient to the day, a sufficiency less elegant than the average aristocrat would have been willing to accept. It took another two years to save enough solid tin to pay for a new window – 50 guineas – but his income from rents was reasonably secure and somehow or other, there was always enough

Figure 30. Frontspiece of 1844 edition of Waterton's Essays, *1st series. (Wakefield MDC. Museums and Arts.)*

money left over to pay for his extended saunters abroad. Enough, too, to indulge one of his favourite little rural pastimes – Christian charity. On 29 February 1844, he sat down at his desk to compose a preface to his second series of Essays:

> This volume which I now present to an indulgent public is an unsolicited donation to the widow of my poor friend, Mr Loudon, whose vast labours in the cause of Science have insured him an imperishable reputation. If this trifling present on my part shall be the medium of conveying one single drop of balm to the wound which it has pleased Heaven lately to inflict on the heart of that excellent lady, my time will have been well employed, and my endeavours amply requited.[35]

As with the first volume, and with the Wanderings, the book went into several substantial editions, so the little present proved a good deal better than 'trifling'. Natural history was still widely thought of as 'idle trifling', unless it was strictly utilitarian, and was usually vested with a strong undertow of sentimental moralising or dilettantism, in self-defence. Gradgrind and Bounderby ruled OK. But populist natural history was still popular, and none was more popular than Charles Waterton.

So the gift was a godsend. 'It is indeed only when we are in trouble that we find out how much real goodness there is in the world', said Jane Loudon, publicly.[36] Her own books, and her husband's, were less popular than Charles Waterton's and the greater part of the debt still hung over her head like the sword of Damocles. She survived her husband by 14 years (she was 24 years younger than he was) and continued the regular visits to Walton Hall until shortly before she died, ever ungrudgingly grateful for the unsolicited donation.

Reviews of the Essays trickled in, over the next couple of years, for the most part critically non-committal or almost completely non-critical. *The Athenaeum* celebrated a nineteenth-century man with the faith and enthusiasm of the thirteenth,[37] and Blackwood's was sure we should never look upon his like again.[38] The *Leeds Intelligencer* (forerunner of the *Yorkshire Post*) was so moderately uncensorious that the Squire wrote to the editor with an invitation to visit Walton Hall. The editor declined the invitation, for the time being, because, he said, he was shortly going on holiday with his family to Scarborough. The rave review of his paper's mildly favourable review clearly nonplussed him. 'I cannot but feel your commendation is greater than the cursory review deserved', he replied, very truly and respectfully.[39] But the Squire's friendships, like his antagonisms, were ever extreme.

One of his very best friends – Captain Jones – the old school chum who climbed the dome of St Peter's with him, in 1817, and who made the joke pair of door knockers at Walton Hall, visited the Hall in the summer of 1844 for the very last time. He was now going blind and would soon suffer complete paralysis. His days of jest and youthful jollity, quips and cranks and wanton wiles, were now over.

Not so the Squire. He was still lithe as a ten-year-old, free as the frolic wind and a stranger to all sexagenarian disabilities. A great boost to his health, during the summer of 1844, took the shape of a three-toed sloth – the first living sloth to appear in Europe.[40] It was introduced into Regent's Park Zoo and obligingly hung upside down from the underside of a branch. Charles rejoiced.

With a new lease of life, in the early autumn of 1844, he and his two dear foster sisters, Eliza and Helen, set their faces once more to the 'smiling south', in quest of new adventures and religious ecstasy. Edmund stayed at college and bided his time.

Chapter 17
1844-1850
Strangers in Paradise

Off they went – Squire, spinsters, footman and maid – 'through fertile lands and magnificent scenery, amongst people whose conduct requires no rural police',[1] to Innsbruck. There, and in Bolzano, they got the necessary letters of introduction to see the chief Catholic tourist attraction of the area – the holy Ecstatica of Caldaro.

Fraulein Maria von Morl – the Ecstatica – had first developed the stigmata of the cross on her hands 12 years earlier, in 1832. Pilgrims descended on her house in their thousands. By 1836, 2,000 a day were fighting for a place in the young girl's bedroom; some of them were hauled up over the banisters and lowered from the window by rope when it was time to leave, such was the crush.[2] After her parents died, she was moved to the nearby Franciscan monastery, where access could be strictly controlled, and it was to this more dignified sanctuary that Charles Waterton and his little band of pilgrims now sought admission.

A three hour journey by post-chaise took them from Bolzano to Caldaro, on All Soul's Eve, 1844.[3] The man they had to find was Pater Capistrani, the Ecstatica's father confessor. (Regularly pecked to death by a sinless duck.) But the day was inconvenient; the Squire's party had to wait for six or seven hours in the convent church before, at a quarter to five, they at last met the Franciscan monk and were taken up the mountain-side to the Ecstatica's cell.

And there she was – kneeling on her bed, her face turned towards a crucifix and 'a well-executed picture of the holy Mother of God' on the wall. Her hands were joined together under her chin as she prayed, whilst a turtle dove cooed from its cage at the other end of the little room, an emblem of innocence and purity.

Father Capistrani brought her out of ecstasy – 'Maria, Maria' – and introduced her to her latest visitors, all the way from Yorkshire, England. The Squire was granted permission to hold her hand, and he felt with his own fingers the ageing cicatrix which had by now closed up and shrunk into little more than a bright red spot on her palm. She relapsed into ecstasy a few more times, was recalled to herself by the father confessor, and gave the travellers some holy pictures as suitable devotional aids and souvenirs. Then, lingeringly, they left, with effusive thanks from the Squire, in Latin, for the good offices of Father Capistrani.

The phenomenon witnessed was apparently no more than religious coma, the 'magnetic affection' witnessed 12 months later by the immortal Anon.[4] The Squire's party was spared the demonics of an earlier stage of her condition when she would vomit pins and needles, splinters of glass, horse-hair, nails, broken knitting needles and bones, all previously swallowed in obedience to some sort of diabolic instruction attendant upon virgin purity. 'Nor did these objects come from her mouth only', said Anon, in reported speech, 'but sometimes presented themselves at different parts of the head, and were with difficulty drawn out by her confessor.' The speech reported was Father Capistrani's own.

All this smacks largely of the 'terrific', the remarkable and interesting events which were then all the rage – the crimes, judgements, providences and calamities of the nineteenth-century *News of the World*, the Penny Dreadfuls. Horrible murders, terrible executions, frightening spectres, radiant apparitions. Fearful plagues, miraculous escapes, wonderful resurrections and holy stigmata. Not to mention eccentric squires.

But Charles Waterton reported none of the unseemly rabble-rousing and restricted himself simply to a faithful account of what he saw, anxious to cast the 'Tyrolean Virgin' in as authentically holy a light as possible – what some might still call rabble-rousing. He saw her as 'a privileged being, exempt from the common law of mortality, and ready, whenever heaven should will it, to take its flight to the everlasting mansions of the blessed in another world.'[5] Conscious of his own reputation for dressing truth, he confidently expected to be taken for 'an incorrigible dupe' by his readers (the very idea!) and eagerly looked forward to the chance to square accounts with them when they, too, reached another world.

The affairs of this world meanwhile interposed. Italian hotels were not exactly mansions of the blessed. In Venice, they arrived at the Hotel d'Europe and found a big fat rat already booked in, staring at the new arrivals, 'as much as to say, "I have capital pickings here, both for myself and for my relatives."'[6] And in Bologna, they stayed at a hotel of 'splendid appearance' but 'serious drawbacks'.

> O ye nasty people of Bologna, of what avail are your gorgeous palaces. your cookery and fruits, whilst your temples to the goddess Cloacina are worse than common pigstyes.[7]

From Bologna, they travelled via Rimini, Pesaro, Ancona and Loretto, to Rome, where they stayed for about seven months. There's a rather sick-looking night heron in the Waterton collection, labelled 'Italian heron. From Rome, 1845'.[8] Apart from that, the Squire's dissecting knife appears to have remained in its sheath. His quill remained in its ink-well.

But there were other English pens in Rome, including the ones which belonged to Charles Dickens and William Makepeace Thackeray. Dickens arrived by the more usual Grand Tourist route[9] such as the Squire's party had followed four years earlier. Thackeray arrived from the south, after jumping ship at Malta, in late October, on his way home from a Mediterranean cruise.[10] He did the round of expatriate Englishmen – the itinerant 'artist banditti' – and became well acquainted with a pious old Catholic gentleman whose surname began with W. Later, this 'good, kind W__' was allocated a paragraph all to himself in *The Newcomes*. Clive Newcome met him in a Roman street and was taken into a church where the Virgin Mary had lately appeared, 'in light and splendour celestial,' to convert a visiting Jewish gentleman to the one true faith.

> My friend bade me look at the picture and, kneeling down beside me, I know prayed with all his honest heart that the truth might shine down upon me too;

> but I saw no glimpse of heaven at all. . . . The good, kind W__ went away, humbly saying that 'such might have happened again if Heaven so willed it.' I could not but feel a kindness and admiration for the good man. I know his works are made to square with his faith, that he dines on a crust, lives as chastely as a hermit, and gives his all to the poor.[11]

This fictional sketch of Charles Waterton is obviously accurate but seriously devalued by its context – all those honest Tom, Dick and Harrys, kind old creatures, dear old souls, worthy gentlemen, capital fellows and good, kind, dear, brave, jovial mortals who jump up and clap their hands with glee at one another's company. 'Why don't you put my uncle into a book?' says the dear, sweet, pretty Miss Ethel, to Pendennis, and answers the question herself – 'He is so good that nobody could make him good enough.'

The good, kind W__ stayed on in Rome for Easter and the Month of Mary and returned home via Civita Vecchia and Marseilles, some time during June.

A couple of days after returning home, he received news of another death in the family – his 50-year-old brother, Edward. Edward had also been returning home – from unknown business in New South Wales – and died at sea of unspecified causes, complicated by sea-sickness. He had on board, when he died, two trunks full of clothes and some suitable little presents for his brother, Charles – 'two living emus, a kangaroo and a squirrel.'[12] Charles considered that they would only serve to increase his sorrow at his brother's death, so he donated them all to the ship's captain who in turn 'made presents of them'[13] to someone who presumably found someone to give them to. What scientific enlightenment and literary gemstones were denied posterity through default of a kangaroo at Walton Hall, defies all serious conjecture. A kangaroo would have suited Walton Hall.

It might also have been appreciated by Charles Darwin, who visited the Hall at about this time. Tantalisingly sketchy impressions of the visit are contained in Darwin's own correspondence with Sir Charles Lyell and Professor Henslow. He'd been to see a farm he'd bought, in Lincolnshire, and afterwards called on the Dean of Manchester, in York.

> I then visited Mr Waterton at Walton Hall, & was exceedingly amused with my visit, & with the man; he is the strangest mixture of extreme kindness, harshness and bigotry that ever I saw.[14]

The synthetic type was shown around the transitional varieties and analogical resemblances of Walton estate, and watched as its divergent character – his 63–year-old host – 'ran down and caught a leveret in a turnip field.' Next day, at early dinner, 'our party consisted of two Catholic priests and two Mulatresses!'[15] – further evidence, to Charles Darwin, that his host was a highly amusing and strange fellow. But that's all we know about the visit. Darwin left Walton Hall and went off to visit the Duke of Devonshire, at Chatsworth House, before going home to Kent, to collate the evidence for extending the duke's pedigree, and the Squire's, beyond Bess of Hardwick and Ashenhold the thane. Walton Hall was a 'fine old house', said Darwin, and Chatsworth wasn't bad either.

Another visitor to Walton Hall at about this time was F. H. Salvin, nephew of the Squire's old friend, Bryan Salvin of Croxdale. On his first visit to the Hall, with his brother, Marmaduke, they were met at the footbridge by the butler, who had a coal

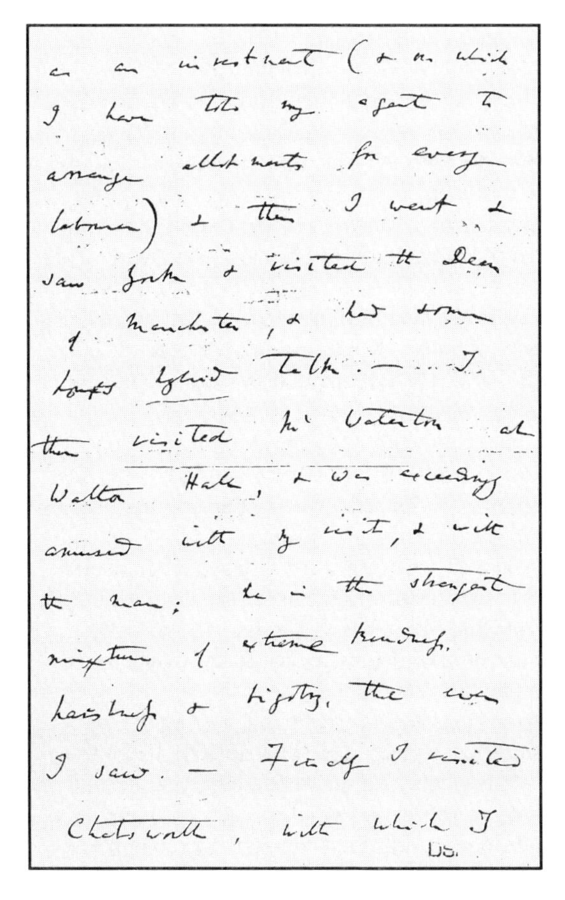
… in contract (& on which I have told my agent to arrange allotments for every labourer) & then I went & saw York & visited the Dean of Manchester, & had some of his splendid talk: — I then visited Mr Waterton at Walton Hall, & was exceedingly amused with my visit, & with the man; he is the strangest mixture of extreme kindness, hairbrainedness & bigotry, the man I saw. — Finally I visited Chatsworth, with which I

Figure 31.Charles Darwin to J.S. Henslow, re Charles Waterton. (Smithsonian Institution Libraries – Special Collections.)

scuttle and brush in his hands.[16]

'Sorry, Sir, Master never allows the collection, nor the grounds, to be seen on Sundays', said the butler.

'But we've come a long way', said Marmaduke, 'and Uncle Bryan went to school with Mr Waterton.'

'Ho, ho, ho', said the butler, revealing his true identity, and he tickled the young men's ribs with the coal brush. Then he scratched his own ear with his big toe (no doubt). Famous people, in those days, were generally more anonymous than now.

FHS became a regular visitor to Walton Hall and on one occasion accidentally broke the glass case which contained the Nondescript, 'which mishap was most good-naturedly forgiven', he said. The case was still intact when Darwin visited the Hall, but his opinion of the potential missing link inside it is not on record. It's a crying shame but the two mulatresses failed to keep diaries.

Eliza's lung trouble was now getting worse, exacerbated by fumes from the new soap works on the edge of Walton village. What she needed, she agreed, was another dose of winter sunshine – southern Spain perhaps, or Madeira. In November, therefore, she left for Madeira, with her sister and brother-in-law, and filled her lungs for the next few months with warm Atlantic sea air, laden with the fumes from sundry winter cocktails of jasmine, daphne, oleander and lemon blossom. The Squire returned home in January, ostensibly because the island offered 'little in the ornithological line',[17] but in fact on account of 'your brother's going thither'.[18] No love lost.

The Portuguese themselves had been leaving Madeira in droves – emigrating to British Guiana. Richard Schomburgk detested them. 'The Portuguese of Guiana are the Jews of Europe', he said,[19] engagingly. But Charles Waterton loved them. He'd lately been awarded a Knighthood of the Order of Dom Miguel of Portugal, approximate equivalent of the British K.G., and the fact that Madeira was as ruthlessly Catholic as Britain was oppressively Protestant, obviously recommended it. But in early January, he left his 'sisters' to their own devices and returned home to attend to the business of a pioneer nature reserve in Yorkshire.

In other circumstances, Walton Hall could have been a suitable lung for the hands of industrial south Yorkshire, but local savages abused the privilege. The Squire returned home to find that the poachers had played havoc with his pheasants and hares whilst his keeper, Warwick, had been 'amusing himself in the alehouse.' 'I do really feel that this is the worst country in the world to live in.'[20]

The gamekeeper from Haw Park was borrowed and four of them – two gamekeepers, a policeman, and a Knight of the Order of Dom Miguel of Portugal – set out to surprise the poachers that night.

> I shot at one fellow but the ball merely grazed him. Luckily, the constable recognised some old and desperate offenders. Nine are now in Wakefield jail. . . .I have been desired by the magistrates to hold myself in readiness for the Lent assizes.[21]

He duly appeared at the Lent assizes and *Punch* got to know about it:

> On the Northern Circuit, at the instance of MR CHARLES WATERTON, the philanthropic naturalist, some men were indicted for trespassing upon his grounds in search of game. MR WATERTON was examined:-
>
> "MR SERJEANT WILKINS asked the witness if a spring gun did not go off?
>
> "Witness – Does not the law of this country forbid the use of spring guns?
>
> "MR SERJEANT WILKINS – That, Sir, is not an answer to the question.
>
> "Witness – If the law forbids the use of spring guns, I want to know whether a man is bound by the law to criminate himself."
>
> We consider this answer quite worthy of the Jesuit's college, and one with which the learned Serjeant very significantly expressed himself 'satisfied.' The witness then wound up by saying – 'There was, indeed, a tremendous explosion!' Artless MR WATERTON! He had no objection to prosecute poachers for breaking the law, but it was not for him to 'criminate himself' by owning to a systematic violation of the statutes. And this is game-law morality.[22]

Whatever might be said of the court's verdict, the verdict of *Punch* was obviously unjust, as the plaintiff was quick to inform one of its main contributors, William Makepeace Thackeray. He fain hoped (as was the wont) that Thackeray had nothing to do with bringing him forward so unhandsomely. He gently reminded him that his invalid sister, Eliza, had been very kind to him in Rome. (When the novelist was short of money.)[23]

But it was all nothing to do with Thackeray anyway. It was probably the work of that 'savage little Robespierre', Douglas Jerrold, the editor's right-hand man.[24] Spring guns were much worse than the crimes they were used against but to caricature Charles Waterton, and Stonyhurst College, as benders of rules for the purpose of oppressing the poor, was clearly ridiculous. When the English shires were full of English squires, good and bad, the Yorkshire Squire was very very good, as Thackeray himself later pointed out in *The Newcomes*. The good, kind W__ did not receive justice from the court report in *Punch*.

When he returned home from court, he supervised completion of a few alterations at Walton Hall and the nearby outbuildings. The grand staircase was replastered and the new 20-foot high window – cost 50 guineas – was put in place on the south-east wall of the house, to light the stairs and replace the window which had been damaged by a storm 12 years earlier.[25] The new window is still in situ, 145 years later.

The pigeon cot had now been replaced with a bigger and better, burglar-proof edifice, two storeys high, the lower storey for use as a tool-shed, and the upper for birds. There were 666 holes for birds[26] but none for criminals, human, vulpine or Hanoverian. There was no purchase for grappling tools on the walls, and the door to the upper storey was flush with the wall, 25 feet or more from the ground.[27] If Charles Waterton couldn't climb it, no-one could. The cot was a great success with the birds, and very nearly fully occupied, from the first night after the old cot was closed down. 'It is allowed to be the finest pigeon-cot in England', said its owner, demurely.[28] Not a ladder in Yorkshire could reach its roof. The building was still standing, its lower floor used as a tractor shed, until shortly before I decided to photograph it, in April 1982.

The old cot had been raided twice whilst the Squire was away in Italy, parent birds taken for the shooting matches which were then a popular sport in the local towns.[29] Poachers climbed to the roof of the cot and put a net over the glover, whilst an accomplice tapped the door, to drive the adult birds into the net. Orphaned nestlings died of starvation.

> Young Mr Draper wants to try a new gun: Squire Goodaim is eager to shew his skill in shooting, before an assembled multitude. . . and Tunley (Smollett's name for an innkeeper) has his eye on an opening to dispose of his beer by the hogshead. Urgent reasons these for destroying kidnapped pigeons![30]

No kidnap was possible with the new cot.

Another shock, when he returned home from Madeira, was to discover that the neighbouring savages had also killed the woodpeckers. A pair of great-spotted woodpeckers had successfully bred in the park, for the first time, in 1844, but none were seen or heard there during the spring of 1846. The whole family had disappeared, presumed killed by savages who should have been stopped by his keeper, Warwick.[31] The estate's mason, John Ogden, was promoted to the post of keeper, and his wages increased in line with the un-codified laws of nineeenth-century feudalism.

Charles Waterton predicted a general decrease in labourers' wages because, he said, agricultural rents would soon have to be lowered, to combat repeal of the Corn Laws. As for the poor factory hands in the nearby 'slave mills', the cotton lords would soon reduce their wages.[32]

But John Ogden's bread, such as it was, was now well-buttered on both sides, and the relationship between that particular slave and his master was set fair to prosper till death did them part.

With a poacher-proof pigeon cot and this new keeper to care for it, the Squire now felt confident enough, in the late summer of 1846, to spend another 'very merry month in Belgium',[33] followed by a visit to the pathetic zoological exhibits in the British Museum. He returned home refreshed, and judged his keeper to be admirably answering the purpose for which he'd been appointed. 'Sixty-one dozens of young ones' were reared during the pigeon cot's first year (over half way to the record 93 dozens in one year, when his father was alive[34]) and J. G. Wood later recorded 'twenty rows' of nests on each wall.[35]

The warm summer of '46 was 'comfortable' to the Squire's cold, venesected frame, and set him up nicely for a Yorkshire winter indoors, by a roaring fire in front of an open window. Emerson visited England, saw the several 'paradises' of the landed gentry, set aside from the nearby roar of industry,[36] but had no occasion to visit the epitome of nineteenth-century paradises at Walton Hall. The next illustrious visitor, in fact, was Edward Charlesworth, the man who had taken over *Loudon's Magazine of Natural History* in 1837. (It survived for three years without Loudon or Waterton before merging with the *Annals of Natural History* which itself soon folded.) Charlesworth had now been appointed to 'a comfortable situation' at York Museum and spent a day at Walton Hall, after Easter, perhaps in quest of a Waterton bequest. He impressed the Squire as 'a civil and obliging person'[37] but the Waterton collection was bound elsewhere than York Museum, when it finally left Walton Hall after the Squire's death.

Another summer at Walton Hall passed unremarked – by doubtful inference unremarkably – and the new Squire/keeper arrangement prospered. The Squire duly negotiated his 65th birthday and confidently expected to remain as strong as a giant until the age of 100, mainly by dint of fresh air, exercise, and disdain of heterogeneous aliment. 'In case of need. . . every day,' he was now taking two grains of sulphate of quinine in the morning, with two drops of diluted vitriolic acid and two thimblefuls of cold water, and each evening, the dose was repeated, 'with marvellously good effect.'[38] Liberal quantities of his own blood were cupped 'when necessary'.

A few doses of vitriol were held in reserve for the soap and vitriol factory on the northern boundary of his land. Trees were being crippled, fish were dying and herons were moving to the southern shore of the lake, further from the source of pollution. 'Friendly warnings' had little effect, and ultimately, the arms of the law were invoked and words became less friendly, as will be seen.

Meanwhile, polluted paradise was administered as 'economically' as ever. "Here are peaceful retreats in ample fields, grottos and refreshing lakes; here are cool valleys and the lowing kine, and soft slumbers beneath the trees", quoted Hobson, from Virgil.[39] Pity about the smell.

The birds of paradise co-existed like lions and lambs in latterday Watchtower publications. Moorhens nested a few feet away from carrion crows, woodpigeons in the same trees as magpies, and geese interbred uninhibitedly. A rook's eggs had been hatched under a carrion crow, a jackdaw's under a magpie, and a magpie's under a jackdaw.[40] In fierce disgust at his reputation as an 'amateur', Charles Waterton pointed to the highly professional success of getting the barn owl, the tawny owl, the heron, the jackdaw, the magpie, the carrion crow, the mallard, the pheasant, the starling, the woodpecker, the great tit, the moorhen, the song thrush and the blackbird to build their nests and take away their young in safety, 'at a stone throw of each other.'[41] And Tommy Pussy took pity on a domestic hen, abandoned by its mother, and nursed it by the kitchen fire like one of his own kittens, allowing the bird to take meat out of his mouth as it pleased.[42]

After 30 years of 'memoranda', 123 species of bird had been recorded at Walton Hall – resident or as birds of passage – including the peregrine, crossbill, grosbeak, 'mountain sparrow' (snow bunting), 'land-rail' (corncrake), 'lesser land-rail' (spotted crake?), 'little land-rail' (little crake?), 'thick-kneed plover' (stone curlew), 'dusky grebe' (black-necked grebe), 'Cape goose' (greylag), black tern, garganey, velvet scoter and common scoter (not common on inland waters).[43]

But much of the habitat which attracted the birds was put in jeopardy by Simpson's soap and vitriol works in Walton village. The grotto was particularly badly affected, as being closest to source. Soil and water were poisoned and the trees became stag-headed 'beacons of desolation', pitiful skeletons 'to warn us advocates of sylvan scenery that we are in horribly bad company.'[44] The Squire had developed a taste for the advocate's fence against bad company, and the parvenu soap boiler was taken to court.

Its a matter for judgment as to who was the richer man in that court room, defendant or plaintiff – middle-class industrialist or upper-class squire. Back in 1824, Charles Willson Peale had been told by 'another source' that Charles Waterton was in receipt of £6,000 a year from his tenants,[45] but since the rents were said to come from an estate in Oxfordshire, it can be assumed that the source was unreliable. The Squire, like many another gentleman of means – Jawleyfords of Jawleyford Courts – was probably less rich than he liked to pretend. Edward Simpson, on the other hand – the horny-handed sire of toil – possibly had more mucky brass under his floorboards than it was policy to admit; in all probability, he was beginning to overtake the Squire financially.

But in a court of law, money mattered less in Victorian England than the divine human right to it, presumed to be conferred by noble birth and primogeniture. Soap-boilers stood no chance against the legal might of pedigree and privilege. Charles Waterton won the case and was awarded £1,100 in costs,[46] to compensate for the imminent destruction of part of his little earthly paradise, and thereby his putative lung for the hands of South Yorkshire.

It was good that he won because, though hands be idle, environmental decency won in the process, and that almost never happened in Victorian England. But it failed to cure the problem. Simpson continued to make soap and money and saw no clean or economical way of doing it. The Squire's warnings became unfriendly:

> Sir,
> The law has most properly condemned you in cost and damages. Still you continue a nuisance which is utterly intolerable at my house whenever the wind blows from the northern quarter.
>
> I take this early opportunity of informing you that unless you immediately do away with the smoke nuisance and cease entirely to pollute the water in my brook with filthy drainage from your works, I shall apply to the law for redress at the next summer assizes held in York.[47]

So the case again went to law – in August 1849 – and this time Simpson's money had sufficient power of attorney to talk big – a whole court-house full of witnesses for the defence was called, including some who might otherwise have appeared for the prosecution. Complainants half dead with respiratory disease found it financially expedient not to complain. False witness won the day.

For Charles Waterton and his long-suffering sisters-in-law, there was just one course of action left open. They bought a piece of riverside land at Thornes, four miles away, and offered it for sale at the kind of knock-down price which Edward Simpson was bound to pick up. Land, power, navigation rights, wharf facilities and a 'residence fit for a nobleman' – Thornes House itself – all for £5,000.[48] Then they went to court for a third time – October 1849 – and this time managed to get an injunction served on the soap-boiler to close down the factory at Walton and to move it, lock, stock and soap-kettle, to the 'prime site' in Thornes, that much closer to the 15,000 inhabitants of Once Merry Wakefield. Simpson procrastinated but by 1853, the Walton soap house had stopped making soap and the World's First Nature Reserve could breathe again.

The manufacture of soap, of course, always has been a filthy business. Its by-products still cause ecological mayhem, hundreds of miles from the point of origin. But where industrial vested interest is concerned, earthly paradise counts for nothing – the powers of darkness are the Lord's anointed. Did I say soap boilers stood no chance against pedigree and privilege? Don't you believe it! Filth and ugliness invariably won in the end.

As in Europe, so in America. George Ord was having exactly the same trouble at exactly the same time.[49] He lived in Southwark, New Jersey, and had a similar soap and candle factory right on the edge of his own garden plot. The owner was ordered by the court to move the factory but continued to make soap as usual, except that the melting of the fat was done elsewhere. The judge then gave it as his opinion that the factory was no longer a nuisance. The order to move was overturned. 'Tell your English jurists of this trick of a Yankee judge, and let them enjoy themselves in a hearty laugh at our expense.'

But Charles Waterton was not disposed to find it funny. His life-work and love of his life, his home and handiwork, had taken a battering, and so had his dear mulatress sister-in-law, Eliza Edmonstone. All in the interests of a manifestly crass form of brutalism called civilization – produce the masses and you have to mass produce.

For the time being, as the two soap operas slowly unfolded, the trees of Walton continued to languish and die and 'a botanist of European reputation' – obviously Jane Loudon – told him that it would take 12 years for the ones that survived to 'recover their pristine health.'[50] The whole dirty episode cast a black and murky cloud over his last years; paradise lost, principle upheld. 'Long is the way and hard that out of hell leads up to light.'[51]

The birds survived, in spite of everything. The 24 Canada geese were 20 too many for the size of the lake; density stress turned them nasty – it happens all the time. Pairs of Canadas flew noisily around the Hall, 'like tipsy women at one of our village fairs.'[52] But the Squire left them unpinioned, in the hope that some would fly away.

All other birds were welcome. The herons repaired their old nests and built new ones, and squabs blown out of their nests were put back in again by the Squire. 'For the greater delectation' of adult herons, a stream was created by cutting a channel between the lake and a little spring at the southern end of the park, within sight of the Hall.[53] The birds took to it like herons to water.

A pair of tawny owls bred in one of the sycamores on the Hall Island, and 'a fine breed of kingfishers' had cause to be grateful to the park wall, but for which, 'their race would be extinct in this depraved and demoralised part of Yorkshire.'[54] A kingfisher fishing stump was provided, gratis, compliments of the management.

Summer lightning, in 1849, helped the soap works to kill some of the trees, but the weather improved in September, in time to welcome another honoured guest to the Hall – the entomologist and artist, William Chapman Hewitson.[55]

A long freeze in the winter brought one of the periodic invasions of 'Bohemian chatterers' (waxwings) to Britain, and a 'lad' who lived nearby – as distinct from a

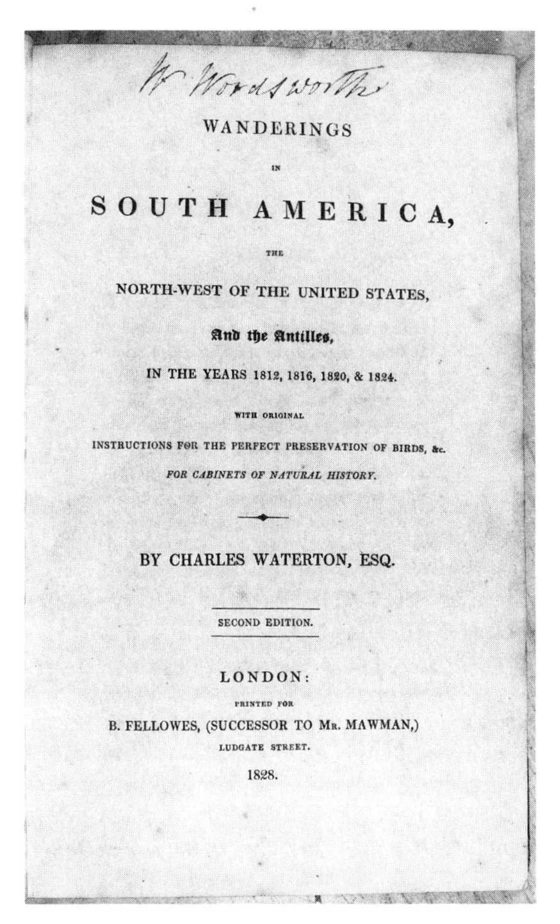

W Wordsworth

WANDERINGS

IN

SOUTH AMERICA,

THE

NORTH-WEST OF THE UNITED STATES,

And the Antilles,

IN THE YEARS 1812, 1816, 1820, & 1824.

WITH ORIGINAL

INSTRUCTIONS FOR THE PERFECT PRESERVATION OF BIRDS, &c.

FOR CABINETS OF NATURAL HISTORY.

BY CHARLES WATERTON, ESQ.

SECOND EDITION.

LONDON:

PRINTED FOR

B. FELLOWES, (SUCCESSOR TO Mr. MAWMAN,)

LUDGATE STREET.

1828.

Figure 32. Wordsworth's copy of Waterton's Wanderings. *(Stonyhurst College Library.)*

neighbouring savage – caught two of them as they fed on whitethorn berries in his garden, and presented them to the Squire for preservation.[56] One of the birds is now in Wakefield Museum.

As the ice on the lake broke up, wildfowl numbers increased, and 1,500 were counted 'within a long gun-shot of the drawing-room window'. The gun-shot, naturally, was hypothetical.

On St George's Day 1850 William Wordsworth died. Not long afterwards, a copy of Waterton's *Wanderings*, with Wordsworth's signature on the flyleaf, found its way to the Arundel Library, at Stonyhurst, but no copy of Wordsworth's works with Waterton's signature on the flyleaf is known to exist.

At about the same time as Wordsworth's death, George Ord went to bed with the 'flu; the Squire sent him a few comforting words on the frequently fatal outcome of the disease. Eliza and Helen went to Bruges, for the Ascension Day procession of Christ's blood through the streets, and Charles went to see Lord Derby's menagerie at Knowsley Park, near Liverpool. 'I think that every handsome species from the tropics has found its way to Knowsley Park', he said.[57] Edward Lear had painted many of them,[58] but no known copy of Lear's work bears Charles Waterton's signature on the flyleaf.

He now returned to Walton Hall, intending to carry on from there to meet up with the ladies in Bruges. But he was delayed by undisclosed 'urgent business' at home – probably the visit of Sir Richard Owen and his wife. The urgent business discussed would certainly have included the abominable stink of soap and the vulture's ability to smell it – Owen had recently preserved a specimen of a turkey buzzard's nose, which the Reverend John Bachman – father-in-law of Audubon's two sons – had refused to look at, on the grounds that the anti-nosarian case had already been proved.[59]

The Owens went home to London and Charles Waterton set off for Bruges, to join Eliza and Helen for the Month of Mary. He almost joined Wordsworth in the next world instead. Which serves as introduction to a curious chapter of accidents.

Chapter 18
1850-1851
Hell and High Water: Incomparable Friends and Shittering Fellows

Squire Waterton's patent cure for influenza, as prescribed to George Ord, was based on the best treatment for yellow fever and back-ache.

> In its early state, it is most effectually subdued by a good blood-letting, followed up by a strong purgative or two. I never knew this to fail. If treated otherwise, influenza becomes exceedingly obstinate and often fatal.[1]

News of the treatment reached Philadelphia long after Ord's recovery by more fatal means. He was denied the pleasure of a fail-safe cure which might have killed him.

The Squire himself was luckier. The same letter contained news of an incident in Dover harbour – later expanded upon in the third volume of Essays – which produced similar symptoms to influenza and was treated accordingly, 'with complete success'.

'Death often spares his victim when far from home, and slays him, at last, close to his own fireside', said the Squire, prophetically. It nearly slew him now in Dover harbour.

He arrived in Dover by night train from London and went in search of his boat to Belgium. 'Italian cloak' over his shoulder, a little portmanteau in one hand and an umbrella in the other, he cast around for a porter to direct him, and found one at last, carrying the luggage for two gentlemen bound for Calais. The Belgian boat was moored slightly ahead of the French steamer and the porter pointed to the little blue light which burned on deck. Charles Waterton made for the light, and a place on deck for his velvet cushion. But the night was black as pitch – darkness drear with dangers compassed round. No moon, no stars, just dense, low cloud.

He set foot on the gangway of the Belgian boat, as he thought, and proceeded to embark for the continent. 'O horrible mistake!'[2] Next he knew, he was head-first in the water, 15 feet below, floundering for his life, whilst his portmanteau, travelling rug, hat and umbrella floated away towards Belgium of their own accord. 'Death now stared me in the face', he said, later; all he could splutter at the time was a feeble cry for help.

There was no time to inflate his Macintosh (which he'd received as a present from Charles Macintosh himself) but he had a yet more infallible preserver of life impended from his neck like a waterwing – 'the miraculous medal of the Blessed Virgin'. With indomitable faith in his dear redeemer, he splashed around, spluttering for help, and found himself, presently, under the paddle-wheel of the boat he was supposed to be aboard. He was about to sink 'for the last time' when a voice called out to him in French. It was a Belgian sailor, summoned by the consolatrix afflictorum round the Squire's neck. He opened the paddle-house door on deck, ran down through the wheel itself, and grabbed the drowning man's hand, to pull him on board, with the help of another sailor who was following close behind. Between them, they got him up on deck, where he sat, 'soaked through and shivering in the midnight blast.' Two police officers turned up and language difficulties were quickly resolved. It was established that he was not a suicidal type of maniac, except by accident, and his request to be taken to the nearest 'respectable' hotel was granted.

So, in due course, the accidental traveller was booked in at the Dover Castle Hotel, kept by a woman who rejoiced in the name of Widow Dyver – as immortally appropriate, in the circumstances, as Madame Van Gulpen had been, at the Dragon d'Or Hotel in Aix-la-Chapelle.

Widow Dyver wanted to call for a doctor. 'Madam,' said the Squire, 'a doctor will not be necessary.' Instead, he rolled himself up in a couple of blankets by the fire and caused a mild sensation by refusing 'inward consolation in the way of cordial.'

Nature rallied during the night, his cloak, umbrella, hat and portmanteau were recovered from the sea, and – plans revised – the afternoon tide took him and his belongings to Calais.[3] In Calais, he sat by a roaring fire, in the coffee room of the Hotel de Paris, and regaled his French host with details of his recent cold bath and various other adventures of accidental travel. Then he caught the overnight train to Flanders.

In Bruges, he met up with his sisters-in-law and began to look forward to the great procession at the end of the month. But soon he began to pay for his recent midnight dip – fever, heats and shiverings, coughs and sneezes and the old 'oppression at the chest' kept him indoors. Things got steadily worse. The procession of the holy blood was imminent but it looked as though he might have to miss what he'd travelled through hell and high water to see. So he took away 25 ounces of his own blood and next morning, 'to make all sure', he took a gentle laxative of 20 grains of jalap, mixed with ten grains of calomel – one of Squire Waterton's pills. 'The operation was crowned with complete success.'

He spent a contented month in Bruges, with his sisters-in-law and 15,000 other pilgrims, and witnessed an Ascension Day procession of the Holy Blood 'of inconceivable splendour', untroubled by any more chest pains, nor by any of the usual Protestant distempers such as pick-pocketing and disorderly conduct.[4] The obligatory pilgrimage to the English nunnery in the Carmerstraat was made, reviving memories of his beloved wife, Anne, but though Eliza and Helen could enter the convent itself, their brother-in-law was naturally confined to the seventeenth-century chapel where he'd been married 21 years earlier. Prayers for his wife's everlasting soul were never far from his lips, and on this occasion, it can safely be assumed there were also the most fervent thanksgivings for his recent deliverance from a watery grave.

As May observances ended in Bruges, Charles Waterton returned home through London and called in once more at the British Museum, where he heard reports of a hybridised 'stoat' in York Museum. Later in the summer, he went on holiday to Scarborough and took the opportunity to call in at York, to examine the animal for himself. It was no hybrid at all, he decided – just a diseased stoat – and he acquitted the taxidermist of anything as disgraceful as taxidermical fraud.[5] A couple of years later, his friend from the Society of Friends, 'Friend Allis', dissected the animal and declared it a new species. The Squire agreed.[6]

Prince Albert was also in York, presiding over an 'Extravagance Culinaire' for local aldermen – the infamous 'Hundred Guinea Dish', concocted in honour of the Great Exhibition – a savage insult to the Jesuitical sensibilities of Charles Waterton. The dish included 5 turtles, 24 capons, 18 turkeys, 18 poulardes, 16 fowls, 40 woodcocks, 100 snipes, three dozen pigeons, 43 partridges, 10 dozen larks, 30 pheasants, 6 plovers, three dozen quails, a few ortolans and a garniture of cock's combs, truffles, mushrooms, crawfish, olives, American asparagus, croustades, sweetbreads, quenelles de volaille, and sauce. Not to mention the spirituous liquors. All this at a time when thousands were dying of starvation in Ireland and two million in England were dependent on public alms. Charles Waterton was furious. The assembled swine at the banquet – including Prince Albert – should be flogged once a week at the cart-tail for a month, and then sent for three months hard labour on bread and water. Albert's portrait should be placed alongside Caligula's, who ordered 600 nightingales to be killed so that he could eat their brains for supper.[7]

> If our well-known bird the owl, sacred to Minerva, had been called upon for an opinion, it would have gravely pronounced that a fox must have presided at the committee, an hyena have been cook, and a stud of asses the consumers of the dish in question.[8]

Nearby Scarborough was less dissolute. It was fast becoming the 'Queen of Watering Places' with the more well-to-do, and soon became as popular with the Squire and his two squiresses as Catholic Bruges or Aix. They usually stayed at Mrs Peacock's, at number 1 The Cliff, though Miss Reid at the Royal was also very good. Mr Roberts' museum had 'great attractions', and Mr Threakstone's establishment for books and newspapers was a daily resort. Mr Wombwell's travelling menagerie, where Miss Blight was the governess, was a great treat.[9] Their bodies are buried in peace, but their names live for ever more.

It was at Wombwell's menagerie, in the autumn of 1850, that he made the acquaintance of a female chimpanzee,[10] recently arrived from West Africa. When Scarborough had seen what a chimpanzee looked like, George Wombwell took it to Wakefield, and the local naturalists received advance notice from C. Waterton. C. Waterton himself stayed with his sisters-in-law in Scarborough, 'the boast of Yorkshire and old Neptune's pride.'

No praise was high enough for Scarborough. Everyone should spend a week or two at the seaside each year, and where better than 'this gay town of Yorkshire's eastern confines', incipient queen of watering places. Its sea breezes were free from the unwholesome vapours from detested 'long chimneys', and the view from Castle Hill was as sublime as any in Europe. But in time, the Squire and his sisters-in-law had to leave the delights of 'pretty, healthy, sweet and enchanting Scarborough' and return to regions of Cimmerian darkness, of blackest midnight born, there to pursue once more the little common occurrences incidental to retirement from a busy world.

On the day that he arrived home, he learned that the chimpanzee from Wombwell's menagerie had just breathed its last in its cage in Wakefield. George Wombwell offered

Figure 33. Walton Park Hotel, formerly Walton Hall.

to donate the dead animal to Walton Hall but since there was already an undertaking to show the animal in Huddersfield, the Squire suggested that Huddersfield should at least have the opportunity to see a dead chimpanzee, and that it could be forwarded to Walton Hall from there – 'the frosty state of the weather was all in its favour.'

So the ex chimp was shown in Huddersfield, and Wombwell then gave it to a carrier to take to its last resting place at Walton Hall. It seemed like the best arrangement.

But the carrier had a cousin in Leeds – 'a fiddler and a soldier by profession.'

Imagine you've never seen a chimpanzee before in your life. Imagine almost nobody in your home town has ever seen a chimpanzee. Imagine there's no television, or cinema, or photographs to tell you what a chimpanzee looks like. Imagine you're suddenly entrusted with the care of a dead chimpanzee, still fresh on account of the frosty weather. Then imagine that your nearby cousin is 'a fiddler and a soldier by profession.' Imagine the fun and games.

The carrier made a bee-line detour to Leeds, and two days later he'd sobered up enough to convey the corpse to Walton Hall. The Squire was not pleased. The delay had lost him five hours 'nocturnal labour' with the dissecting knife and the frosty weather was not quite frosty enough. He set to work immediately and seven weeks later. the chimpanzee was resurrected to eternal life, 'hollow to the very nails', and sitting on a coconut from British Guiana – 'not, by the way, its correct position.'[11] (In case you should think it was.)

The change of air, meanwhile, from pretty, healthy and enchanting Scarborough to regions of Cimmerian darkness, was fraying the Squire's short temper far more than late arrival of the chimpanzee. The atmosphere which in his own recent memory had been pure and clear, was now poisoned. The towns were full of smoke and the trees in neighbouring fields were all dying.[12] Added to which, there was now a hue and cry amongst northern landowners, in fear that the scolytus plague then attacking southern England might any day appear north of the Thames and ruin their fox draws and pheasant coverts.

But the beetles stayed away and the trees crippled by pollution were spared the coup de grace. Many of the trees were still doing well, despite abominable effluvia. One of his standard pear trees suffered, if anything, from an 'overgrown luxuriance'. It needed pruning. Which entailed some radical surgery for the tree surgeon:

The ladder used for the pruning – 'full seven yards long' – was propped against 'a machine of his own invention'[13] which was apparently some sort of wooden scaffold erected around the tree.

Edmund, now 20 years old, well over six feet tall and in need of pruning himself, was home at Walton Hall, newly graduated from Stonyhurst College. Like many a 20 year-old, he began to see flaws in his father's state of mind. He was not impressed with the patent climbing machine. It had no side stays to buttress it and was bound to collapse with a little determined recklessness, which Waterton Senior could be depended upon to provide. Edmund took a long cold look at the contraption, tried to make his father see sense, failed, and left the scene.[14] He refused to answer for what was bound to happen.

Presently, the whole thing gave way and ladder and Squire hit the ground simultaneously; he tried to keep his head up as he fell, but still suffered a recurrence of an old, intermittent disorder, now officially diagnosed as 'a partial concussion of the brain'.[15]

First thought, naturally, when he recovered – or discovered – his senses, was to make himself feel better. His whole side, from foot to shoulder, felt as though it had been pounded in a mill, and his left arm, from shoulder to fingers, was badly bruised.[16] His elbow was dislocated. So he took away a couple of pints of blood from his arm and next day, swallowed the necessary strong laxative of calomel and jalap. That put all to rights and he was able to pursue some more common occurrences incidental to retirement from a busy world.

For his next trick. everyone had to wait until a few days after Christmas. Then he tried the old circus clown routine the one with the disappearing chair.

Unfortunately, it was unrehearsed – unintended in fact. The Squire was feed man, his servant played Puck. Since his fall from the ladder, he'd been unable to wear a coat in the cold weather because he couldn't get his arm into the sleeve, so he draped a Scotch plaid over his shoulder for comfort, and on the day in question, he sat down to dinner in this tartan blanket, with his son and his two sisters-in-law close by. Dinner was rarely more than a 1 p.m. 'snack' for the Squire and when he got up to adjust the blanket on his shoulders, his servant, standing behind him, thought he was finished. He drew away the chair, the Squire sat down again, and a partial concussion of the brain supervened, this time 'to a considerable amount'.

Another 30 ounces of blood were taken from his arm but, nevertheless, 'symptoms of slowly approaching dissolution' became apparent and the Squire anticipated a crisis. He was, after all, 68 years old.

> Having settled with my solicitor all affairs betwixt myself and the world, and with my Father Confessor betwixt myself and my Maker, nothing remained but to receive the final catastrophe with Christian resignation.[17]

But as bad luck would have it, help was at hand. Whilst his agent wrote to Philadelphia to tell George Ord that the end was nigh, the Misses Edmonstone sent off a telegraph to Leeds, to summon their 'incomparable friend and physician', Dr Hobson – now considered the 'sole remaining hope'. Dr Hobson arrived and immediately exerted his 'giant powers'.[18]

The Squire had sacrificed nearly four pints of blood after the two successive accidents and the doctor found him at death's door, twitching spasmodically and drifting, from time to time, into a partial concussion of the brain. It was perfectly clear that insufficient blood had been shed. He sent for 18 medical leeches and supervised their application to the patient's head 'with wonderful precision.' Death chuckled wryly and gave up the ghost.

> Although, after this, I lay insensible, with hiccups and subsultus tendinum for fifteen long hours, I at last opened my eyes and gradually arose – I may remark, from my expected ruins.[19]

So death's faint endeavours were foiled, and life returned from the brink of the tomb. Another old foe, J. J. Audubon, was less resilient. Audubon's mind had been going for some years now and a cerebral haemorrhage in January finished him off completely, just as the Squire was rising from expected ruins at Walton Hall.

Half to rise and half to fall. . . the glory, jest, and riddle of the world. His injured arm was slowly withering and gave him such persistent pain that he was beginning to have nightmares. 'I was eternally fighting wild beasts, with a club in one hand, the other being bound up at my breast. Nine bull-dogs one night attacked me on the high road, some of them having the head of a crocodile.'[20]

After three months of this, he'd had enough. He thought of getting the arm amputated but his keeper, John Ogden, suggested that a better plan would be to try the local bonesetter, Joseph Crowther, 'a stout little man, made like Hercules', as the Squire told George Ord. 'You must have seen him a-fishing here.'[21]

The suggestion was accepted enthusiastically. Chloroform was not yet generally available and ether was still in its infancy as an anaesthetic. Alcohol was out of the question. The operation would have to be performed without a pain-killer. Any opportunity to follow the example of St Ignatius (and J. C. Loudon) was worth improving. Pain was immaterial. 'I dare say I shall have enough of it', said the Squire to the bonesetter. 'You will', came the reply, emphatically.

> 'Ah, Squire,' said he, 'them shittering fellows you have had about you have made sore work of you. You ought to have come to me some months ago and then I could have done your job without much bother. At present, we shall have some trouble.'[22]

It was now Lady Day – 25 March – and three weeks of torture were necessary, 'to prepare the parts.' The Squire sent a horse to Wakefield every day, to fetch the bonesetter, and after a fortnight of 'stretching, pulling, twisting and jerking', the shoulder and wrist began to fall into place. The elbow was more complicated.

On 7 April Edmund reached his 21st birthday and the church bells rang out in Sandal Magna and Wakefield, to mark the occasion. But his father was in no condition to celebrate. The family chose to treat the anniversary as a day of remembrance for Edmund's mother.[23] His father's pain was probably construed as a timely penitence, doubly appropriate in the run-up to Good Friday.

The elbow had been out of joint now since 12 December; it was proving more difficult to 'prepare' than the rest of his arm. But a week or so later, after four months of 'unsupportable' pain, all was ready for the great operation. 'I'll finish you off this afternoon,' said Chiron the Centaur, who clearly had a way with words, and at four o'clock, he took off his coat, rolled up his sleeves, and gave himself a strong anaesthetic of good old Yorkshire ale. The patient made do with soda water and ginger.

Eliza lost courage at the last moment and left the room, unable to bear the sight of sainthood in extremis. But milady's maid, Lucy Barnes, 'bold as a little lioness', stayed on to help. Whilst another accomplice held the Squire down in the chair, Lucy stood by, in case of need. The surgeon took another swig of beer and Lucy gave the patient some more soda water and a quid of ginger. Then Crowther bound his wrist to the Squire's and quickly yanked the arm straight. There was a loud crack and then a pause; the patient thought it was all over – not quite as bad as expected. But the worst was still to come; it usually is. Lucy loosened the buckle which fastened the two wrists together and the final round began.

> Laying hold of the crippled arm just above the elbow with one hand, and below it with the other, he smashed to atoms, by main force, the callous which had formed in the dislocated joint; the elbow itself crackling as though the interior parts of it had consisted of tobacco pipe shanks.[24]

The worst was over. The Squire put a fiver into Crowther's hand – 'by way of extra fee' – the elbow was bound up to set in place, and nine days later –'sound as an acorn' – the risen Lord wrote to his friend, George Ord, to tell him all about it.

For four months and four days the elbow had been dislocated but soon he'd be fit enough to follow his usual pursuits of flood and field without pain, hypothetically the wiser for recent events. (Though you and I know better than that.)

'I shall prefer in future,' he said, 'to mount into trees without the aid of ladders, and should I again have to grope my midnight way along the edge of an unprotected pier, I will bear in mind, at every step, the dismal dip at Dover.'

Chapter 19
1851-1857
Unique Abnormities

The Great Exhibition of the Works of Industry of All Nations opened at the Crystal Palace on May Day 1851. Charles Waterton was not impressed. The whole thing was a waste of time and wedding rings. (200 pawned, he read, so that their owners could pay to see the exhibition.) Professor Owen asked him what he thought of the taxidermy exhibits. 'When the medal is awarded,' came the reply, 'the inscription should read, "To the least bad."' They all deserved to be thrown on the back of the fire.[1]

He also visited the British Museum whilst he was in London, and found no improvement there either. Five years later, Professor Owen himself was appointed superintendent of the BM (Nat. Hist.) and still the standard failed to improve. When the whole department was moved to South Kensington in 1860, the one specimen submitted by Charles Waterton – the ant-eater which had been mauled by dogs at Camouni Creek – was not included. (It's now back in Wakefield.) None of his specimens were ever shown at the Crystal Palace.

In September he was back in London to make the acquaintance of a full-grown orang-utan in Regent's Park Zoo. Meanwhile, his arm was recovering nicely and his general health 'resumed her ancient right',[2] revived by the news that Jonathan had thrashed his surly old grandfather Bull in a yacht race at sea.

The result of the yacht race – won by the schooner America – filled him with patriotic fervour. 'A Yankee has beat them all, hollow, hollow, hollow, by eight miles!'[3] The great exhibitionists of British imperialism – the Law Church and closet naturalists – were humbled. Virtue was vindicated. The America Cup has been contested regularly since 1851 and the Squire's bones dance with glee in their grave to hear that Rule Britannia has never won it – never, never, never.

Very shortly after the boat race, he went to the zoo. The superintendent was a friend of his and the zoo's prize exhibit soon would be. The animal's keeper had obviously been forewarned of the visit and the Squire was granted permission to enter the cage, for a great exhibition to rival the show at Paxton's Palace. The wild man of Borneo and the white man from Yorkshire entered into a minute scientific examination of one another's persons, each equally impressed by the other's eccentricities. The wild man was fascinated by the white man's wrists and the white man was struck by the 'uncommon softness' of the wild man's palms, as fine in texture as the skin of a delicate lady. (Cf the vulture's belly.) The white man opened the wild man's mouth and examined his two fine rows of teeth.

> We then placed our hands around each other's necks; and we kept them there awhile, as though we had really been excited by an impulse of fraternal affection. It were loss of time in me, were I to pen down an account of the many gambols which took place betwixt us, and I might draw too much upon the reader's patience. Suffice it then to say, that the surrounding spectators seemed wonderfully amused at the solemn farce before them.[4]

Satisfied with the result of the experiment – 'a wild animal from the forest may be mollified and ultimately subdued by art' – the Squire took his leave, 'scraping and bowing,' convinced that the orang-utan is a model of good manners and social breeding.

But soon, 'alas', he was obliged to change his mind. 'Scarcely had I retired half a dozen paces from the late scene of action, when an affair occurred which beggars all description. In truth, I can-not describe it: I don't know how to describe it: my pen refuses to describe it.' Then he described it:

> This interesting son of Borneo advanced with slow and solemn gravity to the bars of his prison, and took a position exactly in front of the assembled spectators. The ground upon which he stood was dry; but immediately it became a pool of water, by no means from a pure source. Ladies blushed and hid their faces; whilst gentlemen laughed outright.[5]

The Squire himself, like a Fleet Street editor, was 'scandalised beyond measure.' 'All monkeys are infinitely below us', he concluded – they live in trees. The orang-utan was clearly never intended as a biped, as the briefest of examinations confirmed. To Man alone was granted the sublime privilege of standing upright and using a lavatory.

A New History of the Monkey Family was published in the third volume of Waterton's Essays, six years after the trip to the zoo, and about twelve months before the world première of *Origin of Species*, the material for which had been circulating amongst naturalists for some time. Secretly, Charles Waterton, 27th Lord of Walton Hall, must have been troubled by disclosures that he and the wild man of Borneo might share the same escutcheon.

In the early autumn of 1851, a few days after his trip to the zoo, he was off to Scarborough again, with Eliza and Helen, to survey the new fashion in ladies' knickerbockers.

The knickers were modelled by 'two fair Yankee ladies', touring England as envoys of Amelia Bloomer. They were currently appearing at Scarborough's Mechanics Institute. Where fashion was concerned, Charles Waterton generally preferred Jonathan's daughters to John Bull's. 'But, oh! – ye powers of bad taste and deformity!' It would take old Hogarth's pencil to hit this one off. The top half of her costume would have suited Harlequin and the bottom half Mahomet. That was what he told his English readers.[6] But to his American friend, George Ord, he said:

> I must confess they converted me on the spot, for, when they compared the new fashion with the old, it struck me forcibly that there was much common sense in the first and none in the last.[7]

And it often gets chilly in that region.

Scarborough was always enjoyed best as the summer season ended and parlour fires began to appear in the lodging-houses – roughly coincident with the first woodcocks in the shop windows. As the bathing machines began to be abandoned for the winter, Charles Waterton began to take a fancy to bathe in the sea. He was still enjoying a November dip at Scarborough as late as 1855, in the November of his own life.

Naked bathing, by this time, was beginning to offend Victorian sensibilities, so let's assume, to save the blushes, that he entered the sea in full fig Victorian bathing costume, like a pantomime life-guard. With an occasional urge to scratch the back of his head with his big toe, or to make his elbows meet behind his back, it must have made an interesting sight, as fashionably odd as the fair young Yankee ladies in Mohammedan bloomers. Oh, ye powers of bad taste and deformity!

A letter from George Ord had arrived whilst the family was away in Scarborough. It discussed 'the habitudes of the young cuckoo.'

Edward Jenner was the first to describe a newly-hatched cuckoo throwing birds out of the nest. That was in 1788 (ten years or so before he discovered vaccination).[8] The account was perfectly accurate in every detail but Charles Waterton was not convinced – the legs of a cuckoo could not support the weight of its own body at one day old, let alone the weight of another bird as well. 'The whole narrative is confused', he said.[9]

Now, in 1851, George Ord was writing to his old and dear friend, to plagiarise him, as usual, in the interests of sycophantry. 'Truly there must be something very fascinating in the marvellous, otherwise narratives of this stamp would not be so greedily devoured without the pains of mastication.'[10]

The Squire read the 'welcome' letter, sent back news of the aforegoing events in Scarborough, and settled down for a winter at home, interrupted only by a trip to Stonyhurst College, at Christmas, and again, apparently, early in the New Year.

Not long after his return from Stonyhurst, he was up to his old circus tricks again. He'd arranged for the governor of London Zoo to send him the body of the orang-utan as soon as it became available. The animal now obligingly died but an 'unforeseen event' robbed him of the chance to dissect his former playmate. The governor was not on the premises at the time of death, and the keepers knew nothing of the arrangements for despatch of the corpse. It was 'otherwise disposed of.'

But compensation was soon at hand. Ord sent him a rattlesnake's head through the post, and the same post brought news from Liverpool of of a whole boxful of live rattlesnakes – 34 of them – which had been brought to England by an American blacksmith named Vangordon: good news from a far country.

When he arrived in Liverpool, Vangordon unavoidably fell into 'bad hands' – Watertonian for customs officials. Furthermore, 'this enterprising son of Vulcan' had read Waterton's *Wanderings in South America.* What a capital fellow! The Squire took him under his wing.[11] Dr Hobson was notified and they were all invited to Hobson's house, in Leeds, for the purpose of demonstrating to an invited audience of scientific gentlemen the relative potency of rattlesnake venom and pure curare. It promised to be a good show.

Top of the bill, naturally, was Charles Waterton – 'in his genial element, every inch of him.'[12] There were 28 snakes left by this time, all contained in a box divided into four compartments. As the world's leading authority on snakes (he'd once knocked one out with an uppercut to the jaw) the Squire considered the arrangement unsuitable; his own ant-eater case would serve the purpose much better. He sent it on to Hobson's house and followed, not long afterwards, to personally supervise the experiments and the transfer from one case to the other.[13] But 'where could a mouse be found to hang the bell around the cat's neck?'[14] Silly question.

'Jonathan', said Hobson, was terrified of his own snakes. The Squire, on the other hand, was a perfect model of cool intrepidity. In breathless silence, Hobson lifted the lid of one of the compartments, and the Squire, 'fearlessly but gently' – with utmost composure and Hobson's italics – 'introduced *within the case, his naked hand.*' Then he gently grabbed one of the snakes behind the head and removed it from the box, to an accompaniment of loud hisses and rattles all around his hand. (And a few comments from the snakes too.)

He held the captured snake whilst it bit one of the guinea pigs, then repeated the experiment with other snakes and other victims – guinea pigs, rabbits and pigeons. Tension rose. As he replaced one of the snakes, another rose almost erect out of the box and made a bid to escape. The scientific gentlemen did likewise. They huddled together on the staircase, and in the street outside, like frightened birds. But the Squire and his page were unruffled. Hobson, in charge of the lid, closed it on the half-protruding body, whilst the Squire calmly grasped the snake around the neck and replaced it in the compartment. The faint-hearted scientists were gathered up from the street, to witness the closing stages of the experiment and a bit more derring-do from the man in charge. The curarised victims died within a few minutes, but the ones bitten by the rattlesnakes lasted longer and put up more resistance. 'They sank at last, with a few convulsive struggles'.

By the 'simple process' of placing finger and thumb on each side of the neck,[15] the Squire now lifted each snake out of the box, 'as though selecting the sweetest bon-bon,'[16] and transferred it to the ant-bear case, where it could be examined more minutely by the scientists and seen more easily by future visitors to the travelling show. In their new accommodations, the snakes toured the country with the American blacksmith and helped pay for his extended holiday in Britain.

The main conclusion of the experiment was obvious: murder by curare is more humane than murder by rattlesnake poison, as being much less painful for humans to watch, more civilized. One has to consider these things.

But Hobson drew a supplementary conclusion, less scientific but more to the point: the man who could put his naked hand into a boxful of rattlesnakes was obviously capable of riding an alligator. The Squire was no Baron von Munchausen; he was a bona fide nut-case; his word could be trusted. QED. Dr George Harley, later of Harley Street, confirmed the rattlesnake story as 'truth unadorned', on the basis of an acquaintance with one of the scientific gentlemen who had seen it all with his own eyes.[17]

George Ord received news of all these hepatological goings-on in his sick-bed. He'd got the 'flu again, accompanied this time by a rash.[18] Two Squire Waterton pills and a dose of Epsom salts, said the Squire. It never failed.[19] The lancet should also be used, but 'singular to tell,' it was now falling out of favour amongst the medical profession. 'The result is that two thirds of their patients die of inflammation of the brain, the lungs, the liver and the bowels.'[20]

One victim of modern practice was Robert Carr, the Wakefield solicitor who had been a regular visitor to Walton Hall since the death of his wife, Helen – the Squire's sister – in 1840. Robert Carr shared the new-fangled disdain of the lancet and consequently died, in February 1853, of the inevitable inflammation of his lights. There'd been no child by the marriage, so his estate was left in reversion to Edmund, upon the young man's marriage. He was to receive the whole income from the Carr estate – agricultural rents and so forth – until that time.[21] A new phase in the history of the Waterton family had now begun.

Edmund's father was not a peddler in silver spoons. Like the man who owed much, owned little and gave the rest to the poor, the Squire earned little, ate little, gave his all to the poor and reserved the rest for his son's inheritance – only a slight exaggeration.

Edmund was 'all laugh and activity' at this stage of his life, striding round the corridors of Stonyhurst, on his Christmas visit there, as though he owned the place.[22] But unfortunately, he was already in debt – £800 borrowed through 'a Roman Catholic lawyer'[23] (possibly Carr himself) – and by the time he came into his marriage settlement, nine years later, the debt had risen to £18,000, through a series of wild mortgages on his assets. The ultimate mortgage was on Walton Hall itself.

His father, by now, was not only living like a poor man (it was said that his personal expenses never equalled the wages of one of his own labourers[24]) he was, in fact, by no means a man of great means. By 1859, his income from rents was something like £650-700 a year – about one ninth of what Charles Willson Peale had been led to believe was his income from the same source, 35 years earlier.[25] Consistent with the theory of alternate family fortunes – one generation saving, the next spending – Edmund looked likely to become a definite material hazard to what family fortune remained. For the time being, however, there were more important things to consider.

A few years earlier, a relation of Robert Carr had given to the Squire a couple of feral Egyptian geese – 'great beauties and wonderfully admired.'[26] At about the time of Carr's death, one of the local farmers left a sheep hurdle propped against the park wall and the biggest fox in Yorkshire used it to gain entry. There was a layer of fresh snow on the ground at the time, so the evidence was clear next morning. Remains of the Egyptian gander were found at the base of a sycamore tree, with fox prints all around. A spring-gun was set with no. 3 shot, and soon the fox was 'dead as mutton'.[27] Artless Mr Waterton remained uncriminated.

The snow thawed, spring poked its nose over the wall, and the Squire trimmed his holly hedges, to prepare suitable nesting places for the birds. Local man of letters, Thomas Lister – the postmaster poet from Barnsley – turned up at the Hall one day and found the owner busy breaking his vow with regard to ladders. There he was, said Lister, 'mounted on some high movable steps' – another of his 'contraptions' – preparing nest sites for the birds, to ensure Walton Park had 'music all the year round'[28] – a tantalising claim. The year-round barks and bill-clatterings of the herons – music to any educated ear – now proceeded from a colony of over 30 nests.

Summer declined to show its face in Yorkshire, but the song-birds bred in their little prepared snuggeries and life was punctuated by occurrences yet more common than usual. At the beginning of May, when the dawn chorus was at its best, a morning's minion from the postmaster poet delivered another letter from George Ord which contained an interesting little reference to a mutual friend.

The letter gave news of the American lecture circuit – 'the order of the day' there, as in Britain. Probably on account of his former association with the American Philosophical Society, Ord had recently been introduced to 'an agreeable gentleman' who had made the Squire's acquaintance in Rome. 'He begged to be remembered most kindly to you. His name is Thackery or Thackeray. He also lectures.'[29]

'I knew Thackeray very well in Rome', replied the Squire.[30] 'He is a clever man.'

The first three chapters of Thackeray's new novel – *The Newcomes* – appeared in November, 1853; 30 chapters and 10 monthly numbers later, Charles Waterton appeared in it as the 'good, kind W___', a friend of the old religion.

By the time he'd made his brief appearance, he was back on the continent again, with his two sisters-in-law, Eliza and Helen. They planned to visit the south of France but thought better of it when told about the scale of the cholera epidemic in the towns en-route. So, after another 'delightful' time in Aix-la-Chapelle[31] – unseduced by the heterogeneous aliment or the dissipations of nocturnal gadding – they went on to Brussels, then Ostend, and so back to Yorkshire in late October. At Walton Hall, contingency plans were made for a late autumn return to Scarborough.

There were some ferns at the Hall now – a present from Quaker Allis. Every parlour in the 1850s had its case of ferns, just as in the 1950s every parlour had its three flying ducks on the wall. Charles Waterton wrote to thank Friend Allis from Mrs Peacock's parlour in Scarborough.[32] He was there with the two Misses Edmonstone; Edmund stayed at home, in his fashionable rococo apartment, and borrowed another £2,200 from a Daniel O'Connel Wheble, who had a bit to spare. The Squire had recently made his son a settlement of £200 a year, which was soon mortgaged by Edmund for a further £1,387.10s.5d.[33] Much of the money appears to have been spent on luxurious furnishings and his growing collection of religious rings.

After thanking Quaker Allis for the ferns, Charles Waterton wrote at least two more letters from Mrs Peacock's lounge – one to Dr Hobson and one to George Ord. If either he or the ladies wrote to Edmund, the letter or postcard has not survived.

Letters from George Ord certainly have survived. He now sent news of latest developments in the story of Peale's Museum,[34] of one of its exhibits in particular.

The collection had been sold to the showbiz impresario, Phineas Taylor Barnum, but much of it was destroyed in a fire at Barnum's Theatre. Some of the portraits survived and were pawned to the United States Bank, then sold at auction on 6 October 1854.[35] George Ord bought four of the pictures, including 'Charles Waterton, the English Naturalist and Traveller' which was hung for awhile in Ord's study, 'as perfect as the day it was finished.' A few months later, it was hanging at Walton Hall, which the Squire devoutly hoped would be its last resting place. It has since been to Ushaw College, County Durham, to the National Portrait Gallery in London, and to Stonyhurst College, Lancashire, where it hung on the jamb of an old archway in the main corridor.

On Christmas Eve, the Squire would still join the annual invasion of old boys at Stonyhurst, totally unsuspecting that one day his own portrait would hang there. He clearly stood out like a bird of paradise in a flock of crows. Because, of course, this old bird still wore his 18th-century school uniform, much as it appeared in his portrait, painted 30 years earlier – bright blue swallow-tail coat, gold buttons, stick-up collar and choker, check waistcoat and grey trousers. Air holes were pierced in the sides of his top hat and the crown was half detached, ostensibly to keep his head cool but apparently for talking through. Lewis Carrol obviously had Charles Waterton in mind when he invented the Mad Hatter.

When this vision of boy eternal entered the Academy Room, it would perform a few interesting little acrobatic feats for the students – hand stands, ear scratches, elbow claps, that sort of thing – just to assure them that the most commonplace of men was not only as daft as he looked. It was 'his delight', said the Centenary Record, euphemistically.

He would attend the Christmas Day concert at Stonyhurst and 'not unfrequently, provided a special treat, in the shape of a supper, for the actors.' Then he'd return home to Walton Hall, to face the worst rigours of a Yorkshire winter.

The winter of 1854-55 was one of the worst on record. Flora and fauna were suffering badly. Since Simpson had been indicted, pollution from the soap factory had not perceptibly decreased and hard winters obviously didn't help. The injunction to 'prohibit the continuance of the works' had little immediate effect.

If any divine passion was stirred by the court's judgment, it was visited principally on the judge. Early in the New Year, Mr Hall, the 'arbitrator', was thrown down a 40 foot embankment, in a train derailed by the snow and ice on the track. 'He now lies in Leeds with both arms and legs broken and a large wound in the scalp!' said the Squire,[36] who was still sound as an acorn himself.

His health was sound but his wealth was fragile. Local criminals knew his little ways.

Three of them hired a carriage in Wakefield one day and arrived at Walton Hall under pretence of being close friends of Dr Hobson. One of them professed to be ill and was put to bed. The 27th Lord had no bed of his own to offer, so the 'ungainly animal', who rejoiced in 'broad Yorkshire vernacularisms'[37] was put in a room full of 'tempting trifles unattached to the walls'.

Now the rooms best equipped with tempting trifles were undoubtedly the ones belonging to Edmund, who was not at home. 'Everything therein betokened a taste for ease and luxury', said the Gossip of the Century, Mrs W. Pitt-Byrne,[38] – draperies, tapestries, pillows, cushions, poufs, portieres, fauteuils, girandoles, candlesticks, chandeliers, snuffers, pipes, pipe-racks, cigars, cigar boxes, books and, of course, the growing collection of religious jewellery, most of which was later housed in the Victoria and Albert Museum.

His father's 'bedroom', upstairs in the loft, was stocked with unconsidered trifles hardly worth the snapping up. It was like the cell of Ruggieri, or the den of Faust, said Mrs Pitt-Byrne, appreciatively. A blanket, a few old clothes and a taxidermy apron were usually draped over a rope slung across the room, and a hollow baboon – a trifle light as air – hung from the ceiling in default of chandelier or girandole. Washing facilities were a cracked red pan, on a 'crazy' chair with no back, and a piece of smelly soap in a broken saucer, with a coarse jack-towel hung from a roller behind the door. Bed was a roll of coconut matting or the bare floorboards, and pillow an oblong block of wood polished by his own head.[39]

The ungainly animal was put to bed with his broad Yorkshire vernacularisms, in a room full of tempting trifles which was certainly not the master bedroom up in the loft. It could have been either the special guest room – big enough for a biblical patriarch and all his forefathers to sleep in[40] – or Edmund's sumptuous boudoir in his self-contained, five-star apartment. The man was very large and would need a big, Edmund-size bed to sleep in.

He was big enough, in fact, to have taken a feather bed home with him, if he'd wanted to, but suspicion was aroused just in time by 'certain telegraphic symbols' passing between the men. The plot was thwarted and the would-be thieves sent away in their hired carriage back to Wakefield.

No great fuss was made, no charges preferred, distinction always drawn between crimes committed by the rich – nouveau in particular – and those committed by the 'lower orders'. The Squire was always charitably inclined towards the lower orders, consistent with Jesuit principles.

The only expected reward for charity was gratitude but the recipients were not always as grateful as they should have been, proving what Hobson called 'the impropriety of indulging in promiscuous charity'.[41] Oh dear!

Charles Waterton was properly well disposed towards promiscuous charity and developed his own ways of preventing its exploitation. He'd once given a beggar half a crown for new shoes, only to discover that it was spent on drink. That man was not ungrateful; next time he saw the Squire, he said he'd still got 15 pence left in button park and offered to buy him a pint of heavy to show his gratitude. The offer was turned down.[42]

Clearly, something had to be done; either charitable payments had to stop or a new method had to be found for paying them. So he came to an arrangement with the local cobbler: if any poor beggar needed shoes, the Squire would give him his penknife, to be exchanged at the cobbler's for a new pair. When the cobbler returned the knife, the Squire paid for the shoes.

But still the fraud continued. A barefooted beggar approached him one day, unaware that John Ogden, the keeper, had spotted him hiding a perfectly good pair of shoes in the hedge. The penknife was about to change hands when Ogden turned up with the beggar's own shoes, which were offered, and received, as a direct gift.[43]

A set portion of all earnings at Walton Hall was allocated specifically for relief of the poor – one tenth of total revenues, according to Mrs Pitt-Byrne; the produce from one of his best fields, according to a descendant of one of the servants.[44] The Squire called it 'St Joseph's Acre.'

On Rogation days – the three days before Ascension – the fields were blessed ceremoniously, the processional cross-bearer followed by a priest annointing the soil, and two attendants dressed in scarlet cassock and surplice, then by the Squire in his blue school uniform, then by the family, the domestic servants, some farm labourers and a few villagers, all muttering responses to the litanies chanted by the priest. Sprigs from the Walton Park yew trees were always carried during the procession, just as they had been by generations of 'pious ancestors'.[45]

Adequate supplies requisitioned, it was loaves for all and fishes for the cats; boiled potatoes for the waterhens. But now there was one mouth less to feed, one favourite fewer for widowed squire and maiden aunts to pamper. To who knows what access of grief and prayers, Tommy Pussy passed away, in the year of grace 1855, revered in memory as endeared in life. 'He prayeth best who loveth best, / All things both great and small.'[46]

No stately home is complete without its resident aristocat so, after a decent interval of mourning, a replacement was found during the annual autumn holiday in Scarborough. A suitable aristokitten was acquired, from a suitable alley, and taken home to Walton Hall where it was duly christened Whittington, later abbreviated to 'Whittie'. The newly inducted feline Lord outlived his 27th human counterpart but left no legitimate offspring to continue the male line. Many a tempting trifle was put in jeopardy during Whittie's lifetime but he was always charitably forgiven, on account of his humble origins.

It was also during this holiday in Scarborough that Charles met his new girl-friend, Jenny – a young brown 'chimpanzee' with a big pot belly and a nose which looked as though it had been pressed down by the finger and thumb of some officious midwife at the hour of her birth.[47] From this, and from other little give-aways, it's clear that Jenny was, in fact, the first live gorilla to leave Africa. She was caught on the bank of the Congo river and toured very briefly with Wombwell's menagerie, where she was formally introduced to the Squire and climbed onto his shoulder from time to time, to whisper sweet nothings in his ear. She was a small animal but the description definitely fits a young gorilla, not a chimpanzee.

As with the wild man of the woods in Regent's Park Zoo, so with the little wild woman of the woods in Wombwell's menagerie – Jenny and Charles were clearly fascinated with one another. On their fourth and last date, the Squire mounted the steps into her little prison cell, to bid his last farewell.

> Jenny put her arms around my neck; she looked wistfully at me, and then we both exchanged soft kisses, to the evident surprise and amusement of all the lookers on.
>
> 'Farewell, poor little prisoner,' said I. 'I fear that this cold and gloomy atmosphere of ours will tend to shorten thy days.'
>
> Jenny shook her head, seemingly to say, 'There is nothing here to suit me. The little room is far too hot; the clothes which they force me to wear are quite unsupportable; whilst the food which they give me is not like that upon which I used to feed, when I was healthy and free in my own native woods.' With this we parted – probably for ever.[48]

He took his November dip in the cold, depopulated North Sea and returned home with his new friend, Whittington, to a few more Cimmerian horrors of the Industrial Revolution, Jenny for the moment forgotten.

Cotton mill workers in Lancashire were now on strike; pale, malnourished factory hands were touring the outlying villages, singing for bread. Others were setting fire to the corn stacks which should have been used to make the bread. Many ricks were burned between Walton Hall and Barnsley.[49]

Walton Hall itself was now made over to Edmund, reserving the current squire's personalty (including taxidermy specimens and paintings) with the right to raise £10,000 on the estate and with steadfast trust in God's divine will, as always, to preserve the Home Farm stackyard for relief of the poor and the increasing numbers of waterfowl and finches. Whittie was kept away from the birds as much as possible but soon developed a 'marauding freak'[50] and became as adept at catching leverets as the 27th Lord himself.

Like Jenny the gorilla, Whittie the cat would whisper in the Squire's ear from time to time, particularly if he considered the butler's service at meal times was not up to scratch. There were now three male domestics at Walton Hall, all at a moment's beck and call from his feline lordship.

Jenny the gorilla was not being so well treated. She toured with Wombwell's menagerie, dressed up to titillate as puritanically as possible, and finally breathed her last, 'without any previous symptoms of decay,' on a show ground in the Cheshire town of Warrington. Her keeper, Miss Blight, wrapped her up in a linen shroud and sent her on to Walton Hall, at the end of February 1856. By June, she was sitting in state on the staircase, with various other animals, and still unaccepted for what she was, a young gorilla.

In the spring of 1856, another illustrious visitor arrived at the Hall – Dr George Harley. (Harley Street was named after an ancestor of his – the first Earl of Oxford.)

Harley mixed socially with all the top brass – Kossuth, Mazzini, Faraday, Tyndall, Louis Blanc, Frank Buckland, Thackeray, Dickens, Cruikshank, Landseer, Matthew Arnold, George Eliot, Robert Bell, the Brownings etc. But of all the 'great minds' he met, the deepest impression on his own was made by Charles Waterton, the English traveller and naturalist – 'probably from his possessing a most incongruous mixture of bizarre eccentricity, credulity and unbelief, coupled with a brilliant originality of thought, in a somewhat rough setting of common sense.'[51]

Harley was recommended to the Squire by Mrs Loudon. He'd read all about the latterday Munchausen and privately nursed an 'antipathy'. But soon he was to change his mind. He rapped the great bell-metal mask of tragedy on the door and was shown into the drawing room, to amuse himself for awhile with the Squire's mounted telescope. He was just putting his eye to the glass when his host strode in, crew-cut and shaved like an escaped convict. But the 'beaming smile' and 'speaking eye', the 'winsome expression of truthful sincerity', the 'cordial handshake', won the day.[52] Here was 'a genial, highly cultivated and sympathetic mind' – the most impressive Harley had met so far.

He was shown the taxidermical specimens on the staircase and politely remarked how well they were stuffed.

'Stuffed? Stuffed?! What do you mean – "stuffed"?' A polecat was produced, as from a trouser leg, to demonstrate that the only stuffing used was thin air – no straw, no wire, no spices, nothing. Head and body were pulled apart and the doctor persuaded to poke his fingers inside the animal's head; empty as a Regius professor's. Head and body were then put together again, like screwing the lid back on a box, and George Harley was suitably convinced of air's malleability.

On the upper landing of the stairs, he was introduced to the old 'waggity-waw' clock which oral tradition insisted had once belonged to Sir Thomas More. 'A 17th-century lantern clock – minute hand and pendulum missing', says the British Museum.[53] Sir Thomas More was executed in 1535. ('Minute hands were not thought of until high-pressure engines and high-pressure living came into existence', said the Squire.[54])

Next, Harley was shown into the bedless bedroom in the attic, where he learned the arcane mysteries of the Waterton Method. He studied to such good effect that, by the time the bell sounded for dinner, his fingers were as black as ebony. He washed them in the Squire's own basin and then they both went downstairs, to dine with Eliza and Helen.

'The meal was simple, the conversation delightful, and that funny little dinner party will always be a charming recollection', said Harley, years later. No wine for the host, of course, but he, George Harley, could drink as much as he liked.

In most fine houses, after dinner, the ladies would retire to the drawing room for 'conversazione'; in this one, the gentlemen did. The main reason for Harley's visit was to get fresh supplies of curare for experiments he was conducting into the properties of strychnine. (He subsequently claimed to have discovered that curare was 'a true physiological antidote to strychnine.'[55]) On a little table in the drawing room, the required specimens of curare were laid out – rolled up in balls of beeswax – and Harley spent the evening examining them, until it was time for the ladies to join the gentlemen for a nice cup of tea. Re-enter the butler, followed by Whittie, who was quickly removed to 'a place of safety'.

The Yorkshire tea ceremony, in all polite society, makes great demands on good form and table manners. Amongst other things, it requires a clean table. Breeding will out. Charles Waterton might look like an escaped convict, but there coursed through his veins the blood of several saints and a few characters from Shakespeare. The social proprieties had to be observed. He moved the curare balls to one side, kicked off his slipper, and stood on one leg to polish the table top with the ball of his stockinged foot. Then he prepared to jump over the table, as required by the Walton Hall version of the ceremony. The doctor stationed himself between his host and the table, to dissuade him from going to such unnecessary inconvenience just on his account. The ladies, of course, were both used to it.

Tea-time over, the waggity-waw struck eight and the Squire went to bed in his top floor garret. Harley stayed up until ten o'clock, chatting to the two spinsters. Then he followed the butler upstairs to his own apartment on the first floor – the biggest bedroom he'd ever seen. (It's now used as a conference room.) He flung himself into the armchair by the fire, to marvel at the difference between the Squire's self-denial and the pampering treatment of overnight guests.

George Harley spent several days at Walton Hall and wrote his account of the visit some 20 years later. By the time it was published, Charles Waterton had been dead for 33 years and Harley was a Fellow of the Royal Society and retired professor of medicine at University College, London. He was one of a steady stream of protestant guests at the Hall, which reputedly included the Archbishop of Canterbury. (Harley's family origins were, in fact, Catholic.)

George Harley's literary output was small, but his daughter entered the literary Pantheon, at the turn of the century, with a book about her father, the longest chapter of which was a reprint of his *Reminiscences of Charles Waterton*. Her father had mixed with most of the eminent Victorians, and she herself met some of them as a result – 'Browning was such a gentleman, and his wife such a dear.' But,

> Broadly speaking, the five persons who most fascinated my father, for the reason that they gave him the greatest number of new ideas, were Baron von Liebig,

> the author of the 'Letters on Chemistry'; Professor Sharpey, the secretary of the Royal Society; Catlin, the North American traveller; John Ruskin, the great art critic; and Charles Waterton, the naturalist.[56]

Of whom, obviously, Charles Waterton left the deepest impression.

He was now approaching his 74th birthday and was fit enough to jump over tables and shin up trees like a monkey. He'd bled himself over 160 times and proudly showed off the scars to his friends. His horses were also bled when necessary – half a pail at a time – followed by a dose of calomel and antimony.[57] The tails of the horses remained undocked and their stalls were equipped with inter-connecting windows so that they could chat to one another, after their day's work was over, about the vet's lucky way with horses.

They had plenty to talk about. A young visitor arrived, prepared to trade stories of cool intrepidity with Charles Waterton, a leading authority on the subject. He challenged the Squire to a bare-backed horse-jumping contest, over a bar and rail fence in the park, but Dr Hobson intervened and spoiled the sport.[58] The horses had something juicy to chew in the hay box that night.

Another good subject for gossip was the prodigal son, Edmund, now in Rome, in retreat from his debts. He'd applied for a position in the Vatican household and arranged for a few suitable strings to be pulled by the fond monsignors and cardinals he'd charmed with his piety as a boy. In September 1856 he was still looking for ways to raise a bit more cash to spend in Rome and quickly mortgaged his annuity, his interest in the Robert Carr estate, and the reversion of the Walton Hall estate, for another £700. Two months later, his aunt Eliza bought up the mortgage and advanced him another £2,000, presumably through Prince Torlonia's Bank. By the middle of 1859, he owed his aunt £5,500.[59]

Edmund's references were impeccable, his good character vouched for – in Rome at least. He was appointed chamberlain to Pope Pius IX and in due course received a knighthood of the Order of Christ of Rome,[60] which entitled him to wear the red enamelled gold cross of the order, adorned with the optional precious stones which he would certainly have opted for, *ad majorem Dei gloriam*.

Edmund's Christmas was spent apart from his family but safe in the arms of St Peter. The old platonic triangle spent a quiet Christmas at Walton Hall, happy in the knowledge that at least the dear boy's account with God looked likely to be settled, if nothing else.

Sir James Menteath had sent a couple of 'milk-white geese' to Walton Hall – possibly snow geese – and one of them was killed during Christmas week. 'I think I know the rascal who shot it', said the Squire,[61] bitterly, and speculation was rife in the servant's quarters and stables.

At roughly the same time, a local surgeon learned of a brood of deformed ducklings, of which all but one had been killed out of superstition. He presented the survivor at Dr Hobson's house on a day when Charles Waterton happened to be visiting. It had some webbing missing from between its toes. Hobson naturally gave the duckling to the Squire who released it on Walton Lake and straightway christened it Dr Hobson,[62] doubtless in recognition of some passing resemblance.

Waterton's avowed distaste for freaks and varieties has to be taken with a pinch of salt. The general interest in varieties in nature was a legitimate enquiry into the true affinities of species[63] and the intermediate 'causes' (i.e. origin) of 'productions' (i.e species).[64] Until such causes could be discerned or surmised, all such productions and varieties had to be ascribed to the 'superintending agency of God'. Any implied subversion of that agency obviously upset the Squire. But no-one had a more intense interest in freaks – 'unique abnormities', as Hobson called them – than Charles Waterton.

Albino hedgehogs, crowing hens, rumpless fowls, webless ducks, a sheep with a horn growing from its ear, or a duckling with its head back to front so that it had to turn a somersault to pick up food – they were all welcome at the Walton Hall asylum. He loved to show them off, along with his zoological satires and lusus naturae.

Abnormities and prototypes alike were all relatively safe behind the high park wall; the waterfowl were now considered safe from foxes so it was also considered safe to drain the lake.

There are several reservoirs close to Walton Hall, all of which were there in Waterton's time,[65] but after the lake was drained 'to its last drop', the wintering duck chose to stay in Walton Park. Safe from guns, they gathered in large flocks on the dry ground, their supply of wild seeds not unduly affected and a good deal of animal food – insect larvae, shrimps, snails, mussels, water beetles etc – presumably trapped in or on the exposed mud.

A few more things were also discovered in the mud – broken crockery, glass bottles, a sword blade, a battle spear, two daggers, axe and hammer heads, coins, keys, silver plate, and the small iron swivel gun which had shot one of Cromwell's soldiers in 1644. All flung into the lake, according to reliable hearsay, when the house was searched for arms after the Battle of Culloden.[66]

The dredging increased the depth of the lake, though it was still no more than 18 feet at its deepest. Pondweeds and reed beds soon began to flourish, to make suitable cover, and food, for the birds.

By the middle of May 1857, the Squire had begun work on another great taxidermy project – a three-year-old peacock, provided by Friend Allis. Seven weeks later, he was still working on it and declared it to be 'the most splendid peacock you ever saw',[67] a vast improvement on the scruffy apology for a peacock which 'you' and a good many others had seen at the Great Exhibition.

A picture of Charles Waterton and his peacock – both in spring plumage – is now housed in the Hitchin Museum and Art Gallery, Middlesex. Its colours are fugitive, shadows dark, but its character is undimmed. It was painted by a Samuel Lucas, Senior – member of a celebrated painting family roughly equivalent to the American Peales – and it is easily the most lifelike full-length portrait of Charles Waterton. The peacock is now housed at the Wakefield Museum.

Peafowl were occasionally eaten at Walton Hall – by guests – and some of the birds roosted in one of the four sycamores on the Hall Island,[68] serenading the household to sleep each night like chain saws. But Charles Waterton never got round to writing a detailed account of the peacock, alive or dead.

His little 'book' on the monkey family (20,000 words), with a few discursively interconnected notes on the fox, the dog tribe, the humming-bird, snakes, pigeon cots and pigeon stealers, Scarborough, Aix-la-Chapelle and cannibalism, was now ready for the publisher and appeared a few months later, cloth-bound in demi-octavo, price 3s.6d., as the third and last volume of Waterton's *Essays on Natural History, Chiefly Ornithological.* Within twelve months more, it went into a second printing.

No edition has yet appeared during the decadent dark age of the twentieth century.

Chapter 20
1857-1865
Birds or Boudoirs

'Sublime. . . Horatian. . . educated. . . wit. . . wisdom. . . shrewdness.' *Saturday Review*.

More circumstantially: with regard to the monkey essay, 'These are not very sublime or recondite conclusions', and the short and would-be Horatian epilogue concludes a work from which few would gain much information. Peale's portrait of the author – prefixed to the book – reflects, not unfaithfully, the singular mind of a man who expects educated people to be amused by coarse chatter completely unrelieved by a single flash of wit. 'The expression is strangely compounded of wisdom and silliness, imbecility and shrewdness.'

Thus the 'Saturday Reviler',[1] roughly true to type. *The Gentleman's Magazine* was less kind.

> We candidly confess we have little liking for autobiographies; those more particularly which are intended for publication during the writer's life. In nine cases out of ten inspired by egotism, the work is either redolent of conceit and untruthfulness, or is replete with twaddling details and vapid small talk; of no worldly interest to anyone but the author, or his circle of more intimate friends.[2]

Mr Waterton's book hardly formed an exception to the rule.

Fraser's Magazine – jolly old Tory *Fraser's* – took leave of him more regretfully – regretting that he should have written as provocatively as he had about the established church. 'We still hope to see him less offensive toward those who have chosen a different path to heaven from his own.'[3] 'Informed' sources, said Fraser's, put the first appearance of the brown rat in England at 1737, 23 years after the first Hanoverian king was crowned.[4]

All this must have been religious ratsbane to the author of Waterton's *Essays*, but the bitterest pill of all to swallow would certainly have been the opening paragraph of the very first review:

'Mr Waterton is a favourite writer on natural history,' said *The Leader*, radically. 'He has been an Audubon in his way, and has met with adventures scarcely less wild and picturesque than those of the unparalleled American.'[5] The object of praise swallowed hard and threw up. Subsequent reviews were only slightly less nauseous, though his character emerged relatively unstained – 'imbecile. . . silly. . . arrogant. . .amusing. . . well-read. . . observant. . . skilful . . . flippant. . . credulous. . . incredulous. . . warm-hearted. . . offensive' Mr Waterton.

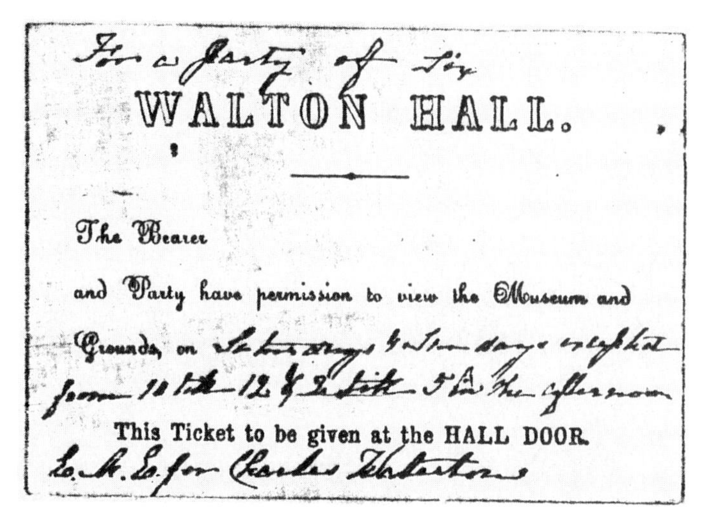
For a Party of Six
WALTON HALL.
The Bearer
and Party have permission to view the Museum and
Grounds, on [illegible]
from 10 till 12 & 2 till 5 in the afternoon
This Ticket to be given at the HALL DOOR.
E. M. E. for Charles Waterton

Figure 34. Ticket of admission to Walton Hall. Eliza's handwriting. (North Yorkshire County Library – York City Archives).

Blackwood's and the *Literary Gazette* were less enchanted by the man but more appreciative of the book; they ignored it. So did the *Quarterly* and the *Edinburgh Review*. So did the *Athenaeum*. Whig, Tory, Radical and Independent, all were equally patronising or in equal degrees dismissive.

Tattered and torn by the cold winds of criticism, he declined to loan any of the paintings from his private collection when Manchester held its winter exhibition in 1858. But bad reviews made no difference whatever to the popularity of Charles Waterton, nor to his little hospitable Garden of Eden, all change Oakenshaw Junction. Enough of the pre-industrial Eden remained to attract a 'vast number' of visitors from far and near.[6] One card of admission was enough to reserve the whole park for the exclusive use of one whole party for one whole day. Applications were rarely refused.

Larger parties – weddings etc – were often accompanied by one of the local brass bands, of which the region had plenty, and dancing was then 'the order of the day,' either out on the naked turf or in one of the classical temples in the grotto. (None of your fancy pump-room stuff – quadrilles and mazurkas; this was the real native McCoy – The Lincolnshire Poacher, The Derby Ram, The Vicar of Bray, The Drunken Sailor. Plus a few home-spun ballads from the penny gaffs in town: She was only the band-leader's daughter but she knew how to score for woodwind and choir.)

If the Squire himself appeared during the proceedings, even in the distance, the band would strike up with a popular little ditty which went something like this: 'King Stephen was a worthy peer; / His breeches cost him but a crown; / He held them sixpence all too dear, / Therefore he called the tailor lown. / Like a fine old English gentleman, / All of the olden time.'

All present would join in the chorus, 'with a manifest warmth of feeling', and then, when the day's entertainments were over, they'd congregate by the footbridge, in front of the Hall, to bid adieu to the fine old English gentleman with 'numberless blessings' and a hearty, feudal rendering of 'God Save the Queen' – enough to move the heart of a republican.

Hobson called them 'mob picnic parties of unlimited numbers and occasionally, even, of dubious cast.'[7] But the Squire hadn't the heart ('could never muster sufficient

Figure 35. The Grotto, c.1860. Picnics and wedding parties were usually free of charge. (Wakefield MDC. Museums and Arts.)

resolution', said Hobson) to refuse anyone. A hundred or more patients from the West Riding Mental Hospital were received each year, with open arms and a like mind. They took their own choir and band with them – 22 of the patients themselves – and happily vied with the Squire himself in light fantastic toes and merry wakes, prisoned souls lapped for the day in Elysium. He danced with them on the lawns, rowed them on the lake, and joined them for dinner in the grotto before they left for Wakefield, all of them wittering gleefully at the manic antics of the funny old nut-case of Walton Hall – 'daft as Dick's hatband', as he sometimes said of himself. 'Condescension', said Hobson, superciliously, 'was a too unlimited pleasure to him.' As was flatulence to the doctor himself.

When no picnickers were using the grotto, the doctor and Squire would use it themselves, beguiling the hours in natural history pursuits for 'upwards of five and twenty years,' said Hobson. Often, the Squire would sit on a little green chair in the grotto – in one of the classical recesses carved into the face of the rock, or in the keeper's sentry box – for all the world a prophet of the cave, 'musing upon many things'. There he'd sit, the livelong summer day, 'listening to the birds', said the Druid,[8] 'even without his hat, under a scorching sun', remonstrated the doctor.[9] In lighter mood, he'd climb to the top of the rock face itself, encouraged by his friends' pleas to stop, and hop along the edge, on one leg, with a drop of about 10 or 15 feet beneath him, which, to a man so well-stricken in years, possibly justified the doctor's description of it as a 'precipice'.

Or, less unkindly, he'd just kick his shoe in the air and scratch the back of his head with his toe. Or walk on his hands, or shin up a tree, or swarm up a pillar, or chase a stray leveret through a turnip field like an adolescent whippet. Anything to get rid of excess energy.

For kicking shoes in the air, he generally wore pumps, one of them slashed with his penknife so that it fitted as loosely as possible. If he saw Dr Hobson's carriage arrive at the lodge gates, he'd rush out onto the footbridge, kick his pump in the air, catch it as it fell, and wag the vestigial tail which a newly-published book about specific origins assured him he'd got.

As a variation on the theme, he'd secrete himself under the entrance hall table, growl like a dog, and bite the doctor's ankle as he walked through the door. Greater love hath no man. Other visitors came in for the same treatment, the same shamefast crisis of response. Correct etiquette was a nervous giggle. Preferred reaction

Figure 36. The Squire's green chair in the precipice wall, where he sat for hours on end 'like a prophet of the cave.' (Wakefield MDC. Museums and Arts.)

was a dog fight.

It was all good practice for the Christmas party, of course. For Christmas 1858, Edmund was home from his two year appointment in Rome, complete with papal brief creating him a Knight of the Order of Christ, and still up to his ears in debt. He was now a full-grown giant of a man – 28 years old, well over 6 feet tall, and increasingly the sensible, spendthrift foil to his father's foolish frugality – a big, handsome, top-of-the-class little goof, and a great disappointment to the family.

Christmas was spent at Stonyhurst, as usual, and the star turn was held in check by his sensible son. When Edmund saw that his father had a shoe on one foot and a slipper on the other, he felt it his filial duty to mention the fact, within earshot of some of the audience; so was spoiled another innocent jape, prepared and rehearsed all year, in the grotto and on the footbridge at Walton Hall.The prodigal son had clearly had enough of his father's embarrassing tricks. He assumed responsibility for the old man's behaviour.

At home also, Edmund was strangely unreceptive to things that really mattered. Back at the Hall, he retired to his fashionable boudoir whilst his father attended to his beloved birds. The wintering duck were flighting in like locusts. By 2 February, there was an 'extra-ordinary show of wildfowl. . . within a half pistol shot of the terrace; a sight not possessed by any other country house throughout the whole extent of Great Britain.'[10] (Until Peter Scott's house at Slimbridge was built.) But Edmund was not impressed. He seldom was.

During the following spring, the Squire managed to cut his head-scratching foot with a 'light hedging bill'. The surgeon from Wakefield prescribed complete rest and repeated applications of cold water to the wound. So the patient stood ankle-deep in cold ditch water and continued to trim his fences with the hedging bill, to the obvious exasperation of his incorrigibly sensible son. Shortly afterwards, the invalid took to his bare floor-boards with a coughing fit and lost so much sleep that he ate his breakfast next morning 'unfeelingly'. It was high time to put things 'peremptorily to rights'.

> I let out above five and twenty ounces of blood. The pulse returned to its normal beat and the cough no longer pained me. I could make my elbows meet behind me, which I can always do when in good fettle.[11]

Soon he was also scratching his head again, with equal self assurance. 'When Mr Waterton was seventy seven years of age,' said Hobson, 'I was witness to his scratching the back part of his head with the big toe of his right foot.'[12] He could also spread his five toes, like fingers, and use his foot as a hand, cockatoo fashion, which was obviously preferable to using his hand for the same purpose.

All good skills for a picnic, as for a Christmas party. Applications for picnics continued to increase each year, either from mobs of dubious cast or from the 'more private and more aristocratic symposia' which naturally appealed more to the refined tastes of Richard Hobson. The resident contortionist was always on hand, whoever the guests, and was laid on free, like the crockery and stabling.

Dubious parties were indubitably best – best suited to the special provisions to be found in the grotto: tea-room, temples, chair swings, Stormfreibüde, secret garden and sequestered nooks, all 'embosomed' in woodland, 'to enhance in the greatest possible degree, its many charms.'[13] Devoted swains gave 'swinging attentions' to buxom wenches, or enjoyed the 'distinguishing peculiarities of the numerous birds in view' – starlings, jackdaws, tawny owls, jays, blackbirds, song thrushes, wood warblers, willow warblers, chiffchaffs, chaffinches, redstarts, robins, bullfinches, nuthatches, wood pigeons, pheasants, and a pair of great tits in the shrubbery. She was only the gardener's daughter but she knew what she was formed by nature to bear.

The weather, sometimes, turned spiteful. In July 1859, a party of about 120 turned up and got caught in a vicious thunderstorm before the day was out. Down came the rain, mixed with hailstones, and 81 panes of glass were broken at the front of the Hall. A woman in one of the carriages held her parasol out of the window, to stop the rain getting in, and the parasol, predictably, was struck by a flash of lightning. Her forefinger and forearm were numbed by the shock but otherwise, she received no ill effects at all[14] – a clear sign, to Richard Hobson, of immunity in the female sex from death by lightning.

The Squire and his sisters-in-law survived the summer storm and went to Aix-la-Chapelle for a few weeks, to take the waters. After four weeks in Aix, they returned home to Walton Hall, where they unpacked their bags and packed them again, for five weeks in Scarborough, out of season – the best time of year to enjoy a good fire.

In Scarborough, they always had everything to their hearts' content – crackling fires, pier concerts, pierrots, travelling shows and lectures, bloomers and harlequinades and monkeys. And, of course, the waters.

This year – 1859 – they also had a hypnotist.

Charles Waterton had no faith in 'the wonderful science of mesmerism'. 'Delusion and humbug', he said. Unbelievable. Completely unmiraculous. Impossible. The mesmeriser asked for a volunteer from the audience and the Squire of Walton Hall delivered himself up for an experiment into the latest quack use of the black arts. But God was with him; the witch doctor was forced to conclude that this was one old Johnny-come-lately who was 'quite out of the reach of mesmeric influence'.[15] He'd seen real miracles with his own eyes; he wasn't going to be fooled by a common trickster.

It was too cold for a walk on the sands, or for a dip in the sea, so, after a couple of visits to the mesmeriser, and everything else to their hearts' content, they took train to Oakenshaw station, then Whitechapel cart to Walton Hall, for the revived annual pantomime season of wintering birds.

The Walton Hall waterhens were now as tame as pet dogs; they marched up to the drawing room windows, in their big green wellies, for a few boiled potatoes direct from the hand. Rumpless fowls and hybrid geese pottered around the lawns and rowdy, pure-bred Canadas circled the lake at dusk whilst peacocks screamed from the sycamores and elm trees like hecklers in the audience. Geese honked, coots tooted, mallard quacked and wigeon whistled. Herons clapped bills in the heronry and geriatric rooks raved randily from the rookery. Flocks of jackdaws snapped overhead, like petulant schoolgirls. Across the moat and down the hill, old man Waterton had a farm, e-i-e-i-o.

The rooks, in truth, deserted the rookery for the first half of the winter, but still took their evening meal on the corn stubble and pasture, safe and sound as almost nowhere else. They'd then fly to their winter roost at Nostell Priory, two miles away, and the resident stage manager on the Hall Island set his watch by their comings and goings. The rest of the cast were full-time members of the company.

When the lake froze, it was often almost covered with birds – 'a startling variety', said Hobson. And when the winter show ended, there was still plenty going on. The whole native fauna at Walton Park – wild, tame and feral – flourished as happily as faith, hope and charity. Starlings swarmed in the starling towers, pigeons prospered in the pigeon cot, titmice tittered in the tit-boxes, barn owls, blackbirds, jackdaws and cats integrated on the watergate, and pheasants philandered in the pheasanteries, safe behind a couple of thick crescents of ornamental yew trees which were in turn encased in a prickly pallisade of forced holly bushes to keep the cattle away from the poisonous berries. Nature smiled and all was gay around.

But pleasure is seldom unalloyed, even in Paradise. Tragedy lurks in every blissful circumstance:

'If you ple-ase Squire, Doctor Hobson is de-ad. . . . I seed him mysen, liggin dee-ad, all on a lump at dam head.'

The Squire, naturally, was overcome with grief. Dr Hobson was an old and much-loved friend. He rushed to the scene of the tragedy, at Nostell Dam, and discovered the doctor, dead on a lump, as reported. He carried him back to Walton Hall, cradled in his arms, and there dissected him, and treated him with bichloride of mercury, until he was presentable enough to be taken home to Leeds at the beginning of July. The Squire, by then, was on his way to Myddelton Lodge, near Ilkley, for a brief stay with his friend, Squire Middleton.[16] He stopped off at the doctor's on the way.

Dr Hobson was presented to Dr Hobson and, according to Hobson himself – Richard Hobson – the Squire was 'as much delighted on the occasion as if he were treading on enchanted ground.'[17] Which just goes to show that bliss lurks in every tragic circumstance, even the death of a web-less duck.

After a slap-up meal in Leeds – hot water and bread and butter – he took the mail coach and omnibus to Ilkley, where he had a 'severe' tea with friends called Moorhead, and arrived at Myddleton Lodge by eight o'clock, his usual bedtime. He was much impressed at Myddleton by the 'calvaire' – a gravel walk in the shape of a crucifix, lined with Georgian flower pots and entered through a rustic arch – 'a judicious work of magic, beggaring all description', he told his sisters-in-law.[18]

He spent a few days at Myddleton, walked on the moor bar t'at, and then returned home through Leeds, where he ordered a fresh cheese at the doctor's, for Eliza and Helen. (He'd forgotten to get one in Wharfedale.) The deformed duck now had pride of place in the doctor's parlour.

His letters were now slightly less legible than usual, on account of an injured thumb. This possibly also accounts for his late entry into the most recent scientific debate – the gorilla warfare fought mainly in the pages of the *Times* and the *Athenaeum*. It repays a little study.

A big hairy beast called Pongo, who walked around with his hands behind his neck, like an offside prop forward, had been chasing elephants and killing humans for 250 years or more, but the first real evidence of his existence – a skull and a skeleton – had only recently been produced.

At about the same time that Charles had an affair with Jenny the Gorilla, Paul Belloni Du Chaillu – a naturalised American who had spent his childhood in French West Africa, was leaving his adopted land to lead an expedition to Gabon, there to confirm his cherished belief that, along with the crested lion of Atlas, the gorilla is the most dangerous animal on Earth.[19] He boxes like a prize fighter, roars like a lion, terrorises lions, leopards and elephants, rapes women, tortures men (nails torn from fingers and toes), and pulls passers-by into the branches of trees to strangle them. Guns are little use; he snaps them in two, like twigs.

Du Chaillu's party eventually came face to face with one of these monsters, who stood up to his full height, glared fiercely, bared his fangs, beat his chest like a big bass drum, and roared.[20] So they shot him.

Civilization never got to see the body; Du Chaillu and his native guides were now in the kingdom of the Fans – fearsome cannibals who seldom say no to a juicy ancestor. The French Americans himself was spared from the cooking-pot by reason of his being curious to the Fans as they and their supper were to him. In due course of time, he produced and appetising pot-boiler of his own – *Adventures in the Great Forest of Equatorial Africa and the Country of the Dwarfs* – inspiration for two loads of belloni from the 20th-century: Tarzan of the Apes and King Kong.

Sir Richard Owen – superintendent of the British Museum Natural History Department – fresh from his attacks on Charles Darwin (in Edinburgh and Oxford) – now raised his banner on behalf of Du Chaillu. On 19 March 1861, Owen delivered a lecture at the Royal Institution, in London, and used the skulls of gorillas brought back by Du Chaillu, to compare their distinctive physiological characters with those of the negro, 'or lowest variety of human life.'[21]

As the most distinguished anatomist in Europe – the highest variety of human life – Owen chose to disparage *Origin of Species* and to lend whole hearted Christian support to the author of *Adventures in Equatorial Africa and the Country of the Dwarfs*. Dr Gray – Keeper of Zoology at the British Museum – took issue with the book, in the letter pages of the *Athenaeum*, and Professor Owen joined forces with John Murray (Du Chaillu's publisher) and Sir Philip Grey Egerton, an eminent deer stalker, in defence of the French-American's story.

Gray claimed that, amongst other things, Du Chaillu 'copies the published figures of well-known animals, and gives them as true representations of what he calls 'new and undescribed' species.'[22] The picture of a gorilla skeleton, reproduced in the book, was copied, said Gray, from a photograph of the skeleton already in the British Museum.[23] Du Chaillu probably never even went into the interior. And the stuffed gorilla he sent to Europe had been shot in the back.[24]

Professor Owen insisted that the beast had been killed in fair combat, face to face in the African jungle; he was supported in this opinion by no less a personage than 'Africanus' himself, latterday Rusticus of Godalming. Gray was 'less civil to a stranger than an Englishman ought to be.'[25]

But a businessman who had known Du Chaillu in Africa – R. B. Walker – now joined the complainants against him and claimed that young gorillas are really quite lovable. Walker had kept one in his factory in Gabon and had actually introduced him to Du Chaillu.[26] Du Chaillu's claims that the gorilla was a homicidal maniac were ridiculous. But Professor Owen appeared again, to say that he'd tested every statement relative to the gorilla in Du Chaillu's book, and that 'those statements, Dr Gray and Mr Waterton will permit me to say, do stand the test.'[27] (Mr Waterton's name introduced, apparently as being the author of *A New History of the Monkey Family*, exposé of Du Chaillu's 'fabricating talents'.)

And this is how the matter stood in October 1861, when, sore thumb or no sore thumb, Mr Waterton – notorious dresser of truth in fiction's garb – was obliged to clear up the argument once and for all.

On 8 October the *Athenaeum* received its first dispatch from Walton Hall and published it in full the same week. If the gorilla was king of the jungle, and every beast flew before it – page 58 of Du Chaillu – how come its short and slender legs were barely able to support its own weight, so that its arms had to be used to sustain its balance – page 434 of Du Chaillu. 'Will Professor Owen, who has raised his seven-fold shield to defend a needy explorer, condescend to explain away this manifold discrepancy?'[28] No reply.

Alright then, perhaps the great anatomist could bring himself to explain how the gorilla managed to 'use its arms as weapons of defence, just as a man or a prize-fighter would' when its 'huge, superimpendent' body needed its own arms for support.[29] (Both quotes from Du Chaillu himself.) No reply.

What it all amounted to then, was this: the gorilla was a 'paragon of perfection' when in a tree, but, like Buffon's sloth, a 'bungled composition of nature' on the ground.[30] Du Chaillu's statements, which stood the test of Professor Owen's 'physiological deductions', were nothing more nor less than romance, 'lamentable blotches on the page of African zoology'.[31] As for apes carrying young girls to the tops of trees, not even he, Charles Waterton, could manage that, and no-one was better qualified. 'Poor hapless damsels!' What happened to their petticoats in the struggle?[32] (Obligatory dress in the African forest.)

But all the abductions, the stranglings of men, the rapine and the torture, Professor Owen accepted as truth unadorned. It was not until one of Du Chaillu's own travelling companions bore witness against him that Owen was obliged to retract. A Gabon missionary reported that he'd just received news from Mongilamba, one of the African hunters who had accompanied Du Chaillu, that most of the gorillas sent to Europe had been killed by others and sold to the explorer on the coast. What's more, said Mongilamba, Du Chaillu never went to some of the places he claimed to have visited and was never more than a three day journey from his own house on the coast.[33] 'M. Du Chaillu falls by the testimony of those whom he publicly called to his aid', said *The Times*.[34]

And that, to all intents and purposes, was that, though Professor Owen was still prepared to sub poena the gorilla in the case against Darwin, if no longer in defence of Du Chaillu. The idea 'foisted upon poor "working men", of their derivation from a gorilla,' was 'unscientific, not to say absurd', said the professor.[35] This, despite anatomical similarities to 'the lowest variety of human race', whom Sir Richard Owen naturally looked up to. Charles Waterton shared his view of Man's derivation but identified the lowest variety differently.

As civil war raged in the national press, and Americans quibbled over secession, Walton Hall pursued the uneven tenor of its ways. 'On a bright afternoon of May, 1861,' Mrs W. Pitt-Byrne arrived at the Hall and stayed for five or six 'dearly remembered' weeks.[36] She thought better of human nature ever after.

She was shown the museum and the master bedroom – the zoological sports and the den of Faust – walked in the park, watched as small birds settled unafraid on the Squire's hand and shoulder, witnessed a rabbit *battue* and the Rogation Day procession to bless the fields, heard Auld Lang Syne played on the 'finger organ', met Whittie the cat and Edmund the son, smiled uncomfortably at a few racy stories ('which in the beginning of the century were quite admissible') and partook of the daily routine of the Hall, measured by the old, one-armed wag-at-the-wall with a voice like a South American bell-bird.

The visitors' day began at eight o'clock, when the breakfast bell sounded. The Squire had already been up for several hours by then and was generally waiting in the breakfast room, toasting a slice of bread by the fire, when the rest of the household appeared. When breakfast was over, he put on his working clothes – canvas coat and wide trousers, worsted stockings, loose-fitting shoes, and a broad slouch hat, like a sombrero – to survey his squiredom.[37] In winter, he wore velveteen or corduroy.

He showed his guest the paternal acres and when conversation floundered, it could usually be accounted for by the fact that he'd merely disappeared up a tree; shinning up trees for the ladies was clearly one of the keener pleasures of old age.

Dinner was always at one o'clock – a 'snack' for the Squire but slightly more substantial for visitors and cats. Whittie was fed with the humans (or vice versa) and the four feral cats in the stable and outhouses were given boiled fish from Wakefield immediately afterwards.

After dinner, if the weather was fine, the Squire would perhaps take his guest on the lake for an hour or two and maybe cross to the place where his tomb was soon to be prepared, on the southern shore. Or he'd take his hedging bill out with him, or an axe, or spade, and do a few odd jobs in the park, returning home with a faggot on his shoulder, 'like a cheery, honest, hard-working son of the soil', much to the amusement of guests and the unconcealed annoyance of his urbane son.

He'd sometimes come in drenched to the skin and sit on a footstool in front of the morning-room fire, 'lost in the cloud of steam.' If water spilled on the floor, or entered through the window, he'd take off his sock, mop up the water, and put the sock back on his foot, as being the most convenient peg to hand. He'd sometimes do this, said Mrs Pitt-Byrne, when about to go in for dinner (i.e. high tea).

High tea was at seven o'clock and the Squire's bedtime one hour later. Queen Victoria herself would have had to adapt to the waggity-waw timetable, if she'd ever been honoured enough to receive an invitation to Walton Hall.

'How if it were the Holy Father?' asked Mrs Pitt-Byrne.

'As soon as his Holiness has set foot on British soil, I'll let ye know', replied the Squire, more in hope than expectation. Pope John Paul at last kissed the tarmac at Heathrow airport on 2 June, 1982 – a few hours before Charles Waterton's 200th birthday.

His 79th birthday occurred during Mrs Pitt-Byrne's dearly remembered holiday at Walton Hall. J. G. Wood also visited the Hall at roughly the same time and, apart from a few details, he generally confirms her account. The grounds were magnificent, he said, their natural advantages seized and improved upon, as nowhere else in England.[38] The date of his first visit is not known but he was still an occasional guest at the Hall as late as 1863. He was clearly made to feel at home there. Protestant clergymen were as welcome in paradise as inmates from the local mental asylum.

Jean Francois Gravelet was not so lucky. But then Jean Francois Gravelet was a tight-rope walker – the world-famous Blondin. He applied for permission to walk on a rope stretched across Walton Lake but was turned down for reasons which should become apparent.

Charles Waterton had held a sprained ankle under Niagara Falls; Blondin had walked over Niagara Falls, several times, on a tight-rope – blindfolded, in a sack, pushing a wheelbarrow, on stilts, and carrying a man on his back. On one occasion, he stopped half way across to cook an omelette, a feat which unquestionably came to be regarded as his chef d'oeuf. The act was clearly a winner.

But 'exhibitions of every description were always at variance with the natural taste and feelings of the Squire', said Richard Hobson,[39] and who could argue with that? If the profits from such a show could have been given to the poor, Hobson was not at all sure that his friend would have refused permission, 'notwithstanding the almost irreparable injury that would have been perpetrated by numerous, motley, probably mischievous, and certainly ungovernable sight-seers.'

Fears were borne out, as it happened, by an event at about the same time in Birmingham – a 'Blondin Riot' in the grounds of another ancestral home, Aston Hall. (Then owned by a private company.) About 20,000 people 'of the better class', said the *Times*, assembled at Aston Park to witness Blondin's 'second ascent', complete with piggy-back and French omelette. But 'a mob of roughs', recurrent scourge of the better class, broke into the park and fought with the police. Four policemen were seriously injured and a police horse had to be put down.[40]

The thought of thousands of similar sight-seers breaching the wall at Walton Park, trampling the turnips and frightening the birds, hardly bore thinking about. As Queen Victoria said of one of her subjects – 'a female' – killed whilst trying to walk a tight-rope at Aston Hall two years after Blondin, she was 'sacrificed to the gratification of the demoralising taste.'[41] Vast numbers of people visited Walton Park, but never 20,000 at a time and certainly never at the sacrificial altar of demoralising taste.

At 79, Charles Waterton was fast approaching the prime of life. Hobson watched him take a 15 yard run at a stout wire fence, 3 ½ feet high, and jump it 'without touching it with either hand or foot.'[42] But shortly after this – in the summer of 1861 – he suffered a recurrence of his old constriction of the chest: 'lung trouble'.

At the end of September, whilst his lungs were still 'tender', he went back to Wharfedale for a few days, to stay with the Middeltons of Myddelton Hall. The 'mountain air' was not entirely friendly to his iron constitution but not to worry; he was still 'quite well'. The Middeltons dutifully killed him with kindness, then took him to see the local doctor, who failed to understand how a man in his 80th year could dare to ease himself of 26 ounces of blood[43] – evidently the chosen treatment for lung trouble. He received kind letters from his sisters-in-law at home and promptly replied with news and views of Wharfedale. He also received letters from their niece, Lydia, who was now staying at Walton Hall, but from Edmund, he received 'not a line'.[44] Relations with his son were slightly strained.

On 2 October Squire Waterton left Squire Middelton and returned home to Walton Hall with his friend and confessor, Canon Browne. The three Edmonstone ladies were waiting for them, with Mrs Pitt-Byrne – there for another short holiday.

The Squire's lungs were still tender. Even Scarborough in November failed to improve them. Nor Stonyhurst at Christmas. He usually made his entry at Stonyhurst either on all fours or on two hands, as befitted the joyfulness of the season, but this Christmas, it was a shadow of his former self which crawled in on hands and knees to entertain the troops. A message was sent to Leeds, imploring Dr Hobson to come quickly – Mr Waterton was dangerously ill.

The doctor set off post-haste but by the time he arrived at Stonyhurst, the patient was fully recovered, clapping his elbows and kicking his shoes in the air with warmest greetings of the season. It had all been an adversity for sweet use; Hobson was to meet the Jesuits. The doctor was formally introduced and he and the fathers got on well together. The Squire was reet glad. His friend could now see for himself that a Jesuit was equipped to convey souls across the Stygian creek 'more judiciously, and with less fear of an upset, than any of those pharisaical parsons who think it damnation to whistle on a Sunday.'[45] But the doctor remained a Protestant, and soon turned out a Pharisee.

Edmund was now £18,000 in debt, mainly to his own aunts. The gilt was off the gingerbread. But by the terms of his marriage settlement, £13,300 was to be immediately disbursed and his two aunts were to be given a £400 rent charge in lieu of the remainder.[46] It was well worth his while to get spliced.

For wife, he chose an Irish colleen named Josephine Ennis, daughter of Sir John Ennis, Bart., MP for Athlone and a governor of the Bank of Ireland.[47] If a father-in-law like that couldn't keep him in check, no-one could. His natural father had given up trying.

The wedding took place in August, by which time the bridegroom's father and the two Misses Edmonstone were several hundred miles away, in Aix-la-Chapelle. George Ord received news of the marriage by way of an envelope with a Dublin postmark, containing two cards inscribed 'Mr & Mrs Edmund Waterton' – nothing more. There was no accompanying letter and nothing in any of the Squire's own letters to explain it. 'Is it possible', thought Ord, 'that he has formed a matrimonial connection without the approbation of his father?'[48] Quite possible. But in fact his father had received a letter from Edmund, with which he expressed himself 'much pleased'.

Whatever the case between father and son, the state of affairs between Squire and doctor had certainly reached a sudden crisis. The two relationships are apparently linked.

In early August, at the Hotel Belle Vue, in Aix, the Squire received a confidential letter from William Longman, the publisher, which evidently contained a few disagreeable facts about Hobson's proposed biography of him, at least some of which was then in Longman's office. The letter was solemnly burned on the back of the fire, as being strictly entre elles-mêmes. Next day, it was answered, with a request that Longman should do the same – 'Please put this letter into the fire when you shall have read it. Now promise me this, and all will be right.'[49]

Ashes raked, a little blaze-up is fanned, but both reputations - doctor's and Squire's – rise phoenix-like from the flames.

From what he knew of the doctor, the Squire considered him 'a good grammarian but utterly deficient in style'. His book, obviously, could never be worthy of the subject.

> I could wish (most confidentially betwixt ourselves) that in your answer to the worthy doctor, you would contrive somehow or other to give it a squeeze or two so that it would never see daylight in print.[50]

When the book eventually did see daylight – grammatically questionable and hyper-abundant in style – the whole family took exception to it, on the shaky ground that it exaggerated the Squire's eccentricities. They'd been conditioned to receive it badly by reservations before it appeared. Hobson was also suspected – unjustly – of sowing dissension between father and son.

Returning home from Aix, after Edmund's wedding was over, the Squire called in at London Zoo, to satisfy himself that the cheetah was a member of the cat family. A young naturalist who believed in fairy stories had sent word to Walton Hall that the cheetah was not a member of the cat family because it had non-retractile claws. Charles Waterton got himself permission to enter the 'leopard's den'.

> He was a noble fine male; and I played my cards so well that he allowed me to examine his forepaws most minutely. I fully satisfied myself that his claws are retractile.[51]

The cheetah is not a member of the cat family; its claws are non-retractile.

In November, the Squire and his dames were back in Scarborough, as usual, and Edmund was back in England with his wife, Josephine. Stonyhurst had not been notified of his return so he received no invitation to the college for Christmas. He would have received a 'pressing' one if they'd known.[52] Josephine, of course, could have stayed at Walton Hall with the ladies. Lydia Edmonstone was now living there almost permanently.

Edmund's father naturally went to Stonyhurst, dressed to the nines and free to play whatever party games he liked, unrestrained by the finnicky nannying of a sensible son. The place, as always, was 'perfect in every department',[53] but Mr Waterton was not always the life and soul. At breakfast, he'd sit crouched over the fire, 'sipping a cup of water with a thimbleful of tea in it, brewed by himself, and looking as miserable as could be.' So said Percy Fitzgerald.[54] But he'd still dish out the 'droll observations. . . in a country tone, as if he was munching walnuts.' And at night, he'd sit up late in the parlour and roll out the traveller's tales till close on midnight – well past his usual bedtime.

Back home, Charles Waterton was now more than ever a squire of dames. Early in the New Year, he was introduced to a mermaid.

The Honourable Stanhope Hawke of Pontefract, another old boy of Stonyhurst, obligingly sent him a kippered mermaid, from the bottom of the deep blue sea, but the Squire indignantly pronounced it to be a compound animal, half monkey, half fish. '"Monstrum horrendum!"'[55] Deception should be left to experts in the field and the Merry Maids of Walton felt bound to agree.

In February, George Ord received a letter from Walton Hall which at last dispelled all apprehension of shot-gun marriages and family strife – it contained a postscript from Edmund himself, 'expressing his connubial happiness'.[56] Josephine had charmed everybody, Squire and dames alike.

The newly-weds went off to Bruges, for a short pilgrimage, which obviously aroused thoughts of a spring wedding 30 years earlier. Josephine was now in the same 'interesting' condition which had marked the end for Anne. Son and daughter-in-law had Charles Waterton's profoundest love and solicitude, come what may.

May came and the Squire went gathering nuts to prove he was sound as an acorn; the young couple went back to County Westmeath. Josephine went into labour.

On 10 June she gave birth to a boy and the dutiful request was made for a religious guarantor at the baptism. Father, grandfather and godfather were a somewhat heavy burden to bear, but the Squire could not find it in his heart to refuse 'our dear Josephine' anything.[57] The boy's religious education was assured.

The sacramental water was sprinkled and the infant diplomatically christened Charles Edmund Maria Joseph Aloysius Pius for short.[58] His godfather longed to know if there was to be a touch of the field naturalist in him, nomenclature notwithstanding.

He will soon begin to shew whether birds or boudoirs are most to his taste.

'Just as the twig is bent, the tree is inclined.'[59]

A fortnight after the boy was born – three weeks after his own 81st birthday – the Squire was back at Myddelton Lodge. A few weeks later still, he went with his sisters-in-law to stay at the home of Mrs Pitt-Byrne, in London. It was some time between these two social visits that he performed the duty of godfather at the Yorkshire baptism of Charles Edmund Maria Joseph Aloysius Pius Waterton.

It can be assumed that they got on well together.

The next significant visitor to Walton Hall arrived shortly after the three residents returned from London. Norman Moore was then a 16-year-old student at Owen's College, Manchester, the same college George Gissing was soon to be expelled from for stealing. (Compare Gissing's life with Waterton's, both of them Wakefield men. Gissing's autobiographical hero – Biffen – 'held it an axiom that fires were unseasonable after the first of May.'[60] Poverty forbad.)

In September 1863, Norman Moore walked over the Pennines from Manchester, to see the famous museum at Walton Hall – a mini version of William MacGillivray's walkathon from Aberdeen to London in 1819.

The boy crossed the bridge, foot-sore and fancy-free, and banged on the Squire's head to announce his arrival. Eventually, he was confronted by the great man himself, who told him to wipe his feet on the door scraper. Then he was admitted through the hallowed portals.

The Squire gave him a catalogue and left him to case the joint at his own leisure, which he proceeded to do with all due reverence and thirst for knowledge. He gazed at the 'gorgeous' cotingas and the 'splendid' jacamars, 'refulgent in gold and metallic green',[61] gawped at the tropical magnificence of cock trogons and tanagers and toucans and toucanets, smiled at the 'taxidermic frolics' of Nondescript and Noctifer, and generally mugged up on all that was apposite to a thorough understanding of art and skin sculpture, such as the prospect of a glittering career in medical science required. When he'd seen most of what he needed to see, on the staircase and in the organ gallery, the butler appeared with a cold lunch on a tray. The youngster ate his meal and asked if he could see Mr Waterton again, to thank him personally.[62]

And that's how the short but fruitful friendship began, between the octogenarian Squire and a teenage admirer. It lasted until the Squire's death and was remembered until Norman Moore's own death, in 1922, by which time he was Sir Norman Moore, Bt., MA, MD, FRCP, Hon FRCPI, LLD, Honorary Fellow of St Catherine's College, Cambridge, Physician to St Bartholomew's Hospital (retired) and President of the Royal College of Physicians. (Though he wasn't a bad lad really.)

One thing Charles Waterton had learned from a long and favoured life – part of it in the tropical hothouse – was that nature's gifts are conditional, strictly on approval. Abuse them and you lose them. Kill the goose and you starve the gander – that sort of thing. Orthodoxy now, but not generally understood in the 19th century.

Walton Hall in the 1860s was more than ever a model of peaceful co-existence, an oasis of life in a desert of destruction. Some of the beneficiaries got confused. A farmyard goose joined the wild Canada geese on the lake, and a Canada gosling with a drooping wing was reared by a domestic barndoor hen and lived with the poultry all its life. It became imprinted twice over – on the hen as its mother and the Squire as its sire. The foster father fed it on bread from his jacket pocket.[63]

Mixed associations of birds, Charles Waterton had long ago advertised (in self-defence) as one of his great contributions to science,[64] though all the scientific research was conducted by the birds themselves. A waterhen nested under a carrion crow's nest;[65] starlings reared their young in the same tree as a kestrel;[66] jackdaws, barn owls and redstarts shared the same nest cavity in an oak tree;[67] stock doves and a barn owl used adjoining holes in the watergate, close to a brood of nightjars;[68] and pheasants continued to breed successfully in the woods, despite the fact that birds of prey were not allowed to be shot.[69]

> I am fully of opinion that there can be plenty of game and plenty of what keepers term varment in the same district. But keepers are merciless savages and they do more harm to our fauna than can possibly be imagined.[70]

Farmers were little better. Hedges remained uncut at Walton Park, so that the berries could grow 'for the blackbird or the poor man'.[71] Fieldfares and redwings obviously joined in the feast. But this winter – 1863-4 – the fieldfares were almost completely absent.

> This I attribute to the want of food, as our blockhead dandy farmers crop their hedges down to the size of gooseberry trees, so that there is neither shelter nor

aliment for them. Add to this the perpetual attacks of gunners upon them to obtain them for the pot.[72]

And you see why Charles Waterton was an ornament to the age.

As the winter visitors left for their arctic breeding grounds, it became more pleasant to take a boat out on the lake, to check on the birds in the reed beds. He was 'an admirable sculler'[73] – you hardly need telling – and delighted his more sycophantic friends with his 'graceful handling of the oars.' He was also good with a sailing boat. (His boat, *Percy*, was convertible.)

On 3 June he rowed the two spinsters to the southern end of the lake – ladies' maid pressed into service – and showed them the stone cross he'd recently had erected between two ancient oak trees, a favourite spot for watching birds. A Latin inscription on the cross was almost complete.

Percy was moored on the shore and the little party advanced in reverence towards the holy shrine, the Squire's emotions clearly on tip-toe; he threw his arms around the cross and pressed his cheek against the cold stone. Then he announced that this was the spot where he intended to be buried, even if he had to do it himself.

'Squire! Oggi è vostro compleanno!' said Eliza, in a foreign language, and the birthday boy smiled and bowed gallantly. The ladies' maid stared respectfully at her own shoes.

The 83rd summer of his age unveiled its annual pageant – eternal in his soul – and Norman Moore popped over during the summer hols, to watch the parade go by. He seemed to bring a slice of good fortune with him every time he called. During one of his visits, the first green woodpecker for about 30 years flew across the lake, laughing as it went. A pair of them nested in the park, a few feet away from a pair of kestrels, and one of them roosted in the starling tower by the watergate during the following winter.

Then there were the sand martins – absent for the past 14 years. The sand martin wall was built in 1850, with 'upwards of fifty' drainpipes. That year, every hole had been tenanted by a pair of martins, but the following year, they failed to turn up and they stayed away until 1864, when Norman Moore charmed them back.[74] Their descendants now use the nature reserve at Fairburn Ings, nine miles away.

As the fledglings began to leave their drainpipes, to fatten themselves up for the long trip south to their winter quarters, the Squire and his sisters-in-law began to prepare for another short trip to Aix-la-Chapelle. By the time they got home, it was to a land parched by drought and summer apparently eternal in his park. The lake had begun to shrink and the last of the sand martins hawked for mosquitos over a beach of baked mud, cracked like crazy paving.

The drought ended, summer's lease expired, and the Squire's congested lungs rallied once more for the oncoming winter. (Though his thumb was still giving him trouble.) Edmund sent news from Ireland of another forthcoming event and three weeks before Christmas 1864, Josephine gave birth to the first of six daughters. Her name was Mary Paula Pia.

The last few months of Charles Waterton's life were clearly preoccupied with concern for his nature reserve and provisions for repair of his slightly shop-soiled reputation. His life-work was on the line. His eternal summer stood fair to fade. He mistrusted his son's ability to provide for his own family, still less for the home and handiworks.

Potential custodians of his work, and of his reputation, were naturally valued. One such was B. Waterhouse-Hawkins, sculptor, painter, palaentologist and general good egg – London residence: Fossil Villa, Belvedere Road, Upper Norwood. (A stone's throw from the Crystal Palace.) At first glance, his credentials were suspect – it was Waterhouse-Hawkins who made the prehistoric stone monsters, embedded in specially imported primeval swamp, for the Great Exhibition. It was also he who made the famous iguanadon which was used to house the leading lights of market zoology for a great zoological banquet in 1853. Head of table on that occasion, suitably seated inside the dinosaur's head, was none other than Professor Sir Richard Owen, highest variety of human life.

But as Rutherford cannot be blamed for the results of splitting the atom, so B. Waterhouse-Hawkins cannot seriously be held to account for the kind of riff-raff who sat in his dinosaur. You just do what you can and hope that it's good enough.

So Charles Waterton had no great reservations about lending some of his natural history specimens to this man, to help him illustrate a course of lectures on taxidermy and thereby, coincidentally, to shed a little lustre on the very slightly tarnished name of Charles Waterton. Hawkins had a nice line in soft soap which inspired the kind of confidence a generous spirit craved. The required specimens of natural history were despatched to London and the lectures were duly delivered, at the Society of Arts in Adelphi Terrace.

The first lecture was given on 12 December 1864, before 'a very crowded and apparently a very intelligent audience.' The artistic works of Charles Waterton Esquire 'received the homage of universal admiration'[75] – a very intelligent audience. Hawkins drew an antelope on the blackboard – 'in its natural form' – and then another, like the sort of deformed object they would find in a costly glass case at the British Museum. Then he showed them the beautiful specimens of natural history so kindly lent by Mr Waterton – 'a stoat and a bird.' The audience crowded around to pay homage and teacher was hard-pressed to prevent them touching the goods which he obviously considered, or implied that he considered, priceless works of art. His were the only hands to hold them.

The second lecture was delivered a week later, with another brief reference to the beauty of true art, and then the truth was sent home to its Patent Office at Walton Hall, Yorkshire. Waterhouse-Hawkins had visited Walton Hall at some time in the recent past, with his wife and daughters. (Daughters were always especially welcome.) The Misses Edmonstone later received a stone bust of their brother-in-law, compliments of Fossil Villa, apparently executed from very personal memory or even, perhaps, from one of the personal sittings the subject endured only for the great giants of artistic genius. The bust now presides over the permanent Charles Waterton Exhibition at Wakefield Museum.

On Christmas Day 1864, Dr George Harley of Harley Street received from Walton Hall the ideal Christmas box for the man who has everything – the shafts of twelve Amerindian blow-pipe arrows.[76] Just what he'd always wanted.

The Squire's Christmas, naturally, was spent at Stonyhurst, his last visit, as it turned out, to the place which had meant so much to him all his life. Edmund was still in Ireland, assembling a quiverful of children – heritage of the Lord. He already signed himself 27th Lord of Walton' (it should have been 28th) and clearly attached more

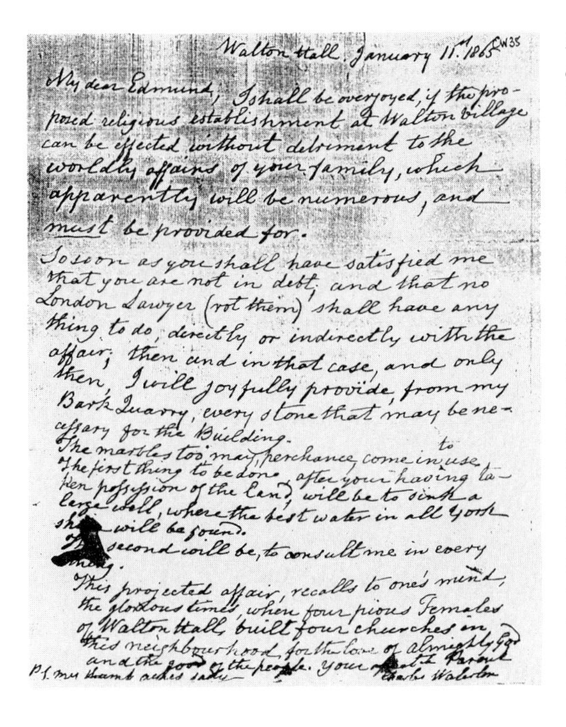

Walton Hall, January 11th 1865 W35

My dear Edmund, I shall be overjoyed, if the proposed religious establishment at Walton village can be effected without detriment to the worldly affairs of your family, which apparently will be numerous, and must be provided for.

So soon as you shall have satisfied me that you are not in debt, and that no London Lawyer (rot them) shall have any thing to do, directly or indirectly with the affair; then and in that case, and only then, I will joyfully provide, from my Bank Quarry, every stone that may be necessary for the Building.

The marbles too, may, perchance come in use.

The first thing to be done, after your having taken possession of the land, will be to sink a large well, where the best water in all York[shire] will be found.

The second will be, to consult me in every thing.

This projected affair, recalls to one's mind, the glorious times, when four pious Females of Walton Hall, built four churches in this neighbourhood, for the love of almighty God and the good of the people. Your affect. Parent

Charles Waterton

P.S. my thumb aches sadly –

Figure 37. January 1865, with a dislocated thumb, to Edmund Waterton. (York City Archives.)

dignity to the title than his subsequent trusteeship justified.

Edmund's intentions were grandiose – money no object. He planned to build some sort of Catholic seminary in Walton village as soon as he got hold of his patrimony or sooner. His father had mixed feelings.

> So soon as you shall have satisfied me that you are not in debt, and that no London lawyer (rot them) shall have anything to do, directly or indirectly, with the affair, then, and in that case, and only then, I will joyfully provide, from my own Bank Quarry [*sic*] every stone that may be necessary for the building. . . .
>
> The first thing to be done, after your having taken possession of the land, will be to sink a large well, where the best water in all Yorkshire will be found.
>
> The second will be, to consult me in everything.[77]

The third would be to look after the birds, of which there were plenty after which to look.

Young Norman Moore was the best chance the birds seemed to have; he was a regular visitor now. In January 1865, from the drawing-room window, he counted 1,640 wild duck – mallard, wigeon, teal and pochard – with 30 coots and 28 Canada geese.[78] And that was just on the lake. Elsewhere in the park, in due season, there was pleasure in the pathless woods and rapture on the lonely shore:

> Walton Hall is twelve miles south of Leeds, and the nightingale breeds here and sings here charmingly.

So wrote the Squire to the Druid, in late January,[79] when the nearest nightingale was a few thousand miles away. Walton Hall was just about the northern limit of the nightingale's range. He sang charmingly – in this, the last outpost of empire – but Edmund's ear was deaf to the call.

Norman Moore was much more appreciative – a surrogate son to the Squire and no mistake. A good friend too, as no prodigal son could ever hope or wish to be. When life's little challenges turn up, to test the manly mettle, it's a pal a chap needs, not a popinjay. Like when a barrel-load of gorilla arrives, in a state of decomposition. What use your fancy boudoirs then? You can't dissect a gorilla in a room full of French bric-a-brac.

The said gorilla was sent from Africa to Manchester in a cask of rum. It arrived at

Figure 38. Walton Hall, c.1860 – later a postcard. The figure on the right is possibly Charles Waterton. (Wakefield MDC. Museums and Arts.)

Manchester Museum in February and was sent from there to Walton Hall, for its ultimate passage into immortality.

A number of essential ingredients are necessary for good dissection and preservation work, as for other kinds of surgery. A good surgical surface is obviously essential – at a height convenient for the most delicate of manipulations. The operating theatre itself should be well-lit and spacious, free of all unnecessary obstructions and bric-a-brac. It should be easy of access – this applies more especially when the subject is large – and some slight concession to cleanliness is also, on occasion, considered useful.

The Squire's bedroom was obviously unsuitable, at least for the early stages, as being right at the top of the house; the corpse was slightly corpulent. And Edmund's boudoir, as aforesaid, was totally good for nothing. The drawing-room might serve – it was close to the main doors and was suitably airy – but the only table in there was too small and frail to take the weight of a full-grown gorilla. The solution was obvious – the operation would have to be performed on the dining-room table, cleared of all cutlery not strictly necessary for the job in hand.

Houseman for the day was Norman Moore – 'one of the most talented and finest youths I ever saw.'[80] The gorilla was rolled in and, with bated breath, the two men opened the barrel to conduct an autopsy. The breath continued to bate. A certain atmosphere prevailed. The air of expectancy suddenly began to smell. In fact, truth to tell, it 'stunk horribly'. The body was badly decomposed and hardly hung together at all – rotten as a London lawyer. The outer skin had peeled away and most of the body hair had gone with it. To all intents, and especially purposes, an ex gorilla.

They toiled manfully at the corpse but finally came to the conclusion that nothing could be saved except the bones.[81] The rest was put back in its barrel of rum and the surgeons laid the operating table for dinner.

After dinner – the customary one o'clock snack – the 'unfortunate' gorilla was packed for its journey back to the museum and a note followed not long afterwards, to explain the nature of the case.

As compensation for losing his ape, the curator at Manchester Museum – a friend of Norman's – was invited to stay at Walton Hall, 'to hear the birds warble'.[82] Only a fool or a philistine could turn down an offer like that. No dead gorilla had yet been born that could sing like a nightingale.

The nightingales turned up allright. On 2 May 1865, as Yorkshire prepared for another long summer, the last shaky entry was made in Charles Waterton's pocket notebook, about 43 years after the first:

> On this night, at 11 o'clock, two nightingales were singing melodiously in the Park at Walton Hall.[83]

Three days later, he wrote to tell Norman Moore about it and asked him to slip over to listen to them. One was singing in Stubbs Piece, south of the lake, and the other in the plover swamp, on the north-east shore.[84] The young man was studying for an examination, as it happened, but a few days after receiving the letter, he arrived at the Hall with his books. The museum curator failed to turn up.

At night, the diligent young student generally sat up late, poring over his books. The Squire suggested that if ever he should be up after midnight, he should pop in for an early morning chat in the attic bedroom-cum-workroom.

> I went accordingly, on May 24th, 1865, and found the dear old wanderer sitting asleep by his fire, wrapped up in a large Italian cloak. His head rested upon his wooden pillow, which was placed on a table, and his thick silvery hair formed a beautiful contrast with the dark colour of the oak.[85]

The Squire woke up and for three quarters of an hour they talked about birds. At breakfast next morning, he was more than usually cheerful. 'That was a very pleasant little confab we had last night: I do not suppose there was such another going on in England at the same time.' Man and boy were boon companions.

After breakfast, they went together to the northern end of the park, in the good ship *Percy*, to supervise work on some little plank bridges close to the stone crucifix where the Squire intended to be buried.

The boat was moored in a little inlet, as usual, whilst the vital work was done. When all was finished, they walked back towards *Percy*, crossing a series of little bridges on the way. As he walked over one of these, just a few yards from the appointed grave,[86] the Squire caught his foot in a bramble and fell on his side. As he fell, he dropped the obligatory log which he was accustomed to carry on his shoulder. It partly broke his fall but obviously failed to cushion it. 'He was greatly shaken and said he thought he was dying', said Moore.[87] It was a less cheery son of the soil than usual who returned home, faggotless, to Walton Hall.

That the injury was serious was obvious – 'internal injuries', said the intern, in retrospect.'Severe concussion of the liver', said the medical consultants.[88] 'Ruptured spleen', said Richard Aldington, years later,[89] a plausible explanation for the pain and discomfort which the fall caused.

A measure of the pain was the fact that he had to ask for help to get back into the boat – the sort of indignity he'd never before suffered. Back at the Hall Island, he managed to disembark and walked into the house unaided. There, he changed his clothes and Norman Moore told Lydia Edmonstone and Alexander Fletcher, who was also staying at the Hall, what had happened. The accidental traveller had suffered his last accident.

Moore rode into Wakefield to notify the surgeon, Mr Horsfall, and on the return journey, he met Lydia, driving into town in the Whitechapel cart, to collect the consecrated oil for extreme unction.

Back in the drawing-room, the Squire was in terrible agony. He announced that he wanted to go to his room but refused any help to get there. Bent almost double with pain, he slowly climbed the stairs and was persuaded to stop half way, to lie down on the sofa in Eliza's room, the nearest approach to lying in a bed for over half a century. Lydia returned with the holy oil and not long afterwards, Mr Horsfall arrived with the leeches. 'Now let us do things neatly', said the dying man to the surgeon, and the leeches were applied in the approved, long-discredited fashion.

Edmund was in Rome, performing his duties as chamberlain to the Pope. A telegram had been sent to Rome, and later in the day an answer arrived, with the benediction of Pope Pius IX himself. It did the patient a power of good.

Next morning, he was feeling a little better and had another pleasant little confab with Norman Moore, mainly about turnips. He looked up from time to time, 'with a gentle smile', and cracked a little joke for his young friend.

In the late afternoon, Mr Horsfall called again, with another doctor, and the patient insisted that they should be paid immediately after the consultation, as always. Dr Hobson was currently out of favour and was himself reported to be ill in bed anyway.

As night fell, the Squire's condition worsened. The rest of the household retired to bed but Norman Moore sat up with his friend. It was clear that the end was now very near. Canon Browne was sent for, from Leeds, to administer the last rites, and the Squire asked for the window to be opened so that he could hear the land-rail call from the mainland. The nightingales were silent. It was the comb and matchbox call of the corncrake which serenaded Charles Waterton's last hours.

By now, the ladies had been stirred from sleep and gathered round the sofa, and gave his blessings to each in turn – to Eliza, Helen and Lydia, to young Masser Charles and Mary Paula Pia, and to his good young Sir Bedivere, Norman Moore. He also left a farewell message for Edmund, who was on his way home from Rome. Then he received the last sacrements, repeating all the responses by heart. In his hand, he clutched a bronze and malachite statue which had avowedly belonged to Henry, the royal cardinal of York, a grandson of King James.

Norman Moore stayed by his side until the end. As with all emotional events, small details etched themselves crystal clear on his menory. It was time to say goodby to the finest of all old English gentlemen, one of the olden times.

> He died at twenty-seven minutes past two in the morning of May 27th, 1865. The window was open. The sky was beginning to grow grey, a few rooks had cawed, the swallows were twittering, the land-rail was craking from the ox-close, and a favourite cock, which he used to call his morning-gun, leaped out from some hollies, and gave his accustomed crow. The ear of his master was deaf to the call. He had obeyed a sublimer summons and woke up to the glories of the eternal world.[90]

There in that great Catholic bird sanctuary in the sky, Charles Waterton's soul resides in perpetual bliss. Heaven on Earth has moved home.

Chapter 21
1865-1995
Sic Transit Gloria Mundi

Mass was said by Canon Browne, in the family chapel on the ground floor. It was scheduled for 6 o'clock in the morning, death or no death.

The funeral was held over until Edmund arrived home from Rome. It took place on the 83rd birthday of the deceased, 3 June 1865 – one week after his death and one year to the day after he'd taken his sisters-in-law to the spot in the park where he wanted to be buried, at the far end of the lake. It was not an extravagant funeral, by Victorian standards, but it was unique and perfectly suited to the man and his life.

It began at 9 o'clock in the morning – late in the day by Waterton time but still too early for many of the invited guests. One notable absentee was Dr Hobson, ill in bed with 'no hope in his recovery',[1] from which condition he soon recovered completely.

Apart from family and friends, and officiating clergy, all and sundry – especially sundry – were perfectly welcome to attend. 83 old people, one for each year of his age, received a parting gift from the Squire, as a last, commemorative act of charity from the other side of the great divide – 83 loaves and 83 sixpences, doled out at the lodge gates before they left.

The family chapel was not big enough to hold all the mourners so the entrance hall, draped in black, was specially consecrated for the occasion. The Squire lay in a polished oak coffin, resting on its catafalque in front of the double front doors. 14 priests were in attendance. Office for the dead was intoned, then Requiem High Mass was sung by the Bishop of Bevereley, with the Squire's cousin, George Waterton, assisting as sub-deacon in proper subjection of high birth to High Church.

The coffin was carried in procession to the watergate and lowered into the Walton Hall coal barge – *Charon's Ferry* – converted for the occasion into a floating bier, 'a dusky barge,' as Tennyson said of another, 'Dark as a funeral scarf from stem to stern.'[2]

The funeral cortège made its way serenely down the lake to the waiting grave, headed by a canal longboat containing the Bishop of Beverley, in mitre and robes, with four Monsignors and 13 priests. It was followed by *Charon's Ferry*, with the coffin, then four boats in formation carrying principal mourners, and finally, towed in the rear, rigged with black bunting and totally empty, the Squire's own boat, *Percy*.

So, to the gentle plash of oars, and the chanting of the office from the leading barge, the great South American wanderer made his final journey. The ladies watched from a seat on the Hall Island, whilst the tenantry and villagers, and other mourners – hundreds of them – walked along the shore, swinging their censers, and disturbing the

Figure 39. 'Funeral boats on Walton Hall Lake.' From Richard Hobson, Charles Waterton, His Home, Habits and Handiwork. *(Reproduced from the Illustrated London News, 17 June 1865).*

herons as never before. One of the walkers was a young man named William Spurr, whose son lived at Walton Manor House until very recently. William Spurr clearly remembered into his own old age the loud animal cries of the herons that day, and the wild, unearthly calls of the Canada geese, as the procession made its way.[3] But the story of flocks of birds gathering, to accompany the body to its grave,[4] is slightly exaggerated. Now if the funeral had been in January, there might have been a whole lakeful of birds.

The coffin was lowered into the vault and the assembled congregation gathered around for the final ceremony. The Benedictus was sung and a cock linnet joined in from the branches of one of the two old oak trees framing the grave. Then the funeral guests returned to the Hall and 83 graduates of textile mill, soap factory, workhouse and mental asylum picked up their dole at the lodge gates before leaving paradise. Charles Waterton had been laid to rest in his own hallowed ground at Walton Hall. He was 'a man who was, perhaps, more thoroughly missed and more truly mourned than any other of his time', said J. G. Wood.[5] The inscription on the gravestone was now finished:

ORATE
PRO ANIMA CAROLI WATERTON
CUJUS FESSA
JUXTA HANC CRUCEM
SEPELIUNTUR OSSA[6]
NATUS 1782 OBIIT 1865

Edmund had planned to put a memorial chapel over the grave but, like most of Edmund's plans, it was not fulfilled. His father had left £10,000 each to Eliza and Helen, executrices of his will, and when he died, there was £23 cash in his room and £585.18s.2d in the bank.[7] No debts. Household bills, doctors' bills, armorial tax and dole had all been paid and remaining effects totalled 'under £14,000'[8] – i.e. over £13,000.

But income from the estate – agricultural rents and so forth – was nowhere near enough to meet the debts of his heir. Edmund had inherited his father's generosity but none of his thrift. The auguries were bad, as any Roman bird-watcher could have pointed out.

Nor was there much to be gained from sale of assets. Edmund was too generous. He gave things away. Priceless family heirlooms made handy little gifts. Norman Moore nobly turned down a first folio edition of *Richard II*[9] – chiefly famous for its reference to Sir Robert Waterton (Act II Scene 1) – and accepted a less extravagant memento in its place, to help write his sorrow on the bosom of the earth. The two ladies recognised the danger and arranged with Alexander Fletcher for the taxidermy specimens to be sent away for 'safe custody'. This was in November 1865, by which time the ladies themselves had moved to a house in Scarborough. The place chosen for the collection was St Cuthbert's College, Ushaw.[10] (Formerly Tudhoe.) Fletcher, like the dead man himself, was an old boy of the school.[11] The whole exhibition eventually moved to Alston Hall, in Lancashire – ancestral home of Edmund's second wife – then to Stonyhurst, and then to the City Museum in Wakefield. Edmund's own collection of religious jewellery was eventually sold to the South Kensington Museum (now the Victoria and Albert) for an undisclosed sum. That collection, too, is now in Wakefield Museum.

Home, Habits and Handiwork, meanwhile, launched itself on the world's wide stage in Leeds, and there was very little in it to substantiate claims that Hobson had made mischief between father and son. Congratulations were conferred on the book by the heads of two of the leading catholic families in the land – F. H. Salvin and Sir James Stuart Menteath, who was current representative of the Royal House of Stuart. Salvin sent a couple of little items for the next edition of the book and expressed his delight that the biography of the most wonderful man he'd ever seen had been so well received.[12] (It was almost totally ignored.)

Menteath, likewise, was pleased with the book's reception and he, also, offered a couple of reminiscences for the next edition, culled from his own personal communication with the great man.[13] Both of these letters were happily reproduced in full when Charles Waterton reappeared, hide, heels and hair, in the new, enlarged, second edition of *Home, Habits and Handiwork*, published in 1867.

But the book's keen reception failed to extend to Walton Hall. Despite Hobson's claim that all the facts had been verified by Waterton himself – 'in his 80th year' – Eliza and Helen knew that the Squire had disapproved in advance; they therefore disapproved in their turn, by proxy. Their own opinions of the 'correct grammar' of the book, and the 'deficiences of style' – its ocular inspections and abominable effluvia, plethoric conditions and amphitheatrical configurations, its finny squadrons disporting in the lake and feathered songsters constructing fabrications at the apical extremities of boughs – have to remain a closed book. But the same little peculiarities of style which now endear the author, the same verified facts which immortalise the subject, seem to have offended the artless Misses Edmonstone. The all time epic badness of the book apparently failed to register – 'so bad', enthused H. J. Massingham, fifty-odd years later, that, like high game, it was 'as palatable to the literary digestion as the freshest art.'[14]

The complaint that Hobson had sown dissension between Edmund and his father was apparently waived by the ladies. Eliza's complaint now was against Edmund himself. On behalf of her sister, Helen, and her niece, Lydia, she trembled for the future of their long-established home. Edmund ran away from his debts and became an absentee landlord. His debts exceeded his income, and his aunts' inheritance was less than was owed them by the legatee's son. The upkeep of Walton Hall was obviously expensive.

A formal Bill of Complaint was filed against Edmund on 3 March 1866, amended later in the same year, and again in January 1867. The plaintiff – his sometime besotted aunt – prayed in chancery that, amongst other considerations, an account might be taken of what was due to her for principal and interest on the mortgage securities, and that Edmund might be restrained by an injunction of the court from 'felling, cutting, or disposing of any of the timber or timber-like trees now standing, or from digging for stone or coal under the surface.' Also, 'some proper person' should be appointed, if necessary, to collect rents.[15]

Rent arrears accumulated but amounted to very little anyway. Soon, Walton Hall itself had a lodger, and by 1869, it had a tenant – Charles George Fane. Valuable possessions – feather beds, mahogany tables, rosewood card tables, music stand, Venetian glass chandelier and sundry baroque accoutrements of a fashionable French boudoir – together with a 'round disc in the inner hall, from a tree planted by Squire Waterton' – were left in Charles Fane's trust until April 1870.[16]

The Hall and part of the parkland were then leased to Edward Hailstone, a solicitor, late of Horton Hall, near Bradford.[17] Hailstone, 'a fellow of several learned societies', promptly filled the house with priceless books and objets d'art, and lived there for about 20 years.

The freehold, meanwhile, was put up for sale. It was the only way Edmund could afford new debts – he'd now been officially declared bankrupt. In 1878, after 700 years as the family seat, Walton Hall was sold, with all of its 259 acres, for £114,000 – the price inflated by recent discovery of coal seams and the prospect of mining within the grounds.[18] The headstock of Sharlston colliery, with its flat-topped spoil heap and derelict NCB offices, still lours over the park wall, but the grounds of the ancestral paradise itself were never mined. The nature reserve, nevertheless, soon became a thing of the past. The new owner of Walton Hall had no grounds to complain about encroaching industry, nor any inclination so to do, for Edmund had sold out to industrial pillage personified, to the local embodiment of uncritical Victorian optimism, one of nature's congenital rapists. He'd sold out to the spawn of the industrial devil himself, son of soap-boiler. I've said it before, and I'll say it again, with little fear of contradiction:

Sic transit gloria mundi.

The tithes of Walton also moved to Edward Simpson, who thereby became lay rector of Sandal Magna, along with Sir Thomas Pilkington of Haw Park and the Church of England vicar.[19] But the Simpsons were unable to live at Walton Hall until Edward Hailstone's lease expired, in 1891, by which time Hailstone was dead.

Edward Hailstone had kept the Squire's memory fresh and green for awhile and honoured his traditions as though they were his own – 'even that of hospitality', said Mrs Pitt-Byrne. But she re-visited the Hall, in the autumn of 1891, after the Simpsons had moved in, and found the place much altered.

She arrived prepared for change and somehow predisposed to disapprove. The new industrial bourgeoisie were everywhere taking over. In some places, it didn't make much difference; here, it obviously did. The parvenu soap-boiler was 'an uncultivated and unappreciative successor.' So was his wife.

> Mrs Soap-boiler was evidently indifferent to the Squire's friends, and did not care to hear about him. . . . She had not even enough manners to ask me into the house.[20]

Instead, a servant was deputed to show her around the estate, with no especial orders to be other than distantly correct. 'There was nothing of the venerable family butler about him,' said Mrs Pitt-Byrne, snootily, 'but he was very well for the mushroom flunkey of a nouveau riche.'

The flunkey took the visitors in tow – Mrs Pitt-Byrne and her ten-year-old grand-daughter – and marched them around the lake to the Squire's grave. They found it badly neglected and thickly overgrown with moss, the surrounding woods silent and dreary as the tomb itself. The heronry was already defunct and few small birds sang, or flew to greet them, as in the days when the Squire was still alive. The flunkey was 'too stolid and indifferent' to offer any explanation. (He could have pointed out that it was autumn.)

Suddenly, the silence was shattered by a shout: Mrs Pitt-Byrne demonstrating the echo stone. The guide was unsurprisingly astonished; he'd never heard a lady shout before. The soap-boiler probably also heard it, from the house, and must have heard the clap of thunder which followed not long afterwards. Its difficult to say which would have been the more unnerving.

When the rain came, it was accompanied by sudden gusts of wind which blew the umbrellas inside out. Mrs Pitt-Byrne and her grand-daughter were soon soaked to the skin. But – '"like master, like man"' – not so much as a shed was offered for shelter. When they reached the footbridge, no drying facilities were offered by the manufacturers of soap and the two pilgrims were not invited into the house. Nor was a pony and trap made available for the journey back to the station. The Gossip of the Century and her little, rain-sodden grand-daughter trudged miserably through the mud to Oakenshaw Junction, in the pouring rain.

Charles Waterton turned in his grave.

Before Mrs Pitt-Byrne's account was published, in 1898, a few more contributions to the Waterton corpus burst on the scene with a dull thud. Moore's edition of the *Essays* appeared, praising the dead man to the skies, and J. G. Wood's reissue of the *Wanderings* did likewise. Two booklets by James Simson damned him to eternal fire in Hell. The Simpsons naturally identified with Simson – a rose by any other name.

Moore's work came out as early as 1871, when he was still only 24 years old. It made the kind of extravagant claims which were soon to prove a liability. Charles Waterton was a 'profound naturalist', it said , whose observations were 'so accurate that they delight the profoundest philosopher.' He'd never in his life been convicted of an error and his essays deserved to rank as the first of his own time and in no age surpassed. Etcetera.[21] Simson felt he owed it to the world to prove otherwise.

Charles Waterton was not a man of 'acute intellect', said Simson, nor a 'profound naturalist'; his observations were not accurate and he certainly had been convicted of error. He was

> testy and easily riled, as well as spiteful and revengeful, self-engrossed and illogical, and in the highest degree pragmatical and dogmatical, presumptuous and arrogant, in matters with which he was little conversant.[22]

He should not be allowed to occupy a position 'beyond what he is entitled to.'

James Simson was chiefly unknown as the editor of *Simson's History of the Gypsies*. But any gypsy in his own soul was strictly of the urban commuter variety. It was social etiquette and bureaucracy which appealed to James Simson – custom house procedure for instance; Charles Waterton's lack of respect for it was 'in the highest degree unreasonable'.[23] Accidental travel was an offensive peculiarity. And as for writing in the forest, that was roughly on a par with reading in trees – extremely offensive. Walking barefoot in the forest was clearly in the same category: 'Even the Negroes go shod on Sundays.'

Then, when J. G. Wood's effort appeared, in 1879, the second instalment from Simson was published, exclusively for 'the lovers of truth'. *Wanderings in South America* was

> an exceedingly ill arranged, rambling, and wandering account of his adventures and observations, mixed with many simpering sentimentalisms, trifling egotisms, and pedantic quotations of no earthly use to a large part of his readers; peculiarities seldom or never met with in a character that is judicious and manly, or really amiable.[24]

Charles Waterton's social position, and his privileged education, should have guarded him against peculiarities of character, but his writings were 'in every way offensive to good taste and sound criticism.' They were 'poorly put together, and sometimes sadly mixed with extraneous matter, showing the want of a well-trained and scientific mind.' He had insufficient talent to make use of his opportunities, and his labours were meagre, considering they extended over 70 years.

> It is to be hoped that the worship of no such nondescript rural deity will be established at Walton Hall, by the Pagani connected with the Moores and the Woods, for even the county police of literature should not permit it.[25]

Edward Simpson, JP, Lord of the Manor of Walton, sat in judgement on county court proceedings and on this occasion, found in favour of the prosecution.

Joseph Hatton and Mrs Pitt-Byrne joined the pagani, to salvage something of the dead man's reputation, but enough damage had already been done to partially de-sanctify the rural deity. In the twentieth century, his popularity revived steadily but one strange notion persists: that his reputation for eccentricity somehow gets in the way of his just deserts.

Walton Hall, in the twentieth century, enjoyed mixed fortunes. The soap-boilers, at

Figure 40. Frontspiece of Hobson's Home Habits and Handiwork. *2nd edition, 1866. (Wakefield MDC. Museums and Arts.)*

first, had done pretty well for themselves. Edward Simpson – begetter of abominable effluvia – died in 1873. His son, Edward, became Lord of the Manor and he sent his own son, Edward, to Harrow and Oxford,[26] whilst his wife ran a public coffee and reading room in Walton village.[27]

The soap-house continued in business at Thornes but gradually ran into difficulties which meant that the family were obliged to let the Hall to tenants for a time, whilst they themselves went back to their old home in the village – Thornhill House. Edward the Third married into a well-to-do local family and was soon able to return to the Hall, where he stayed until his death at the beginning of the First World War. His two daughters and his son, Edward the Fourth, remained in occupation until the Second World War, when Walton Hall was again vacated, though it remained the family seat for a few years more.[28]

After the War, it was used as a private maternity home for a few years, whilst the grotto area and the walled garden, complete with sand-martin quarry and 33 drainpipes, remained in the possession of the Home Farm, by now a separate concern. The maternity home died young – complications at birth – and the Hall then fell into a state of serious neglect[29] until, in 1979, a local businessman carried out the necessary repairs and turned it into a sporting club and conference centre, with 5-star restaurant, guest bedrooms, snooker room and disco.

By the summer of 1981, a pair of Australian black swans and a mandarin drake had been introduced, and a red and yellow macaw lived in the sports shop with the receptionist. On the lake itself, you could fish, sail, swim, wind-surf or water-ski, or dress up as a pirate and join in the Jolly Roger race. 2,000 people watched the launch of a new swimming pool, with hydro bath and jacuzzi, later towed into position at the back of the Hall by speedboat.[30]

On 19 April 1982, I took a Kodak Instamatic to the sand-martin wall, in time to snap the dumper truck which had just knocked it down. Plinths of the two classical temples in the grotto were still there, though a line of sheep fence now passed through the middle of one of them. The 'precipice' wall had partly collapsed, and the little one-roomed cottage at one corner of it was a heap of stone slabs and rubble. The stream was lost in a tunnel of bramble and barberry.

In June 1982, Pope John Paul II arrived in Britain, to celebrate Charles Waterton's 200th birthday, and the Guild of Taxidermists held a day-long symposium at Walton Hall, with guest speakers, buffet lunch, Grand Celebration Dinner, bird-mount competition and a visit to the Bi-centenary Exhibition at Wakefield Museum. The Yorkshire Society of Anaesthetists held a similar symposium two days later.[31]

The Bi-centenary Exhibition was a great success and on 15 April, 1988, a permanent Charles Waterton Exhibition was officially opened by the local MP, David Hinchliffe. As J. G. Wood said of the same collection, when it was housed at St Cuthbert's College, Ushaw, 'It is worth travelling the length of England to see.'

Walton Hall itself is now the Waterton Park Hotel. In the summer of 1988, the old stable block was sold to a firm of property developers and in 1989, one of the partners bought the land itself. In 1990, Wakefield City Council commissioned a group of land use consultants to study the possibilities for making some kind of nature reserve based on Charles Waterton's original. It now looks likely that a scheme proposed by the International Centre for Landscaping Ecology (based at Loughborough University) will be adopted as part of the design for an 18-hole golf-course, 'monitored' by Wakefield Naturalists' Society. Also included in the proposals is a luxury housing development on the edge of Walton village. The scheme as a whole has the support of English Nature but it has met with opposition from local residents.[32] The most acceptable alternative is clearly out of the question – careful neglect, like happy poverty, costs money.

When Edmund sold Walton Hall, he bought a house in what was then a small village called Deeping St James, near Stamford, Lincolnshire. He was happy to believe, or to believe that he believed, that Deeping Waterton Hall was part of a more ancient possession of the family than Walton Hall itself. There was also a fair bit of change out of £114,000.

But Edmund fooled nobody, not even himself. In modern terms, he was young and downwardly mobile. Lincolnshire, as any Yorkshireman will tell you, is a terrible come-down from Yorkshire. Almost anywhere is infra dig after Walton Hall.

He continued as privy chamberlain to the Pope, but much of his life from now on was spent in south Lincolnshire with his steadily growing family and a new generation of debts. Josephine gave him another four children – six in all – and died of exhaustion in 1879. He then married again – Ellen Mercer, of Alston Hall, near Preston – and sired another two daughters by her, before his own death, after a long illness – possibly GPI – in 1887. He was buried with Josephine, in the family chapel at Deeping Waterton,[33] 85 miles from his parents in Yorkshire.

Edmund left to the world a collection of religious finger rings, a seminal essay on the rings of popes and cardinals,[34] a few essays on the devotion of the blessed virgin in

England, and seven surviving children. (The eighth died in infancy.)

Masser Charles – his firstborn son – married in 1890 and fathered four sons of his own in the last five years of his life; he died in 1906, aged 44. His great grandson, Charles Alec Oliver Waterton, born in 1947, continues the male line to the present day. He lives in Western Australia.

Deeping Waterton Hall, meanwhile, has virtually ceased to exist. It was sold to the Xaverian Brothers early in the 20th century and was known for many years, as it was probably known before Edmund bought it, simply as the Old Manor House. The chapel, with the tombs of Edmund and Josephine inside, remained in the Waterton family until the 1950s. Of the rest, all that now remains is the old stable block by the river, rafters bare to the sky and tufts of charlock and deadnettle sprouting from the eaves. The estate, like many another old manor house and grounds, with all demesnes attached, has been ruthlessly and systematically swamped by the coops and incubator pens of countless other broods, no more nor less productive than the Watertons – new, invasive, open-plan, bungaloid testimonials to indiscriminate procreation.

In the Waterton Arms, they still talk about the family, with all the rude authority of wishful thinking.

'Edmund? Oh yes, he was a right little tearaway. His old man was a bit of a lad as well. Whatsisname?'

Notes and References

Part One

A White Man From Yorkshire

CHAPTER 1: BLESS THE SQUIRE . . .

1. Charles Waterton, 'Living in the Tropics'. (Unpublished.) Brit. Mus. (Nat. Hist.), f.9. Transcribed by Norman Moore, 1867.
2. Charles Waterton, *Essays on Natural History, Chiefly Ornithological*, Series 3. (Longmans & Co., London, 1857), 'Cannibalism', p. 230.
3. James Simson, *Charles Waterton, Naturalist* (Baillikre, Tyndall & Co., London, 1880), p. 2.
4. Charles Waterton's notebook. Wakefield Museum.
5. Charles Waterton to George Ord, 4/3/1836. In Norman Moore (ed.), Waterton's *Essays on Natural History* (Warne, London, 1871), Appendix.
6. 'Short Communications', Charles Waterton. *Loudon's Magazine of Natural History*, Vol. VIII, No. 53 (Sept. 1835), p. 516.
7. Charles Waterton, *An Ornithological Letter to William Swainson, Esq., F.R.S &c. &c.* (Richard Nichols, Wakefield, 1837), pp. 5-6.
8. Richard Hobson, *Charles Waterton, His Home, Habits and Handiwork.* (2nd edn., London, 1867), p. 344.
9. *ibid.*, pp.329-31.
10. Charles Waterton, Preface to *Essays on Natural History*, Series 1. (Longmans, London,1838), pp. xiii-xiv.
11. Charles Waterton to Richard Hobson, 4/11/1854. In Hobson, *op. cit.*, pp. 282-3.
12. Lord Tennyson, *Locksley Hall*, 1.117.
13. W. Montagu, *Ornithological Dictionary of British Birds*, 2nd edn. Edited by James Rennie. (London, 1831).
14. Charles Waterton, *Wanderings in South America, the North-West of the United States, and the Antilles, in the years 1812, 1816, 1820 and 1824* (O.U.P. edn, 1873), p. 48.
15. Charles Waterton, *An Ornithological Letter to William Swainson*, pp. 6-7.
16. Thomas More, *Utopia*. (Everyman edn), pp. 102 & 123.
17. 'Hints to Ornithologists', Charles Waterton. Essays, Series 1, p. 311.
18. Charles Waterton, *A Letter on the Reformation Occasioned by the Attack of the Reformation Society of Wakefield on the Roman Catholic Faith* (Richard Nichols, Wakefield, 1838), p. 9.
19. Charles Waterton, Preface to *Essays on Natural History*, Series 2 (Longmans, London, 1844), p. xcvii.
20. *Essays*, Series 3, p. xix.
21. Hobson, p. 302.
22. Charles Waterton to Eliza and Helen Edmonstone, 17/11/1843 (American Philosophical Society, Philadelphia.).
23. Rev J.G. Wood (ed.), Waterton's *Wanderings in South America* (London, 1879).
24. Harleian Society MS No. 381, f. 171. (Normanby is on the other side of the river.)

25. In Shakespeare, *Richard II*, 2. i. 277-88, Sir Robert Waterton was travelling from Brittany, with Bolingbroke, but in fact, he was already in Yorkshire at the time and greeted him on the quay at Ravenspurgh. (Now under the sea.)
26. *ibid.*, 3. iii. 9-12.
27. Robert Somerville, *History of the Duchy of Lancaster* (London, 1953), Vol. 1, pp.184-5.
28. H Ellis, *Original Letters Illustrative of English History* (London, 1824), i. 2.
29. 'St Helen's Church, Sandal Magna', J. W. Walker. *Yorkshire Archaeological Journal*, Vol. XXXII, No. 93 (1916).
30. Rev. W. B. Stonehouse, *The History and Topography of the Isle of Axholme* (London, 1839), p. 453.
31. David Waterton-Anderson, personal communication.
32. Francis Blomefield, *An Essay Towards a Topographical History of the County of Norfolk* (London, 1807), Vol. VI, p. 178.
33. Quoted in Christopher Morris, *The Tudors* (Fontana, London, 1966), p. 156.
34. Charles Waterton, *A Letter on the Church of England by Law Established; Occasioned by Parson Gregg's Unprovoked Attack on the Catholic Religion, at Willis's Rooms, Wakefield* (Richard Nichols, Wakefield, 1837), p. 7.
35. Charles Waterton, Preface to *Essays on Natural History*, Series 3 (Longmans, London, 1857).
36. *Burke's Peerage*. Bedingfeld.
37. *Essays*, Series 2, p. xxvi.
38. Kathleen Meyer, *How to Shit in the Woods: An Environmentally Sound Approach to a Lost Art* (Berkeley, California, 1989).
39. 'Remarks on Professor Rennie's Edition of Montagu's Ornithological Dictionary', Charles Waterton. *Loudon's Mag. of Nat. Hist.*, Vol. IV (Nov. 1831), p. 517.
40. William Swainson to John James Audubon, January 1828. *The Auk*, Vol. XXII (July 1905), pp. 248-58.
41. *Dublin Review*, Vol. VIII (May 1840), pp. 321-2.
42. Charles Darwin to J.S. Henslow, 28/10/1845 (Smithsonian Institution Libraries – Special Collections – Washington, D.C.).
43. 'A Naturalist's Home', Unsigned. *Chambers' Edinburgh Journal*, Series 4, No. 158 (Jan. 1867), pp. 14-16.

CHAPTER 2: ORGAN BLOWER AND FOOTBALL MAKER

1. Charles Waterton, *Essays on Natural History*, Series 2. 'The New Chimney-Sweeping Act' (Longmans, London, 1844), p. 85.
2. *Essays*, Series 3. 'Notes on the Dog Tribe' (Longmans, London, 1857), pp. 177-180.
3. Charles Waterton to Bryan Salvin, 29/8/1815 (Durham Record Office, Londonderry Estate Archives, D/Sa/C 196.).
4. Preface to *Essays*, Series 1. (Longmans, London, 1838), p. xx.
5. Norman Moore, 'A Life of the Author'. From Charles Waterton, *Essays on Natural History*, ed. Norman Moore (Warne, London, 1871).
6. Preface to *Essays*, Series 1, p. xxii.
7. Joseph Hatton, *Old Lamps and New – An After-Dinner Chat* (Hutchinson, 1889), p. 324.
8. Moore, *op. cit.*
9. Preface to *Essays*, pp. xxvi-xxvii.
10. Edmund Selous (ed.), Introduction to *Wanderings in South America* (Everyman, London and New York, 1925).
11. Essays, Series 1. 'Notes on the History and Habits of the Brown or Grey Rat', p. 210.
12. Preface to *Essays*, p. xxxvi.
13. *Stonyhurst College Centenary Record, 1894* (Arundel Library, Stonyhurst).
14. Preface to *Essays*, p. xxiii.
15. William Cowper to Mary Unwin.

16. William Cobbett, *Autobiography*, ed. W. Reitzel (London, 1933), p. 17.
17. Lord Byron to Elizabeth Pigot, 26/10/1807 (Nottingham Public Libraries.).
18. Preface to *Essays*, pp. xxv-xxvi.
19. *Stonyhurst Centenary Record.*
20. Preface to *Essays*, p. xxiv.
21. *ibid.*, pp. xxvii-xxviii.

CHAPTER 3: BLACK VOMIT

1. *History of the County of Yorkshire*, Vol. II. (Dawsons of Pall Mall, 1912).
2. Mrs W. Pitt-Byrne, *Gossip of the Century* (Downey & Co., Leeds, 1899), Vol. IV.
3. Francis Carter, *A Journey from Gibraltar to Malaga, with a View of that Garrison and its Environs*, etc. (London, 1777). Cited in *Gentleman's Magazine*, 1783, pt. ii, pp. 716 & 843.
4. *Review of J.G Wood's edition of Waterton's Wanderings*, 'A.R.W.'. Nature, 24 April 1879.
5. *Essays*, Series 3 (1857), 'A New History of the Monkey Family', pp. 14-18.
6. Preface to *Essays*, Series 1 (1838), p. xxxiv.
7. *ibid.*
8. *ibid.*, p. xxxv.
9. *ibid.*, pp. xxxvi-xxxvii.
10. *ibid.*
11. William Mark, Consular Papers (Public Records Office).
12. William Mark, *At Sea with Nelson* (Sampson Low, London, 1929), p.234.
13. Preface to *Essays*, *op. cit.*, p. xxxix.
14. *ibid.*, pp. xl-xli.

CHAPTER 4: OUTLAWS AND IN-LAWS

1. Joseph Hatton, *Old Lamps and New* (Hutchinson, London, 1889), p. 321. Based on conversation with H. H. Dixon ('The Druid').
2. Charles Waterton to John Wells, Mayor of Nottingham, 27/4/1839. Transcribed, with an unfriendly annotation, by 'W.C'. (Sir Richard Owen correspondence, vol. 26, no. 204. British Museum (Natural History).)
3. Preface to *Essays*, Series 1, p. xlii.
4. *ibid.*
5. 'Charles Waterton in Demerara', James Rodway. *Timehri*, May 1915, pp. 257-65.
6. James Simson, *Charles Waterton, Naturalist* (Baillikre, Tyndall & Co., London, 1880), p. 27.
7. Wanderings in South America (O.U.P. edn, 1973), p. 64.
8. James Rodway, *Guiana: British, Dutch, and French* (Unwin, London, 1912), p. 114.
9. Simson, *op. cit.*, pp. 27-8.
10. *Essays*, Series 3, 'Notes on the Dog Tribe', pp. 159-62.
11. *ibid.*, 'A New History of the Monkey Family', pp.23-5.
12. *ibid.*, p. 38.
13. Norman Moore, 'A Life of the Author'. From Charles Waterton, *Essays on Natural History*, ed. Norman Moore. (Warne, London, 1871), p. 45.
14. Preface to *Essays*, Series 1, pp. xlix-li.
15. *ibid.*, pp. lii-liii.
16. Guiana Chronicle, 1/4/1809.
17. Preface to *Essays*, *op. cit.*, pp. lviii-lix.
18. J. G. Wood, Preface to *Wanderings in South America* (Longmans, London, 1879).

CHAPTER 5: SCENES OF NO ORDINARY VARIETY

1. 'Experiments and Observations on the Different Modes in which Death is Produced by Certain Vegetable Poisons', B. C. Brodie. *Philosophical Transactions*, 101, 178 (1811). Cited in 'Waterton and Wourali', W. D. A. Smith. Br.J.Anaesth (1983), 55, pp. 221-5.

2. Norman Douglas, *Experiments* (Chapman & Hall, London, 1925), p. 184.
3. 'The Use of Curare in General Anaesthesia', H. R. Griffith and G. E. Johnson. *Anesthesiology*, 3 : 481 (1942).
4. 'Life Before and After Curare', F. F. Folden. In Bowman, Dennison and Feldman, *Neuromuscular Blocking Agents: Past, Present and Future* (Excerpta Medica, Amsterdam [1990], pp. 5-14) and Philip Smith, *Arrows of Mercy* (Doubleday, Toronto & New York, 1969), p. 125.
5. Sir Walter Ralegh, *The Discoverie of the Large, Rich and Bewtiful Empire of Guiana*, etc. 1596.
6. Charles Waterton to Rev. Charles Wright, Jan. 1813. (Stonyhurst College Archives).
7. *Wanderings in South America* (O.U.P. edn, 1973), p. 3.
8. *ibid.*, pp. 6 & 8-9.
9. Vincent Roth, *A Plea for the Better Conservation and Protection of Wild Life in British Guiana*. (Daily Chronicle Ltd., Georgetown, 1949), p. 3.
10. *Wanderings*, p. 9.
11. *ibid.*, p. 12.
12. C. L. Burland, *Men Without Machines* (Aldus Books, 1965), pp. 89-90.
13. Simon, Sirtema and Hancock, *Expedition Sketch Map*, 1810-11 (Royal Geographical Society).
14. *Othello*, 1. iii. 144-5.
15. *Wanderings*,p. 23.
16. Richard Aldington, *The Strange Life of Charles Waterton* (Duell, Sloan & Pearce, New York, 1949), p. 58. and Gilbert Phelps, *Squire Waterton* (EP Publishing, Wakefield, 1976), p. 50.
17. Alexander von Humboldt, *Personal Narrative of Travels to the Equinoctial Regions* (London & Paris, 1814), Vol. 3.
18. 'Retrospective Criticism', Charles Waterton. *Loudon's Mag. of Nat. Hist.*, Vol. VI, No. 35 (Sept. 1835), pp. 464-8.
19. Humboldt, *op. cit.*, Vol. VIII, p. 153.
20. Richard Schomburgk, *Travels in British Guiana, 1840-1844*. Translated and edited by Walter E. Roth (Georgetown, 1922), p. 350.
21. Francis T. Buckland, *Curiosities of Natural History* (British Book Centre, New York, 1975), pp. 198-9. (First published, London, 1857).
22. Mrs Alec Tweedie (ed.), *George Harley, F.R.S. The Life of a London Physician* (Scientific Press, London, 1899), pp .277-8.
23. Charles Waterton to John Wells, 27/4/1839 (Sir Richard Owen Correspondence, Brit. Mus. (N.H.), Vol. 26, No. 204).
24. *Wanderings*, p.40.
25. A. R. Aitkenhead & G. Smith, *Textbook of Anaesthesia* (Churchill Livingstone, Edinburgh, 1990).
26. 'Retrospective Criticism', pp. 467-8.
27. *Wanderings*, p. 45.
28. Charles Waterton to John Wells, *op. cit.*
29. *Essays*, Series 1, 'Notes on the Habits of the Chegoe of Guiana, Better Known by the Name of Jigger, and Instances of its Effects on Man and Dogs'. (Longmans, London, 1838), pp. 239-44.
30. Charles Waterton to Rev. Charles Wright, Jan. 1813 (Stonyhurst College).
31. *Essays*, Series 1, 'Notes on the History and Habits of the Brown or Grey Rat', p. 214.
32. James Barnes, *Naval Actions of the War of 1812* (Harper, New York, 1896), pp. 105-7.

CHAPTER 6: EMPIRE POSSESSED: TIGHT SHOES, TIGHT STAYS AND CRAVATS

1. Preface to *Essays*, Series 1, p. lxvii.
2. *Wanderings in South America* (O.U.P. edn., 1973), p. 47.

3. J. G. Wood, Preface to *Wanderings* (Macmillan, 1879).
4. J. G. Wood, *Out of Doors* (Longmans, London, 1895), 'The Home of a Naturalist'. (First appeared in Cornhill.)
5. *Essays*, Series 1. 'The Barn Owl and the Benefits it Confers on Man'.
6. Richard Hobson, *Charles Waterton, His Home, Habits and Handiwork* (London, 1867 edn), pp. 16-17.
7. *Essays*, Series 1. 'Notes on the History and Habits of the Brown or Grey Rat', pp. 215-6.
8. Joseph Hatton, *Old Lamps and New* (Hutchinson, London, 1889), p. 329.
9. Hobson, *op. cit.*, pp. 9-10.
10. 'Game Law Morality', Unsigned. *Punch*, Vol. 10, p. 144 (Mar. 1846).
11. *Essays*, Series 2. 'On Beauty in the Animal Creation', pp.161-2.
12. 'Waterton's Home', Unsigned. *Bentley's Miscellany*, Vol. 60 (1866), p. 268.
13. *Essays*, Series 2. 'Tight Shoes, Tight Stays, and Cravats'.
14. 'An Old Friend with a New Face', Thomas Hughes. *Macmillan's Magazine*, Vol. 39, pp. 330-1 (Feb. 1879).
15. Preface to *Essays*, Series 1, p. lxviii.
16. *Bentley's Miscellany*, *op. cit.*, p. 269.
17. Charles Waterton's Notebook. Wakefield Museum.
18. *Essays*, Series 1, 'Notes on the Habits of the Windhover Hawk', p. 259.
19. Robert Waterton had received the king's writ to arrest the widow and son of the slain Harry Hotspur (son of the Earl of Northumberland) after the Battle of Shrewsbury. He was then sent to Alnwick Castle to negotiate for the lives of the two captives and, not surprisingly, was taken prisoner himself. His brother, John Waterton stood hostage for him. *Yorkshire Archaeological Journal*, Vol. XXX (1924), p. 359.
20. Charles Waterton to Bryan Salvin, 29/8/1815. Durham Record Office, Londonderry Estate Archives, D/Sa/C 196.

CHAPTER 7: IN QUEST OF THE FEATHERED TRIBE

1. Preface to Essays, Series 1, p. lxx.
2. Sir Joseph Banks to Charles Waterton, 9/3/1816 (Banks Correspondence, *Brit. Mus. (N.H.), Dept of Biology*, ff. 116-7).
3. Robert Southey, *History of Brazil* (London, 1810-19), Vol. 2.
4. *Wanderings in South America* (O.U.P. edn., 1973), pp. 58-9.
5. *Essays*, Series 1, 'Phaeton, or the Tropic-Bird', pp. 284-91.
6. Alexander von Humboldt to Wilhelm von Humboldt, Oct. 1799. 'in the evening, on the coast of Cayenne, I found the cold uncomfortable at 15 degrees centigrade.'
7. *Essays*, Series 1, 'On Preserving Insects Selected for Cabinets', p. 74.
8. 'A Life of the Author', Norman Moore. *Wanderings in South America* (Warne, London, 1871).
9. *Essays*, Series 3, 'On Snakes', p. 265.
10. *ibid.*, 'The Humming Bird', p. 125.
11. Review of Wanderings, Sydney Smith. *Edinburgh Review*, Vol. XLIII (Feb. 1826), pp. 302-3.
12. *Wanderings*, p. 76.
13. Charles Waterton to George Ord, 17/8/1846 (Sir Norman Moore papers). In R. A. Irwin, *Letters of Charles Waterton* (Rockliff, London, 1955).
14. *Wanderings*, p. 71.
15. 'Life History of the White-tailed Trogon', Alexander F. Skutch. *Ibis*, Vol. 104 (1962), pp. 301-13.
16. Richard Schomburgk, *Travels in British Guiana*, 1840-1844. (Georgetown, 1922), Vol. 3, p. 685.
17. Smith.
18. 'Account of the Habits of the Turkey Buzzard (*Vultur aura*) particularly with the view of exploding the opinion generally entertained of its extraordinary power of smelling', John J. Audubon. *Edinb. New Phil. J.*, Vol. 2, pp. 172-184 (Oct. 1826 – April 1827).
19. See Chapter XVI – *Foul Falls for the Devil.*
20. Charles Waterton Bi-centenary Catalogue. (Wakefield Museum), p. 8.

21. *Essays*, Series 1, 'Phaeton, or the Tropic-Bird', pp. 284-91.

CHAPTER 8: DELICATE MANIPULATIONS

1. 'On Preserving Birds for Cabinets of Natural History', Charles Waterton. In *Wanderings in South America* (O.U.P. edn), p. 190.
2. 'Reminiscences of Charles Waterton', Dr George Harley. *Selborne Magazine*, Nov. 1889. Reproduced in Mrs Alec Tweedie, *George Harley, F.R.S. The Life of a London Physician.* (London, 1899), pp. 270-1.
3. Charles Waterton to T. Allis, 5/5/1857 and 'Saturday evening', 1857, with annotation by 'E.P.' (nèe Allis), Nov. 1898: 'My father sent C.W. a fine peacock in full plumage – killed on purpose – over which C. Waterton spent many hours in daily manipulation.' North Yorkshire County Library. NRA 9202.
4. *Essays*, Series 2, 'The Ivy'.
5. Mrs W. Pitt-Byrne, *Gossip of the Century* (Downey, Leeds, 1899), IV. 13.
6. Samuel Johnson, *The Idler* (1760.)
7. Charles Waterton to Norman Moore, 2/10/1861 (Sir Alan Moore, Bt., Hancox, Sussex.).
8. Preface to *Essays*, Series 1, p. lxxi.
9. Charles Waterton to Pope Pius VII, dated Dec 1817 but obviously written earlier (Sir Alan Moore, Bt., Hancox). Quoted in Charles Waterton, Shane Leslie. Dublin Review, No.446 (1949), pp.59-60.
10. *Stonyhurst Centenary Record*, 1894 (Stonyhurst College).
11. Preface to *Essays*, Series 1, p. lxxii.
12. *ibid.*
13. *Essays*, Series 2, 'The Ivy'.
14. City of Wakefield M.D. Archives, Goodchild Loan MSS.
15. *Wanderings*, p. 92.
16. Charles Waterton to Sir Joseph Banks, 22/12/1819 (Banks correspondence – 7th Lord Brabourne, ff. 171-2.).

CHAPTER 9: PRESENCE OF MIND AND VIGOROUS EXERTIONS

1. *Burke's Peerage, Baronetage and Knightage, 1859.* Edmonstone.
2. Richard Schomburgk, *Travels in British Guiana, 1840-1844* (Georgetown), pp. 20-21.
3. Edward Walford, *The County Families of the United Kingdom* (London, 1868 edn), Edmonstone,Wilbraham, and Skelmersdale.
4. *Essays*, Series 1, 'On Snakes, Their Fangs, and Their Mode of Procuring Food', p. 206-7.
5. *Wanderings* (O.U.P. edn), p. 89.
6. Charles Darwin, *Autobiography*, ed. Gavin de Beer (O.U.P., 1974), pp. 27-8.
7. *Wanderings*, p. 92.
8. Charles Waterton, 'Living in the Tropics'. (Unpublished). Transcribed by Norman Moore. *Brit. Mus.(N.H.)*, ff. 2-4.
9. Comte de Buffon, *Histoire Naturelle.*
10. *Wanderings*, p. 97.
11. *ibid.*, p. 95.
12. Review of *Wanderings*, Sidney Smith. *Edinburgh Review*, Vol. xliii (Feb. 1826), pp. 306-7.
13. W.H. Bates, *The Naturalist on the River Amazons* (Everyman, London and New York, 1969), p. 209. (First published 1863.)
14. *Wanderings*, pp. 97-8.
15. 'Mr Audubon Again' (Charles Waterton's reply to 'R.B.'). *Loudon's Mag. of Nat. Hist.*, Vol. VI, No. 35 (Sept. 1833), p. 468.
16. *Wanderings*, p. 129.
17. *Essays*, Series 2, 'On Beauty in the Animal Creation', p. 159.
18. *Wanderings*, p. 95.
19. Charles Waterton Bicentenary Catalogue. Exhibit no. 63. Wakefield Museum.

20. Bates, *op. cit.*, p. 98.
21. C. Barrington-Brown, *Canoe and Camp Life in British Guiana* (Edward Stanford, London, 1876), pp. 119-20.
22. William Beebe, *Jungle Peace* (Witherby, London, 1919), p. 271.
23. *Wanderings*, p. 103.
24. 'The Squire'. Richard Aldington (Unsigned). *Times Literary Supplement*, 29/12/1932.
25. *Wanderings*, p. 102.
26. *ibid.*, pp. 102-3.
27. *Essays*, Series 1, 'Notes on the Habits of the Chegoe of Guiana. . . ', p. 241.
28. *Essays*, Series 3, 'On Snakes', pp. 273-4.
29. *Essays*, Series 1, 'On the Faculty of Scent in the Vulture', p. 19.
30. *Wanderings*, p. 122.
31. W. B. Alexander, *Birds of the Ocean* (Putnam, New York, 1954), and H. G. Alexander, personal communication.
32. *Wanderings*, p. 120.
33. Edith Sitwell, *The English Eccentrics* (Dobson, London, 1933), pp. 272-4.
34. Francis T. Buckland, *Curiosities of Natural History* (London, 1857), pp. 202-3. The 'Cobra di Capello'.
35. *Wanderings*, pp. 131-2.
36. *ibid.*, p. 133.
37. *ibid.*, pp. 133-4.
38. 'Article 2' (Unsigned), *Quarterly Review*, Vol. 33 (Mar. 1826), p. 323.
39. Pliny the Elder, *Historia Naturalis*, lib. viii, cap. 25. Quoted in 'An Attempt to Ascertain the Animals Designated in the Scriptures by the Names Leviathan and Behemoth', Thomas Thompson. *Loudon's Mag. of Nat. Hist.*, Vol. VIII, p. 196 (April 1835).
40. Herodotus, *Euterpe*, ch. 70. Quoted in Thompson.
41. *Essays*, Series 2, 'The Cayman', p. 46.
42. William Swainson, *Fishes*. (Lardner's Cabinet Cyclopedia, 1836), Vol. 2, p. 111.
43. J. G. Wood, Preface to *Wanderings* (1879 edn).
44. James Simson, *Charles Waterton, Naturalist* (Baillikre, Tyndall & Co., 1880), p. 30.
45. R. E. Raspe, *The Adventures of Baron Munchausen* (Methuen, 1987), p. 21. (First part first published, Oxford, 1785. Enlarged edn, 1786).
46. James Buchanan to Charles Waterton, 8/7/1861 (North Yorkshire County Library – York Library Local Studies Collection. CW28/NRA 9540).
47. *Wanderings*, p. 137.

CHAPTER 10: TIN, TAX AND TAXIDERMY

1. Charles Waterton to H.M. Commissioners of Customs, May 1821.
2. 'Charles Waterton', Norman Moore. *Dictionary of National Biography*.
3. Thomas S. Woodcock to Charles Waterton, 9/7/1847 (North Yorkshire County Library, CW20/NRA 9540).
4. Treasury Chambers to Commissioners of Customs, 18/5/1821. Quoted in preface to *Essays*, Series 1, p. lxxvii.
5. *Essays*, Series 1, p. lxxviii.
6. Vivien Noakes, *Edward Lear* (Collins, 1968), *passim*.
7. James Boswell, *Life of Johnson* (Penguin edn, 1979), p. 296 (18/4/1783).
8. Richard Hobson, *Charles Waterton, His Home, Habits and Handiwork*.
9. City of Wakefield M.D. Archives. Goodchild Loan MSS.
10. *ibid.*
11. *ibid.*
12. *Essays*, Series 3, 'Notes on the Dog Tribe', p. 169.
13. *ibid.*, p. 170.

14. Charles Waterton's Notebook. Wakefield Museum.
15. Hobson, *op. cit.*
16. *Essays*, Series 1, 'Defence Against Animals of the Feline and Canine Tribes', p. 114.
17. Charles Waterton to Eliza and Helen Edmonstone, 2/11/1843, in re Father Morris, S.J., Stonyhurst College. (Am. Phil. Soc.)
18. Notebook, *op. cit.*
19. Alexander Wilson, *American Ornithology* (Philadelphia, 1808), Vol. 1.
20. Charles Willson Peale to Thomas Jefferson, 4/7/1824 (Library of Congress, Washington D.C. Peale/Sellers Papers, Smithsonian Institution).
21. Goodchild Loan MSS.

CHAPTER 11: SEX AND THE SINGLE SQUIRE

1. *Wanderings in South America* (O.U.P. edn., 1973), p. 157.
2. *ibid.*, p. 144.
3. Mrs Frances Trollope, *Domestic Manners of the Americans* (4th edn, London, 1832), p. 239.
4. Charles Dickens, *American Notes* (Hazell, Watson & Viney, London), p. 107. (First published, Chapman & Hall, 1842).
5. Charles Waterton to George Ord, 20/9/1831. In R.A. Irwin, *Letters of Charles Waterton* (Rockliff, 1955).
6. William Cobbett, *Life and Adventures of Peter Porcupine* (Philadelphia, 1796), p. 14.
7. James Fenimore Cooper, *England, with Sketches of Society in the Metropolis* (London, 1837), Vol. III, p. 175.
8. *Wanderings*, p.158.
9. *ibid.*, p. 148.
10. Lorenz Hart, 'Manhattan'. From *Pal Joey*.
11. Charles Willson Peale, *Autobiography*. Transcribed by Horace Wells Sellers (Peale Family Papers, Smithsonian Institution), p. 476.
12. Alexander Wilson, *American Ornithology; or The Natural History of the Birds of the United States*. (Edinburgh, 1832 edn), Vol. 2, pp. 89-93.
13. I. M. Spry (ed.), *The Papers of the Palliser Expedition, 1857-1860* (The Champlain Society, Toronto, 1968).
14. 'Charles Waterton (1782-1865): Curare and a Canadian National Park', J.R. Maltby. *Canadian Anaesthetists' Society Journal*, Vol. 29, No. 3 (May 1982), pp. 195-200.
15. *Wanderings*, p. 156.
16. Trollope, op. cit., p. 211.
17. *Wanderings*, p. 156.
18. *The Ballad of the Caps, 1656*. 'The Monmouth Cap, the Saylor's thrum, / And that wherein the tradesmen come, / The Physick, Lawe, the Cap divine, / And that which crowns the Muses nine, / The Cap that Fools do countenance, / The goodly Cap of Maintenance, / And any Cap what ere it be, / Is still the sign of some degree.' See Alice Morse Earle, *Two Centuries of Costume in America, 1620-1820*. (Macmillan, New York, 1903).
19. *Wanderings*, p. 157.
20. Timothy Dwight, *Travels in New England and New York* (New Haven, 1822), Letter 5, Vol. 3, p. 333.
21. William Cobbett, *Political Register*, XXVI (1814).
22. Charles Coleman Sellers, *Charles Willson Peale* (American Philosophical Society), Vol. 1, p. 4.
23. Charles Willson Peale to Thomas Jefferson, 2/7/1824 (Library of Congress, Washington DC. Peale/Sellers Papers, Smithsonian Institution.).
24. *Essays*, Series 1,' Notes on the History and Habits of the Brown or Grey Rat', pp. 217-8.
25. Charles Coleman Sellers, Mr Peale's Museum: *Charles Willson Peale and the First Popular Museum of Natural Science and Art*. (W.W. Norton, New York, 1980), pp. 248-9.
26. Charles Willson Peale, *Autobiography, passim.*

27. Peale to Jefferson, *op. cit.*
28. Charles Waterton to George Ord, 4/7/1833 (Smithsonian Institution Libraries – Special Collections.).
29. Rembrandt Peale to T. R. Peale, 13/4/1833 (Baltimore City Life Museums), and Lillian B. Miller, *In Pursuit of Fame. Rembrandt Peale, 1778-1860.* (Univ. of Washington Press, 1992), p. 218.
30. Charles Waterton to George Ord, 16/2/1832 (Smithsonian Institution). At Dr Mease's, Audubon also met the painter, Thomas Sully, who introduced him to Charles Lucien Bonaparte. See Alexander B. Adams, *John James Audubon.* (Gollancz, 1967), Ch. XVI.
31. Audubon's Journal, 24/4/1824 'from want of knowledge of the habits of birds in a wild state, he represented them as if seated for a portrait, instead of with their own lively ways when seeking their natural food or pleasure.' Quoted in Lucy Bakewell Audubon, *The Life of John James Audubon, the Naturalist.* (Putnam, New York, 1902), pp. 100-108.
32. William Cobbett to Rachel Smithers, 6/7/1794. In Lewis Melville, *Life and Letters of William Cobbett in England and America.* (Bodley Head, 1913), Vol. 1, p. 85ff.
33. Preface to *Essays*, Series 2, pp. xxxiii-xxxiv.
34. Review of *Wanderings in South America*, Unsigned. Quarterly Review, Vol. 33 (Mar. 1826), p. 332.
35. Review of *Wanderings in South America*. Sidney Smith. Edinburgh Review, Vol. 43 (1826), pp. 307-8.
36. *Essays*, Series 1, 'Notes on the Habits of the Chegoe of Guiana . . .' , p. 242.

CHAPTER 12: GATHERING CLOUDS: THE HAPPIEST MAN ALIVE

1. Rev. T. Dixon to Richard Hobson, undated. Cited in Hobson, *Charles Waterton, His Home, Habits and Handiwork* (1867 edn), p. 343.
2. *The Times*, London. 30/8/1825.
3. *The Monthly Review*, Vol. CVIII (Sept. 1825), pp. 66-77.
4. *Literary Gazette and Journal of Belles Lettres, Arts, Sciences &c.*, No. 468 (7 Jan , 1826), pp. 4-8; No.470 (21 Jan, 1826), pp. 39-40; and No. 471 (28 Jan, 1826), pp. 56-7.
5. *ibid.*, No. 471.
6. *Quarterly Review*, Vol, 33 (Mar. 1826), pp. 314-32.
7. *British Critic*, Vol. II (April 1826), pp. 93-105.
8. *London Magazine*, 1826.
9. *Edinburgh Review*, Vol. xliii (Feb. 1826), pp. 299-315.
10. *Blackwood's Magazine*, June 1826, p. 661.
11. Charles Waterton to Alfred Ellis Esq. (2nd son of Sir Richard Ellis of Belgrave Hall, Leics.) In R. A. Irwin, *Letters of Charles Waterton* (London, 1955).
12. J. G. Wood, *Out of Doors* (Longmans, 1895).
13. Charles Darwin, *Autobiography*, ed. Gavin de Beer. (O.U.P., 1974), p. 28.
14. *ibid.*, pp. 27-8.
15. Audubon's Journal, 5/11/1826. Quoted in F.H. Herrick, *Audubon the Naturalist.* (New York, 1917), Vol. II, p. 91.
16. 'Account of the Habits of the Turkey Buzzard (Vultur Aura). . . ', John J. Audubon. *New Edinburgh Philosophical Journal*, Vol. 2 (Oct 1826 – April 1827), pp. 172-184.
17. Charles Waterton to Charles Edmonstone, 17/1/1827. (Smithsonian Institution). Reproduced in Irwin, op. cit., p. 15.
18. Mrs W. Pitt-Byrne, *Gossip of the Century* (London, 1899).
19. 'Notes on the Rattlesnake (Crotalus horridus); in a Letter Addressed to Thomas Stuart Traill, M.D. &c.', John J. Audubon. *New Edinburgh Philosophical Journal*, Vol. 3 (April - July 1827), pp. 21-30.
20. Anne Edmonstone to Mrs Edmonstone, undated (Am.Phil.Soc.).
21. Anne Edmonstone to Mr (Charles) Edmonstone, undated (Am.Phil.Soc.).
22. Anne Waterton to Eliza Edmonstone, 13/10/1829 (Am.Phil.Soc.).

23. Anne Edmonstone to Mrs Edmonstone, undated (Am.Phil.Soc.).
24. Charles Waterton, *An Ornithological Letter to William Swainson, Esq*. Richard Nichols, Wakefield, 1837), p. 4.
25. 'Some Account of the Work Now Publishing By M. Audubon, Entitled The Birds of America', William Swainson. *Loudon's Magazine of Natural History*, Vol. I, No. 1 (May 1828), p. 43.
26. Swainson to Audubon, Jan. 1828. *The Auk*, Vol. XXI (July 1905). Hitherto unpublished letter.
27. Signed 'A. Nondescript, April 22nd 1829.' Inscribed, in handwriting of Captain J. Wood of Sandal, on endpaper of Edmund Waterton's copy of *Wanderings in South America.* (Dr J. Lines, Wisbech, Cambs.).
28. 'To Ignota', Charles Waterton. *ibid.*
29. Anne Edmonstone to Helen Edmonstone, 20/4/1829 (CW Bicentenary Exhibition, Wakefield Museum. J. C. M. Daniel).
30. Anne Waterton to Helen Edmonstone, undated (Am.Phil.Soc.).
31. *ibid.*
32. Charles Waterton to George Ord, Ghent, 20/5/1829 (Am.Phil.Soc.).
33. Eliza Edmonstone to Anne Waterton, 23/6/1829 (J. C. M. Daniel).
34. 'A Life of the Author'. From Norman Moore's edition of Waterton's *Essays*. (Warne, London, 1871).

CHAPTER 13: SEVERED CORD

1. C. Barrington-Brown, *Canoe and Camp Life in British Guiana* (Edward Stanford, London, 1876), p. 24.
2. Helen Edmonstone to Anne Waterton, Bruges, Sept. 1829. (Am.Phil.Soc.).
3. Anne Waterton to Eliza Edmonstone, 18/10/1829 (Am.Phil.Soc.).
4. Eliza Edmonstone to Anne Edmonstone, 23/6/1829 (J. C. M. Daniel).
5. Sir George Head, *A Home Tour Through the Manufacturing Districts of England in the Summer of 1835.* (Cass Library of Industrial Classics, London, 1968), p. 169.
6. Anne to Eliza, *op. cit.*
7. *ibid.*
8. Anne Waterton to Mrs Helen Carr, undated (Am.Phil.Soc.).
9. Anne to Eliza, *op. cit.*
10. Anne to Eliza, undated, but probably Nov. 1829 (Am.Phil.Soc.).
11. Edward Jones, *Design for Knockers on Hall Doors of Walton Hall* (Am.Phil.Soc.).
12. 'A Naturalist's Home',unsigned. *Chamber's Journal*, 4th Series, No. 158 (5 Jan 1867), p. 15.
13. 'A Life of the Author', Norman Moore. From Waterton's *Essays*, ed. Norman Moore (Warne, 1871), pp. 125-6.
14. Joseph Hatton, *Old Lamps and New* (Hutchinson, 1889), pp. 322-3.
15. Charles Waterton Exhibition, Wakefield Museum.
16. Waterton Exhibition, Wakefield Museum.
17. 'Taxidermy', J. G. Wood. In Charles Waterton, *Wanderings in South America* (1879 edn.)
18. Anne Waterton to Mrs Robert Carr, undated (Am.Phil.Soc.).
19. Anne Waterton to Mrs Robert Carr, dated by unknown hand '1830, April 30', the day of Anne's funeral.
20. Charles Waterton to Josephine Waterton, ne Ennis (Edmund's first wife), 5/7/1863.
21. 'J.W.' (Captain Wood), 'To Charles Waterton Esq of Walton Hall, on the Death of his Wife'. (Am.Phil.Soc.) The quoted lines are the third and sixth stanzas of a handwritten, seven-stanza poem, dated 'Sandal, April 30th, 1830'.
22. Charles Waterton to George Ord, June 1830 (Am.Phil.Soc., Film, 1298).
23. Mrs Alec Tweedie (ed.), *George Harley, F.R.S. The Life of a London Physician* (London, 1899), p. 273.

Notes and References
Part Two
The Fine Old English Gentleman

CHAPTER 14: FOUL FALLS FOR THE DEVIL

1. Charles Waterton, *Essays on Natural History*, Series 1 (Longmans, London, 1838), pp. lxxiii-lxxiv.
2. The saints appear to be Vladimir the Great, St Anne of Russia, the Holy Martyrs Boris and Gleb, King Stephen of Hungary, Queen Margaret of Scotland and Mathilde of Germany. He could also have claimed indirect descent from Count Humbert III of Savoy, King Ferdinand III of Castille and King Louis IX of France, all of whom were canonised, but most of it seems to hinge on an alleged 15th-century marriage between John Waterton and Elizabeth Clifford.
3. Mrs Alec Tweedie (ed.), *George Harley, F.R.S. The Life of a London Physician* (London, 1899), p. 273.
4. Quoted in 'The Squire', Unsigned (Richard Aldington), *Times Literary Supplement*, 29/12/1932.
5. Charles Waterton to George Ord, June 1830 (Am.Phil.Soc., Film, 1298).
6. 'Prefatory Remarks', Charles Waterton. *Catalogue of Pictures at Walton Hall* (Wakefield, 1865), p. 4. (Wakefield Museum).
7. Sir Norman Moore, 'Charles Waterton'. *Dictionary of National Biography*.
8. Revolution in Belgium', Charles Waterton. *The Examiner*, Vol. XXIII, No. 1184 (10/10/1830), p. 645.
9. *Essays*, Series 1, p. lxxiv.
10. 'Revolution in Belgium', p. 646.
11. Charles Waterton to George Ord, 7/5/1831 and 20/9/1831 (Am.Phil.Soc.).
12. J. J. Audubon, *Ornithological Biography*, Vol.II, 'The Passenger Pigeon'. (Adam & Charles Black, Edinburgh, 1834).
13. *Wanderings in South America* (O.U.P. edn, 1973), p. 9.
14. Howard Corning (ed), *Journal of John James Audubon Made During His Trip to New Orleans in 1820-1821* (Boston, 1929).
15. Alexander Wilson, *American Ornithology* edited by Professor Robert Jameson (Constable's Miscellany of Original and Selected Publications, Edinburgh, 1831), Vol. IV, pp. 239-362.
16. From J. J. Audubon, *Birds of America* (George R. Lockwood, New York, 1839), pp.53-5.
17. Power of Smell Ascribed to the Vulture, Perceval Hunter. Loudon's Mag. of Nat. Hist. Vol. III, No. 15 (Sept. 1830), p. 449.
18. 'Remarks on Professor Rennie's Edition of Montagu's Ornithological Dictionary', Charles Waterton. *Loudon's Mag. of Nat. Hist.*, Vol. IV, No. 22 (Nov. 1831), p. 517. Reproduced in *Essays*, Series 1, pp. 1-6.
19. W. Montagu, *Ornithological Dictionary of British Birds*. 2nd edn. James Rennie (ed.) (London, 1831).

20. 'Eggs Containing Chicks Not to be Successfully Hatched if Suffered to Cool', Professor Rennie. *Loudon's Mag. of Nat. Hist.*, Vol. V, No. 23 (Jan. 1832), pp. 101-2.
21. 'On the Faculty of Scent in the Vulture', Charles Waterton. *Loudon's Mag. of Nat. Hist.*, Vol. V, No. 24 (March 1832), pp. 233-43. Reproduced in *Essays*, Series 1, pp.17-29.
22. William Cobbett, *Political Register*, VI (29/9/1804), p. 451.
23. *ibid.*, XIII (26/3/1808), p. 486.
24. Charles Waterton, *Notice to the Electors of Manchester* (Richard Nichols, Wakefield, 28/3/1832). One copy survives – Wakefield Museum.
25. Mrs Frances Trollope, *Domestic Manners of the Americans.* (4th edn, London, 1832), p. 295.
26. Charles Waterton to George Ord, 1/10/1832. In Irwin, op. cit.
27. Philip Gosse, *The Squire of Walton Hall.* (Cassell, 1940), p. 233. Primary source unknown.
28. 'Retrospective Criticism' (i.e. Letters to Editor), Perceval Hunter, *Loudon's Mag. of Nat. Hist.*, Vol. VI, No. 31 (Jan. 1833), p. 84.
29. 'On the Biography of Birds' (Retrospective Criticism), Charles Waterton. *Loudon's Mag. of Nat. Hist.,* Vol. VI, No. 33 (May 1833), p. 215.
30. George Sim, 'William MacGillivray' (Unpublished MS. City of Aberdeen Arts and Recreation Division – Library Services).
31. 'Retrospective Criticism', V. G. Audubon. *Loudon's Mag. of Nat. Hist.*, Vol. VI, No.34 (July 1833), p. 369. and 'Observations on Mr Waterton's Attacks on Mr Audubon' R.B. *Loudon's Mag of Nat Hist.*, Vol. VI, No. 34 (July 1833), p. 371. and 'Reply to Mr Audubon Jr., Charles Waterton'. *Loudon's Mag of Nat Hist.*, Vol. VI, No. 35 (Sept. 1833), p. 464 and 'Mr Swainson in Reply to Mr Waterton', William Swainson. *Loudon's Mag of Nat Hist.*, Vol. VI, No. 36 (Nov. 1833), p. 550 and 'Retrospective Criticism', Charles Waterton. *Loudon's Mag of Nat Hist.*, Vol. VII, No. 37 (Jan. 1834), p. 66-7.
32. 'Reply to Audubon Jr.', Charles Waterton *ibid.*, p. 67 and see 'The Romance of the Rattlesnake', Thomas P. Jones. *Franklin Journal*, Vol. II (Aug. 1828), p. 144 and *Mechanics' Magazine*, No. 447 (3/3/1832), pp. 403-5.
33. 'Aerial Encounter of the Eagle and the Vulture', Charles Waterton. *Loudon's Mag of Nat Hist.*, Vol. VII, No. 37 (Jan. 1834), p.69. Reproduced in *Essays*, Series 1, p. 122. and Corning. *op.cit.*
34. J. J. Audubon, *Ornithological Biography* (Adam & Charles Black, Edinburgh, 1834), Vol. III, p. 248. and 'Audubon's Humming-bird', Charles Waterton. *Loudon's Mag of Nat Hist.*, Vol. VII, No. 37, pp. 71-2 and *Essays*, Series 1, pp.125-7.
35. 'The Virginian Partridge', Charles Waterton. *Loudon's Mag. of Nat. Hist.*, Vol. VII, No. 37, p. 74 and *Essays*, Series 1, p. 128.
36. 'Remarks in Defence of Mr Audubon', Rev. John Bachman. *Loudon's Mag of Nat Hist.*, Vol. VII, No. 38 (Mar. 1834), pp. 164-72.
37. *ibid.*
38. 'The Vulture's Nose', Charles Waterton *ibid.*, No. 39 (May 1834), p. 277.
39. Some Account of Walton Hall, the Seat of Charles Waterton Esq., James Stuart Menteath Esq. *ibid.*, Vol. VIII, No. 43 (Jan. 1835), p. 31.
40. 'The New Chimney Sweeping Act', Charles Waterton. *Essays*, Series 2 (Longmans, London 1844), p. 85.
41. 'Notes of a Visit to the Haunts of the Guillemot and Facts on its Habits', Charles Waterton. *Loudon's Mag of Nat Hist.*, Vol. VIII, No. 47 (Mar. 1835), p. 165.
42. 'Notes on a Visit to the Haunts of the Cormorant and Facts on its Habits', Charles Waterton *ibid.*, p. 169.
43. 'On Snakes, Their Fangs, and Their Mode of Procuring Food', Charles Waterton. *ibid.*, No. 56 (Dec. 1835), p. 663. and *Essays*, Series 1, pp. 204-5.
44. 'On Beauty in the Animal Creation', Charles Waterton. *Essays*, Series 2, p. 158.
45. 'On the Geology and Natural History of the North-Eastern Extremity of the Allegheny

Mountain Range in Pennsylvania', Richard C Taylor. *Loudon's Mag. of Nat. Hist.*, Vol. VIII, No. 54 (Oct. 1835), pp. 529-41.
46. 'On Snakes. . . . ' pp. 207-8.
47. George Ord to Charles Waterton, 9/1/1836 (Am.Phil.Soc.). Extract in *Loudon's Mag of Nat Hist.*, Vol. IX, No. 64 (Aug. 1836), pp. 416-7.
48. 'Review of Audubon's Ornithological Biography' (Vol. II), Professor Robert Jameson. *New Edinb. Phil. J.*, Vol. X (Jan. 1835) and *Loudon's Mag of Nat Hist.*, Vol. VIII,No. 47 (Mar. 1835).
49. Charles Waterton, *A Letter to James Jameson, Esq* (Richard Nichols, Wakefield, 1835). Wakefield Museum. Reprinted in 'Quarrels of Zoologists', Unsigned. *Fraser's Magazine*, Vol. XI (Mar. 1835), pp. 325-9. Quoted in 'Reviews', Unsigned. *Loudon's Mag of Nat Hist.*, Vol. VIII, No. 47 (Mar. 1835), pp. 190-2.
50. Charles Waterton, *A Second Letter to Robert Jameson, Esq*. Wakefield Museum.
51. Charles Waterton to Neville Wood, 20/5/1835 (Pierpont Morgan Library, New York. MA 22761).
52. Charles Waterton, Letter to the Rev J P Simpson, Curate of All Saints Church, Wakefield. (Richard Nichols, Wakefield, 1833), pp. 9-10. Wakefield Museum.
53. *ibid.*, p. 11.
54. 'Notices of the Affinities, Habits, and Certain Localities of the Dipper (Cinclus aquaticus)', Rev. Francis Orpen Morris. *Loudon's Mag. of Nat. Hist.*, Vol. VIII, No. 51 (July 1835), pp. 374-6.
55. 'On Snakes', Charles Waterton. *Essays*, Series 3 (Longmans, London, 1857), p. 270. and 'Short Communications', Charles Waterton. *Loudon's Mag of Nat Hist.*, Vol. VIII, No. 53 (Sept. 1835), p. 516.
56. 'On Birds Dressing Their Feathers with Matter Secerned From a Gland', Rev. F. O. Morris. *Loudon's Mag of Nat Hist.*, Vol. VIII, No. 55 (Nov. 1835), p. 637.
57. T. Allis to Charles Waterton, 11/11/1836 (North Yorkshire County Library. NRA9540).
58. Charles Waterton to T. Allis, 16/11/1836 (*ibid*).
59. William Swainson, *On the Natural History and Classification of Birds* (Lardner's Cabinet Cyclopaedia, London, 1836), Vol. I, p. 48.
60. George Ord to Charles Waterton, 18/5/1836. Quoted in Charles Waterton, *An Ornithological Letter to William Swainson, Esq* (Richard Nichols, Wakefield, 1837), pp. 4-5.
61. 'Sense of Smell in Carrion Birds', Rev F. O. Morris. *The Naturalist*, Vol. II, No. 7 (April 1837), pp. 34-5.
62. 'The Habits of the King of the Vultures (Sarcoramphus papa)', Richard Schomburgk. *Edinb.J.Nat.Hist.*, Vol. II (May 1839), pp. 14-15.
63. Richard Owen, *Proceedings of the Zoological Society of London* (London, 1837), Pt. V, pp. 33 5. See also 'Vultures', Unsigned. *The Spectator*, Vol. 83, p. 491 (7/10/1899).
64. 'Olfactory Sense in the Turkey Vulture (Cathartes aura)', C. L. Hopkins. *Auk*, Vol. 5 (1888), p. 248.
65. 'Food Detection by the Turkey Vulture', I. Sayles. *Auk*, Vol. 6 (1889), p. 51.
66. 'Notes From Connecticut', Louis B. Bishop. *Auk*, Vol. 38, No. 4 (Oct 1921), p. 585.
67. W. B. Alexander, *Birds of the Ocean* (Putnam's, New York, 1954).
68. 'Observations of Social Behaviour in Turkey Vultures', Howard H. Vogel Jr. *Auk*, Vol. 67 (April 1950), pp. 210-16.
69. 'Feeding Habits of the Black Vulture', E. A. McIlhenny. *Auk*, Vol. 56, No. 4 (Oct 1939), pp. 472-4.
70. Roger Tory Peterson and James Fisher, *Wild America* (Houghton Mifflin, Boston, 1955).
71. 'Proceedure for Studying Olfactory Discrimination in Pigeons', Wolfgang J. Michelson. *Science*, Vol. 130 (1959), p. 630 and C. B. Ferster and B. F. Skinner, Schedules of Reinforcement. (Appleton-Century Crofts, New York, 1957).
72. 'Anatomical Evidence for Olfactory Function in Some Species of Bird', Betsy Garrett

Bang. *Nature*, Vol. 188, No. 4750 (12/11/1960), p. 548.
73. 'Indication of the Sense of Smell in the Turkey Vulture from Feeding Tests', O. T. Owre and P. O. Northington. *Am.Midl.Nat.*, Vol. 66 (1961), pp. 200-205.
74. 'The Role of Olfaction in Food Location by the Turkey Vulture', K. E. Stager. *Contributions in Science*, 81 (1964). Los Angeles County Museum.
75. Francis Hobart Herrick, *Audubon the Naturalist* (New York, 1917), Vol. II, p. 83.
76. Terence Michael Shortt, *Wild Birds of the Americas* (Houghton Mifflin, Boston, 1977), p.72.
77. Frederick II of Hohenstaufen, *De Arte Venande cum Avibus* (The Art of Hunting with Birds). 1245.
78. 'Does the King Vulture', *Sarcoramphus papa*, 'Use a Sense of Smell to Locate Food', David C. Houston. *Ibis*, Vol. 126 (1984), pp. 67-9 and 'Report of Birds of Prey Conference', Philip Burton. *B.T.O. News*, May 1984.
79. Charles Darwin's Notebook, Callao, Peru, 19/7/1835. Quoted in Nora Barlow (ed.), *Charles Darwin and the Voyage of the Beagle* (New York, 1946), p.244. 'Smelling properties discussed of Carrion Crows, Hawks, Magazine of Natural History.'
80. 'Further Remarks on the Affinities of the Feathered Race; and Upon the Nature of Specific Distinctions', Edward Blyth. *Loudon's Mag of Nat Hist.*, Vol. X, No. 70 (Feb 1837), p. 135.
81. Charles Darwin. Quoted in Professor Loren Eiseley, *Darwin and the Mysterious Mr X* (Dent, 1979), pp. 46-7. (Eiseley's 'mystery' is contrived, 'cryptic' clues completely unconcealed).
82. Charles Waterton, *To the Rev Disney Robinson*, 11/10/1836. Wakefield Museum.
83. Charles Waterton, *A Letter on the Church of England by Law Established, Occasioned by Parson Gregg's Unprovoked Attack on the Catholic Religion* (Richard Nichols, 1837), p. 3. Wakefield Museum.
84. Swainson, *op. cit.*, Ch. 1: 'An Enumeration of the Chief Works on Ornithology, With Critical and Explanatory Remarks', p. 211.
85. Charles Waterton, *An Ornithological Letter to William Swainson, Esq.*
86. *Dictionary of Scientific Biography*. 'William Swainson'.
87. Charles Waterton, *An Ornithological Letter...*, pp. 13-14.
88. Charles Lucien Bonaparte to William Swainson, 19/7/1838 (Swainson Correspondence. Linnaean Society, London).
89. Charles Waterton, *The Law Church* (Richard Nichols, 28/6/1837). Wakefield Museum.
90. Jane Loudon, *The Lady's Country Companion, or How to Enjoy a Country Life Rationally* (London, 1845), p. 331.
91. Unsigned, *Literary Gazette*, No. 1109 (21/4/1838), pp. 244-5.
92. Unsigned, *The Athenaeum*, No. 549 5/5/1838), p. 321.
93. Unsigned, *Edinburgh Journal of Natural History and Physical Sciences*, No. 38 (July 1838), p. 152.
94. 'Progress of Publication', Unsigned. *The Spectator*, Vol. XI, No. 511 (14/4/1838), p.355.
95. 'Waterton the Wanderer', Unsigned. *Chambers' Edinburgh Journal*, No. 492 (3/7/1841), p. 188.
96. *Gentleman's Magazine*, Vol. 163 (May 1838), p. 526.
97. *Essays*, Series 1, pp. xiii-xiv.
98. Charles Waterton, *A Letter on the Reformation, Occasioned by the Attack of the ReformationSociety of Wakefield* (Richard Nichols, 15/6/1838), p. 11. Wakefield Museum.
99. 'Waterton and Wourali', W. D. A. Smith. *British Journal of Anaesthesia*, Vol. 55 (1983), pp. 223-4. *Proceedings of a Symposium Held to Commemorate the Bicentenary of Charles Waterton* (at Walton Hall, 5/6/1982).
100. 'The Ass Wouralia', Unsigned but written by Charles Waterton (in third person) with accompanying letter. *St James's Chronicle*, 23/2/1839.
101. *Essays*, Series 1, pp. xiv-xvi.

102. See K. Bryn Thomas, *Curare, its History and Usage* (Pitman Medical Publishing, London, 1964), p. 87 *et seq.*
103. Charles Waterton to John Wells, 27/4/1839 (Anonymous transcription). Sir Richard Owen Correspondence, Brit. Mus. (N.H.), 26:204.
104. 'Wourali: Analysis and Bioassay', J. Cooke, A. Cawood, A. Crossley and F. R. Ellis. Br.J.Anaesth., Vol. 55, p. 225. Proceedings of bicentenary symposium, *op. cit.*
105. 'Notes on the History and Habits of the Brown or Grey Rat', Charles Waterton. Essays, Series 1, pp. 210-218.
106. *Essays*, Series 1, p. xiv.
107. *ibid.*, p. xxx.

CHAPTER 15: HETEROGENEOUS ALIMENT

1. Charles Waterton, *Essays on Natural History*, Series 2 (1844), p. xxxi.
2. *ibid.*, p. xxxiii.
3. *ibid.*, p. xxxvi.
4. *ibid.*, p. xlvi.
5. *ibid.*, p. xlviii.
6. *ibid.*, pp.xlviii-xlix.
7. 'Aix-la-Chapelle', Charles Waterton. *Essays*, Series 3 (1857), p. 136.
8. *Tristra*, 1.1.101.
9. *Essays,* Series 2, p. l.
10. Charles Waterton to George Ord, 25/3/1847. In R. A. Irwin, *Letters of Charles Waterton.* (Rockliff, 1955), pp. 62-3.
11. 'Waterton as an Ethologist', F. Richard Ellis. *Newsletter* (Yorkshire Society of Anaesthetists), Waterton Bicentenary Issue, 1981.
12. Charles Waterton to George Ord, 25/3/1847, *op. cit.*
13. 'Aix-la-Chapelle', p. 135.
14. *Essays*, Series 2, p. liii.
15. *ibid.*, p. lvii.
16. Charles Waterton to George Ord, Rome, 4/8/1840. In Irwin, *op. cit.*
17. *Essays*, Series 2, p. lix.
18. Ovid, *Metamorphoses*, XIII, 16-18.
19. Charles Waterton to George Ord, 4/8/1840, *op. cit.*
20. Essays, Series 2, p. lxvii.
21. *ibid.*, pp. lxix-lxx.
22. *ibid.*, pp. lxxii-lxxiv.
23. 'The Civetta, or Little Italian Owl', Charles Waterton. *Essays*, Series 2, p. 17.
24. Anglo-Italian Society for the Protection of Animals, Baker Street, London.
25. 'Flower Gardens and Song Birds', Charles Waterton. *Essays*, Series 2, p. 9.
26. 'King David's Solitary Sparrow.' In Octagonal Case of Birds, with 'three trogons, six orioles and three cuckoos.' Wakefield Museum.
27. 'The Roller, Called Pica Marina in Italy', Charles Waterton. *Essays*, Series 2, pp. 26-31.
28. *Essays*, Series 2, p. lxxiv.
29. *ibid.*, pp. lxxv-lxxvi.
30. *ibid.*, pp. lxxvi-lxxvii.
31. James Simson, *Charles Waterton, Naturalist* (Baillere, Tyndall & Co., London, 1880), p. 6.
32. Edward Lear to Anne Lear, May 1838. Quoted in Vivien Noakes, *Edward Lear* (Collins, 1968), pp. 55-6.
33. *Essays*, Series 2, pp. lxxviii-lxxix.
34. Charles Waterton to George Ord, 21/9/1840 (Am.Phil.Soc. Film).

35. *Essays*, Series 2, pp. lxxxii-lxxxiii.
36. *ibid.*, p. lxxxiv.
37. *ibid.*, p.lxxxvii.
38. 'The Victorian Setting', from David Elliston Allen, *The Naturalist in Britain.* (W. H. Allen, 1972). and 'William Swainson's Ornithological Collections', Phil Parkinson. *Archives of Natural History* (Soc. for the Hist. of Nat. Hist.), Vol. 15 (1988), No. 1,pp.77-8.
39. *Essays*, Series 2, p. civ.
40. *ibid.*, p. xciv.
41. Virgil, *Aeneid*, 1.83. 'qua data porta, ruunt et terras turbine perflant.'
42. Charles Waterton to George Ord, 7/6/1841. In Irwin, *op. cit.*
43. *Essays*, Series 2, pp. cxi-cxii.
44. *ibid.*, p. cxiv.
45. 'The Civetta. . . ', pp. 17-18.
46. *Essays*, Series 2, p. cxxv.
47. Charles Waterton to George Ord, 6/7/1841. In Irwin, *op. cit.*
48. 'The Civetta. . . ', p. 20.
49. Charles Waterton to Father Singleton, 10/12/1841. (Smithsonian Institution Libraries – Special Collections. MSS 153A).
50. 'The Canada or Cravat Goose', Charles Waterton. *Essays*, Series 2, pp. 114-5.
51. *ibid.*, p. 119.
52. *Essays*, Series 2, p. cxxxii.
53. Charles Waterton to Father Singleton, *op. cit.*

CHAPTER 16: FAITH, HOPE AND CHARITY

1. Charles Waterton, 'Essays on Natural History', Series 2 (Longmans, London, 1844), pp. cxxvii-cxxx.
2. Percy Fitzgerald, 'Stonyhurst Memories'. (Bentley's Miscellany, 1895), *passim.*
3. George Ord to Charles Lucien Bonaparte, 10/9/1841. (Am.Phil.Soc.)
4. George Ord to Charles Waterton, 18/9/1841. (Am.Phil.Soc.)
5. *Essays*, Series 2, p. cxxxii.
6. Richard Hobson, *Charles Waterton, His Home, Habits and Handiwork* (2nd edn, 1867), p. 159.
7. Charles Waterton to Father Singleton, 10/12/1841 (Smithsonian Institution – Special Collections. MSS 153A).
8. J. H. Newman, *Tracts for the Times*, No. 90. 'Remarks on Certain Passages in the Thirty-nine Articles.' A. W. Evans (ed.). (Constable, London, 1933), p. 98. First published, Oxford, 1841.
9. Mgr T. G. Glover to Charles Waterton, Rome, June 1842 (North Yorkshire County Library – York City Archives. NRA9202/CW6).
10. *Essays*, Series 2, pp. cxxxii-cxxxiii.
11. 'The Civetta, or Little Italian Owl', Charles Waterton. *ibid.*, pp. 17-18.
12. 'Ringing Recovery Circumstances of Small Birds of Prey', David E. Glue. Bird Study, Vol. 18, No. 3 (Sept. 1971), p. 143.
13. Mgr T. G. Glover to Charles Waterton, *op. cit.*
14. *Essays*, Series 2, pp. cxxxiii-cxxxv.
15. 'Thunder Storm at Walton Hall', Charles Waterton. *ibid.*, pp. 24-6.
16. 'The British Contribution to Bird Protection', Phyllis Barclay-Smith. *Ibis*, Vol. 101 (1959), p. 118.
17. Mgr T. G. Glover to Charles Waterton, 17/4/1843 (North Yorkshire County Library. NRA9202/CW9).
18. Father A. Barrow to Charles Waterton, 16/9/1843 (North Yorkshire County Library. NRA9202/CW10).

19. 'Anecdote of a Combat Betwixt Two Hares', Charles Waterton. *Essays*, Series 2, pp. 88-90.
20. Charles Waterton to Eliza and Helen Edmonstone, 19/11/1843 (Am.Phil.Soc.).
21. 'The Wren, the Hedge Sparrow, and the Robin', Charles Waterton. *Essays*, Series 2, pp. 95-6.
22. Charles Waterton to Eliza and Helen Edmonstone, 8/11/1843. In Irwin, *op. cit.*
23. Fitzgerald, *op. cit.*, p. 43.
24. Father A. Barrow to Charles Waterton, 11/3/1844 (North Yorkshire County Library. NRA9202/CW8).
25. Edmund Waterton to Helen Edmonstone, 2/12/1843 (North Yorkshire County Library. NRA9202/CW12).
26. Fitzgerald, *op. cit.*, p. 78.
27. Bea Howe, *Lady With Green Fingers*; *The Life of Jane Loudon* (Country Life Ltd., 1961), pp. 91-2.
28. Hobson, *op. cit.*, p. 187.
29. Elizabeth Burton, *Early Victorians at Home* (Longmans, London, 1971), p. 289.
30. Letter from J. W. Loudon. *Athenaeum*, No. 850 (10/2/1844), p. 140.
31. *ibid.*
32. Charles Waterton to Eliza and Helen Edmonstone (at Funchal, Madeira), 27/2/1846. In Irwin, *op. cit.*
33. 'Waste Land – Why Not Improved?' Unsigned. *Chambers' Edinburgh Journal*, No. 6, new series (10/2/1844), p. 82.
34. 'Waste Lands', Charles Waterton. *Essays*, Series 2, pp. 105-6.
35. *Essays*, Series 2. Dedication.
36. *Athenaeum*, No. 856 (23/3/1844), p. 276.
37. Review of Waterton's Second Series of Essays, Unsigned. *Athenaeum*, No. 874 (27/7 1844), p. 692.
38. Review of Waterton's Second Series of Essays, Unsigned. *Blackwood's*, Vol. LVIII, No. 354 (Sept. 1845), p. 300.
39. Chris Kennisley to Charles Waterton, 20/8/1844 (North Yorks County Library. NRA9202/CW13).
40. 'On Beauty in the Animal Creation', Charles Waterton. *Essays*, Series 2, p. 164. and *Blackwood's*, *op. cit.*, pp. 299-300.

CHAPTER 17: STRANGERS IN PARADISE

1. Charles Waterton, *Essays on Natural History*, Series 3 (Longmans, London, 1857), p. ii.
2. 'A Pilgrimage to Caldaro', Anon. *Dublin University Magazine*, Vol. 25 (March 1845), pp. 305-18.
3. *Essays*, Series 3, pp. iv-v.
4. 'A Pilgrimage to Caldaro', p. 312.
5. *Essays*, Series 3, pp. xi-xiii.
6. *ibid.*, pp. xiv-xv.
7. *ibid.*, p. xix.
8. Charles Waterton Exhibition, Wakefield Museum.
9. Charles Dickens, *Pictures From Italy* (Revised edn, 1868), 'Rome'.
10. Mr M. A. Titmarsh (Thackeray), *Notes of a Journey from Cornhill to Grand Cairo* (London, 1846).
11. W. M. Thackeray, *The Newcomes* (Everyman edn, 1969), pp. 383-4. First published in monthly parts, 1853-5.
12. Charles Waterton to George Ord, 24/7/1845. In R. A. Irwin, *Letters of Charles Waterton* (Rockliff, 1955).
13. B. Dailey to Charles Waterton, 12/7/1845 (North Yorks County Library. NRA9202/CW15).

14. Charles Darwin to J S Henslow, 28/10/1845. (Smithsonian Institution - Dibner Collection. MSS 405A).
15. Charles Darwin to C. Lyell, 8/10/1845. In Francis Darwin (ed), *The Life and Letters of Charles Darwin* (John Murray, 1888), Vol. 1, pp. 343-4.
16. F. H. Salvin to Richard Hobson, 20/10/1866. Reproduced in Hobson, *Charles Waterton, His Home, Habits and Handiwork* (2nd edn., 1867), pp. 122-5.
17. Charles Waterton, *Essays on Natural History*, Series 3., p. xxvii.
18. Charles Waterton to Helen Edmonstone, 4/4/1846 (Am.Phil.Soc.).
19. Richard Schomburgk, *Travels in British Guiana, 1840-1844* tr. Walter E. Roth (Georgetown, 1922), Vol. 1, pp. 25-7.
20. Charles Waterton to Eliza and Helen Edmonstone, 22/1/1846 (Am.Phil.Soc.).
21. Charles Waterton to Eliza and Helen Edmonstone, 13/1/1846 (Am.Phil.Soc.).
22. 'Game Law Morality', Unsigned. *Punch*, Vol. 10 (Mar. 1846), p. 144.
23. Charles Waterton to W. M. Thackeray, 28/3/1846 (Am.Phil.Soc.).
24. W. M. Thackeray to his mother, 1842. Quoted in Susan & Asa Briggs (eds), Cap and Bell (MacDonald, 1972), p.xii.
25. Charles Waterton to Eliza Edmonstone (at Dalston, London), 4/4/1846 (Am.Phil.Soc.).
26. Charles Waterton to Eliza & Helen Edmonstone (at Funchal, Madeira), 27/2/1846. In Irwin, *op. cit.*
27. J. G. Wood, Preface to Waterton's *Wanderings in South America* (London, 1879).
28. Charles Waterton to George Ord, 12/11/1846. In Irwin, *op. cit.*
29. Charles Waterton to George Ord, 17/8/1846 *ibid.*
30. 'Pigeon Cots and Pigeon Stealers', Charles Waterton. *Essays*, Series 3, p. 112
31. Charles Waterton to George Ord, 17/8/1846. *Op. cit.*
32. Charles Waterton to George Ord, 29/6/1846. In Irwin, *op. cit.*
33. Charles Waterton to George Ord, 17/8/1846. *Op. cit.*
34. 'Pigeon Cots and Pigeon Stealers', *op. cit.*
35. Wood, *op. cit.*
36. R. W. Emerson, *English Traits* (Houghton Mifflin, Boston, 1856), 'Aristocracy'.
37. Charles Waterton to George Ord, 25/3/1847. In Irwin, *op. cit.*
38. Charles Waterton to George Ord, 24/10/1847. In Irwin, *op. cit.*
39. Hobson, *op. cit.*, p. 66.
40. 'Some Account of Walton Hall, the Seat of Charles Waterton Esq.', James Stuart Menteath. *Loudon's Mag of Nat Hist.*, Vol. VIII (Jan 1835), pp. 33-5.
41. Charles Waterton, *An Ornithological Letter to William Swainson Esq.* (Richard Nichols, Wakefield, 1837), p. 4.
42. Charles Waterton to George Ord, 14/11/1848. In Irwin, *op. cit.*
43. Hobson, *op. cit.*, pp. 111-2.
44. *Scarbro*', Charles Waterton. *Essays*, Series 3, pp. 195-6.
45. Charles Willson Peale, *Autobiography* transcribed by Horace Wells Sellers. (Peale Family Papers, National Portrait Gallery, Smithsonian Institution), p. 475.
46. Hobson, *op. cit.*, pp. 53-5.
47. Charles Waterton to Edward T. Simpson, 19/2/1849 (Goodchild Loan MSS. Wakefield District Archives).
48. Goodchild Loan MSS.
49. George Ord to Charles Waterton, 23/9/1848, 18/2/1850 and 8/7/1850. (Am.Phil.Soc.).
50. 'Scarbro'', *op. cit.*, pp. 195-6.
51. Milton, *Paradise Lost*, II, 432.
52. Charles Waterton to George Ord, 6/3/1849. In Irwin, *op. cit.*
53. Mrs W. Pitt-Byrne, *Gossip of the Century* (Downey & Co., Leeds, 1899), Vol. IV, Ch. XIII. And J. G. Wood, *Out of Doors.*

54. Charles Waterton to George Ord, 27/7/1849. In Irwin, *op. cit.*
55. Charles Waterton to George Ord, 21/9/1849 *ibid.*
56. Charles Waterton to George Ord, 6/2/1850 *ibid.*
57. Charles Waterton to George Ord, 11/5/1850 *ibid.*
58. Edward Lear (Text by J. E. Gray), *Gleanings from the Menagerie and Aviary at Knowsley* (Privately published, 1846).
59. George Ord to Charles Waterton, 6/10/1850 (Am.Phil.Soc.).

CHAPTER 18: HELL AND HIGH WATER

1. Charles Waterton to George Ord, 20/6/1850. In R. A. Irwin, *Letters of Charles Waterton* (Rockliff, 1955).
2. Charles Waterton, *Essays on Natural History*, Series 3 (Longmans, London, 1857), p. xxviii.
3. *ibid.*, pp. xxviii-xxxii.
4. Charles Waterton to George Ord, 20/6/1850. *Op. cit.*
5. Charles Waterton to George Ord, 9/11/1850. In Irwin, *op. cit.*
6. Charles Waterton to George Ord, 31/5/1851 *ibid.*
7. Charles Waterton to George Ord, 9/11/1850. *Op. cit.*
8. 'Cannibalism', Charles Waterton. *Essays*, Series 3, pp. 222-4.
9. 'Scarbro' ', Charles Waterton. *ibid.*, pp. 197-8.
10. 'A New History of the Monkey Family', Charles Waterton *ibid.*, p. 53.
11. *ibid.*, p. 55.
12. 'Scarbro' ', pp. 195-6.
13. J. G. Wood, Preface to Waterton's *Wanderings in South America* (London, 1879).
14. *ibid.*
15. *Essays*, Series 3, *op. cit.*, p. xxxv.
16. Charles Waterton to George Ord, 25/4/1851. In Irwin, *op. cit.*
17. *Essays*, Series 3, p. xxxvi.
18. *ibid.*
19. *ibid.*, pp. xxxvi-xxxvii.
20. *ibid.*
21. Charles Waterton to George Ord, 25/4/1851. *Op. cit.*
22. *ibid.*
23. John Goodchild, *Edmund Waterton* (Goodchild Loan MSS.).
24. *Essays*, Series 3, pp. xxxix-xl.

CHAPTER 19: UNIQUE ABNORMITIES

1. Charles Waterton to George Ord, 31/5/1851. In R. A. Irwin, *Letters of Charles Waterton* (Rockliff, London, 1955).
2. Charles Waterton, *Essays on Natural History*, Series 3 (Longmans, London, 1857), p. xl.
3. Charles Waterton to George Ord, 28/8/1851. In Irwin, *op. cit.*
4. 'A New History of the Monkey Family', Charles Waterton. *Essays*, Series 3, p. 58.
5. *ibid.*, pp. 60-1.
6. 'Scarbro' ', Charles Waterton. *Essays*, Series 3, p. 199.
7. Charles Waterton to George Ord, 26/11/1851. In Irwin, *op. cit.*
8. *Philosophical Transactions*, Vol. 78 (1788), Pt. 2.
9. 'The Wren, the Hedge Sparrow, and the Robin', Charles Waterton. *Essays*, Series 2 (1844), pp. 96-100.
10. George Ord to Charles Waterton, 13/10/1851 (Am.Phil.Soc.).
11. Charles Waterton to George Ord, 4/11/1852. In Irwin.
12. Richard Hobson, Charles Waterton, *His Home, Habits and Handiwork* (2nd edn, 1867), p. 68.
13. 'On Snakes', Charles Waterton. *Essays*, Series 3, pp. 281-6.

14. Charles Waterton to George Ord, 4/11/1852. In Irwin, *op. cit.*
15. *ibid.*
16. Hobson, *op. cit.*, pp. 70-1.
17. Mrs Alec Tweedie (ed.), *George Harley, F.R.S. The Life of a London Physician* (London, 1899), p. 280.
18. George Ord to Charles Waterton, 12/12/1852 (Am.Phil.Soc.).
19. Charles Waterton to George Ord, 23/3/1853. In Irwin, *op. cit.*
20. *ibid.*
21. Goodchild Loan MSS., Wakefield District Archives.
22. Charles Waterton to Eliza Edmonstone, 29/12/1852 (Am.Phil.Soc.).
23. Goodchild Loan MSS.
24. *The Squire*, Unsigned (Richard Aldington). *Times Literary Supplement*, 29/12/1932.
25. Charles Willson Peale, *Autobiography* transcribed by Horace Wells Sellers (Peale Family Papers, Smithsonian Institution), p. 475.
26. Charles Waterton to George Ord, 23/3/1853. *Op. cit.*
27. *The Fox*, Charles Waterton. *Essays*, Series 3, p. 260.
28. Gilbert Phelps, *Squire Waterton* (E.P., Wakefield, 1976), p. 143.
29. George Ord to Charles Waterton, 17/4/1853 (Am.Phil.Soc.).
30. Charles Waterton to George Ord, 12/6/1853. In Irwin, *op. cit.*
31. *Essays*, Series 3, p. xli.
32. Charles Waterton to T. Allis, 31/10/1854. (North Yorks County Library – York City Archives. NRA9540).
33. Goodchild Loan MSS.
34. George Ord to Charles Waterton, 7/12/1854 (Am.Phil.Soc.).
35. Charles Coleman Sellers, *Mr Peale's Museum* (Norton, New York, 1980), pp. 312-3. Ord paid $12 for the painting.
36. Charles Waterton to George Ord, 11/1/1855. In Irwin, *op. cit.*
37. Hobson, *op. cit.*, pp. 102-3.
38. Pitt-Byrne, *op. cit.*
39. *ibid* and Tweedie, *op. cit.*, pp. 269-73.
40. Pitt-Byrne, *op. cit.*
41. Hobson, *op. cit.*, p. 299.
42. *ibid.*, p. 205.
43. *ibid.*, pp. 206-7.
44. Mr W. B. Spurr of Walton Manor House, whose father had known Charles Waterton. See G. Bernard Wood, *Yorkshire Villages* (Hale, 1971).
45. 'The Yew Tree', Charles Waterton. *Essays*, Series 3, pp. 64-5.
46. S. T. Coleridge, 'Ancient Mariner'.
47. 'A New History of the Monkey Family', pp. 64-6.
48. *ibid.*, pp. 66-7.
49. Charles Waterton to George Ord, 4/12/1855. In Irwin, *op. cit.*
50. Hobson, *op. cit.*, p. 130.
51. 'Reminiscences of Charles Waterton', Dr George Harley, F.R.S. *Selborne Magazine*, Feb 1889. Reproduced in Tweedie, *op. cit.*
52. *Selborne Magazine*, March 1889.
53. Waterton Bicentenary Exhibition Catalogue (Wakefield Museum, 1982), p. 4.
54. *Selborne Magazine*, Aug. 1889.
55. 'Notes of Lectures on the Physiological Action of Strychnine', Dr George Harley. *Lancet*, 1856.
56. Tweedie, *op. cit.*, p. 249.
57. Hobson, *op. cit.*, p. 230.

58. *ibid.*, p. 235.
59. Goodchild Loan MSS.
60. *Dilecto Filio Edmundo Waterton de Walton Nostro Cubiculario Intimo* Pius PP.IX. 6/3/1858 (Am.Phil.Soc.).
61. Charles Waterton to George Ord, 19/12/1856. In Irwin, *op. cit.*
62. Hobson, *op. cit.*, p. 250.
63. 'Varieties of Animals, with Observations on the Marked Seasonal and Other Changes which Naturally Take Place in Various British Species, and which Do Not Constitute Varieties', Edward Blyth. *Loudon's Mag. of Nat. Hist.*, Vol. VIII, No. 45 (Jan 1835), pp.40-53 and 'Observations on the Various Seasonal and Other External Changes which Regularly Take Place in Birds', Edward Blyth. *ibid.*, Vol. IX, No. 64 (Aug 1836), pp. 393-409 and Further Remarks on the Affinities of the Feathered Race, and 'Upon the Nature of Specific Distinctions', Edward Blyth. *ibid.*, Vol.IX, No.66 (Oct 1836), pp. 505-14. Etc.
64. 'Remarks on Varieties', Unsigned (probably William MacGillivray). *Edinb. J. of Nat. Hist. & Physical Sciences*, Oct., Nov., & Dec. 1837, pp. 116, 119-20 & 124.
65. One-inch Ordnance Survey, First Edition. *Doncaster* (Sheet 87), 1840-41.
66. Wood, *op. cit.*, and Hobson, *op. cit.*, p. 172. (CW's father had discovered the cannonball which had reportedly broken the Cromwellian soldier's leg as he carried a keg of ale on his shoulder to the beseiging troops. Cannon and ball were a perfect match.)
67. Charles Waterton to Geoffrey Wentworth Esq (of Woolley Park), 1/7/1857 (Yorkshire Archaeological Society).
68. 'Waterton's Home', Unsigned. *Bentley's Miscellany*, Vol. 60 (1866), p. 264.

CHAPTER 20: BIRDS OR BOUDOIRS

1. *The Saturday Review of Politics, Literature, Science and Art*, Vol. VI (3/10/1857), pp. 307-8.
2. *Gentleman's Magazine*, Vol. 203 (Nov 1857), p. 543.
3. *Fraser's Magazine*, Vol. LVI, No. 336 (Dec. 1857), p. 649.
4. *ibid.*, p. 638.
5. *The Leader*, No. 388 (29/8/1857), p. 836.
6. Charles Waterton to Edward Jupp, 7/3/1858 (Smithsonian Institution Libraries – Special Collections. MSS 153A).
7. Richard Hobson, *Charles Waterton, His Home, Habits and Handiwork* (2nd edn, 1867), p. 247.
8. Joseph Hatton, *Old Lamps and New* (Hutchinson, 1889), p. 328.
9. Hobson, *op. cit.*, p. 37.
10. Charles Waterton to George Ord, 2/2/1859. In Irwin, *op. cit.*
11. Charles Waterton to Richard Hobson, May 1859. In Hobson, *op. cit.*, p. 218.
12. Hobson, *op. cit.*, p. 66.
13. *ibid.*, pp. 34-5.
14. Charles Waterton to George Ord, 26/7/1859. In Irwin, *op. cit.*
15. Charles Waterton to George Ord, 20/12/1859 *ibid.*
16. Charles Waterton to Eliza & Helen Edmonstone, 3/7/1860 (Am.Phil.Soc.).
17. Hobson, *op. cit.*, p. 250.
18. Charles Waterton to Eliza & Helen Edmonstone, 3/7/1860. *Op. cit.*
19. Michel Vaucaire, *Paul Du Chaillu: Gorilla Hunter*, tr. Emily Pepper Watts (Harper, New York, 1930), p. 50.
20. Paul Du Chaillu, *Adventures in the Great Forest of Equatorial Africa and the Country of the Dwarfs* (Harper, New York, 1861).
21. *The Athenaeum*, No. 1743 (23/3/1861).
22. *The Athenaeum*, No. 1753 (1/6/1861), p.728.
23. *ibid.*, No. 1758 (6/7/1861), p. 20.

24. *The Times*, 17/9/1861.
25. *ibid.*, 25/9/1861.
26. *Morning Advertiser*, 16/9/1861. Reprinted in *The Athenaeum*, No. 1769 (21/9/1861), p. 373.
27. *The Athenaeum*, No. 1771 (5/10/1861), p. 445.
28. *ibid.*, No. 1772 (12/10/1861), pp. 478-9.
29. *ibid.*, No. 1773 (19/10/1861), p. 509.
30. *ibid.*, No. 1774 (26/10/1861), pp. 543-4.
31. 'A New History of the Monkey Family', Charles Waterton. *Essays*, Series 3, p. 41.
32. *ibid.*, pp. 41-2.
33. *The Athenaeum*, No. 1781 (14/12/1861), pp. 806-7. In re Rev. W. Walker to P. L. Simmonds, Nov. 1861.
34. *The Times* and *The Athenaeum*, No. 1784 (4/1/1862).
35. Richard Owen, *On the Anatomy of Vertebrates*, Vol. 3, p. 797.
36. Mrs W. Pitt-Byrne, *Gossip of the Century* (Downey, 1899), Vol. IV, Ch. 13.
37. *ibid* and Hobson, *op. cit.*, pp. 245-6.
38. J. G. Wood, *Out of Doors* (Longmans, 1895), 'The Home of a Naturalist'.
39. Hobson, *op. cit.*, pp. 64-5.
40. *The Times*, 18/9/1861.
41. Queen Victoria to Mayor of Birmingham, July 1863 (Birmingham City Archives).
42. Hobson, *op. cit.*, p. 142.
43. Charles Waterton to Eliza Edmonstone, 1/10/1861 (Am.Phil.Soc.).
44. Charles Waterton to Helen Edmonstone, 1/10/1861 (Am.Phil.Soc.).
45. Charles Waterton to Richard Hobson, Christmas 1861. In Hobson, *op. cit.*, p. 302.
46. Goodchild Loan MSS. Wakefield District Archives.
47. Edward Walford, *The County Families of the United Kingdom* (London, 1868), *Ennis*, of Ballinahown, Westmeath.
48. George Ord to Charles Waterton, 11/3/1863 (Am.Phil.Soc.).
49. Charles Waterton to William Longman, 8/8/1862 (Goodchild Loan MSS).
50. *ibid.*
51. Charles Waterton to George Ord, 11/9/1862. In Irwin, *op. cit.*
52. Charles Waterton to Eliza Edmonstone, 29/12/1862 (Am.Phil.Soc.)
53. *ibid.*
54. Percy Fitzgerald, *Stonyhurst Memories* (London, 1895).
55. Charles Waterton to R. C. Yarborough Esq., the Hon Stanhope Hawke, 16/1/1863 (Am.Phil.Soc.).
56. Charles Waterton to George Ord, 21/2/1863 (Am.Phil.Soc. Film) and George Ord to Charles Waterton, 11/3/1863 (Am.Phil.Soc.).
57. Charles Waterton to Edmund Waterton, 12/6/1863 (North Yorks County Library – York City Archives. NRA9202/CW31).
58. *Burke's Landed Gentry* (18th edn, 1972), Vol. 3. 'Waterton, formerly of Deeping Waterton'.
59. Charles Waterton to Edmund Waterton, 21/6/1863 (North Yorks County Library. NRA9202/CW32).
60. George Gissing, *New Grub Street.*
61. A rather doubtful jacamar had recently been reported shot, during a snow storm, on the Montreal off Wicklow Heads, Ireland. 'Its appearance amongst us has floored me out.' Charles Waterton to Lieut Col Drummond, 3/2/1860 (Perth and Kinross District Archives. MS 34 Bundle 37).
62. Norman Moore (ed.), *Waterton's Essays on Natural History* (Warne, 1871), 'A Life of the Author', pp. 126-7.
63. Charles Waterton to George Ord, 25/1/1864. In Irwin, *op. cit.*
64. Charles Waterton, *An Ornithological Letter to William Swainson Esq.* (Richard Nichols, Wakefield, 1837), p. 5.

65. Hobson, *op. cit.*, p. 49.
66. *Illustrated London News*, 17/6/1865, p. 583.
67. Hobson, *op. cit.*, pp. 45-6.
68. *Illustrated London News*, *op. cit.*
69. 'Some Account of Walton Hall, the Seat of Charles Waterton Esq.', James Stuart Menteath. *Loudon's Mag of Nat Hist.*, Vol. VIII (Jan 1835), p. 33.
70. Charles Waterton to Alfred Ellis Esq., Feb. 1864. In Irwin, *op. cit.*
71. Hatton, *op. cit.*, p. 329.
72. Charles Waterton to Alfred Ellis, *op. cit.*
73. 'Waterton's Home', Unsigned. *Bentley's Miscellany*, Vol. 60 (1866), p. 328.
74. Moore's edn of *Essays*, p. 125.
75. B Waterhouse Hawkins to Charles Waterton, 13/12/1864 (North Yorks County Library. NRA9202/CW34).
76. 'Reminiscences of Charles Waterton', Dr George Harley. *Selborne Magazine*, Nov 1889. Reprinted in Mrs Alec Tweedie, George Harley, F.R.S. (London, 1899), p. 278.
77. Charles Waterton to Edmund Waterton, 11/1/1865 (North Yorks County Library. NRA9202).
78. *Dictionary of National Biography*, Vol. 59.
79. Charles Waterton to Henry Dixon, 22/1/1865. In Hobson, *op. cit.*, p. 331.
80. Charles Waterton to Keeper of Zoology, Manchester Museum, 27/3/1865. In Irwin, *op. cit.*
81. Charles Waterton to George Ord, 27/2/1865 (Am.Phil.Soc.).
82. Charles Waterton to Manchester Museum, *op. cit.*
83. Charles Waterton's Notebook. Wakefield Museum.
84. Moore's edn of *Essays*, p.131.
85. *ibid.*
86. Wood's preface to *Wanderings* (1879 edn).
87. *Dictionary of National Biography*, *op. cit.*
88. Canon R. A. Browne to George Ord, 27/5/1865 (Am.Phil.Soc.).
89. Richard Aldington, *The Strange Life of Charles Waterton* (New York, 1949), p. 214.
90. Moore's edn of *Essays*, p. 132.

CHAPTER 21: SIC TRANSIT GLORIA MUNDI

1. J. A. Rhodes to Edmund Waterton, Pontefract, 4/6/1865. (North Yorks County Library, NRA9202/CW36). Mr Rhodes, a friend of Hobson, had also been invited to the funeral. 'The hour, however, was so early, & the place of interment so distant, that I did not find it possible to be there in Time.'
2. *Idylls of the King: The Passing of Arthur*, 361-2.
3. G. Bernard Wood, *Yorkshire Villages* (Hale, 1971)
4. 'Funeral of the Late Mr Charles Waterton', Unsigned. *Illustrated London News*, 17/6/1865, p.583.
5. J. G. Wood, Preface to Waterton's *Wanderings in South America* (1879 edn).
6. 'Pray for the soul of Charles Waterton, whose wearied bones lie near this cross.'
7. John Goodchild, 'Edmund Waterton'. (Goodchild Loan MSS. Wakefield District Archives).
8. *Calender of the Grants of Probate and Letters of Administration*, 1865, p. 148.
9. 'Charles Waterton', Shane Leslie. *Dublin Review*, No. 446 (1949), p. 56.
10. Robert Tate to the Misses Edmonstone, St Cuthbert's College, Ushaw, 14/12/1868 (Goodchild Loan MSS). 'I hereby declare that I received the said curiosities, and I undertake on my own part, and on that of my successors, to return them, within six months from the date of demand.'
11. Edward Walford, *The County Families of the United Kingdom* (London, 1868), *Fletcher*, of Dunans, Argyll.
12. F. H. Salvin to Richard Hobson, 20/10/1866. In Hobson, *Charles Waterton, His Home, Habits and Handiwork* (2nd edn, 1867), pp. 122-5.

13. Sir James Stuart Menteath to Richard Hobson, 25/10/1866 *ibid.*, pp. 187 & 199.
14. H. J. Massingham, *Some Birds of the Countryside: The Art of Nature* (Unwin, 1921), p. 198.
15. Bill of Complaint, E. Edmonstone v. Waterton, 8/1/1867 (Goodchild Loan MSS).
16. 'Inventory of Furniture, China, Glass and Other Effects at Walton Castle'. (Goodchild Loan MSS). 'Castle' was Edmund's idea.
17. Walford's *County Families*. (1864, 1878, 1880 & 1890 edns), *Hailstone*.
18. 'Walton Hall', John Goodchild. *British Journal of Anaesthesia*, Vol. 55 (1983), p. 228.
19. 'St Helen's Church, Sandal Magna', J. W. Walker. Yorkshire Archaeological Journal, Vol. XXII, No. 93 (1916), p. 9.
20. Mrs W. Pitt-Byrne, *Gossip of the Century* (Downey, Leeds, 1899), Vol. IV, Ch. 13.
21. Norman Moore (ed.), Waterton's *Essays on Natural History* (Warne, 1871), 'A Life of the Author'.
22. James Simson, *Contributions to Natural History, and Papers on Other Subjects* (London & Edinburgh, 1875. New York, 1878), p. 47.
23. James Simson, *Charles Waterton, Naturalist* (Baillikre, Tyndall & Co., London, 1880), p. 16 et seq.
24. *ibid.*, pp. 23-4.
25. *ibid.*, p. 40.
26. Walford's *County Families*, 1880 & 1920 edns. 'Simpson of Walton Hall'.
27. *Kelly's Directory of the West Riding of Yorkshire* (1893), Part 1. 'Walton'.
28. *Burke's Landed Gentry* (17th edn, 1952), 'Simpson of Walton Hall'. (Seat – Walton Hall. Residence – Thornhill House, Walton.'
29. It was saved from demolition only because the Hall Island is considered a valuable Saxon dwelling site. The prospect of a 5-year archaeological survey deterred the developers. David Waterton-Anderson, personal communication.
30. *Squire* (Magazine of Walton Hall Country Club), No. 4 (Summer, 1981).
31. 'Proceedings of a Symposium Held to Commemorate the Bicentenary of Charles Waterton, F Richard Ellis et al'. *Br.J.Anaesth.*, Vol. 55 (1983), pp. 221-3.
32. *Wakefield Express*, 3/4/1992 and 10/4/1992.
33. *Burke's Landed Gentry* (18th edn., 1972), Vol. 3. 'Waterton, formerly of Deeping Waterton'.
34. 'Rings of Popes and Cardinals', E. Waterton. *Archaeologia* (London), No. 40 (c1877),

Bibliography

The following should serve as a guide for anyone who wants to search further. Most sources referred to in the text are recorded in detail in the Notes and References section.

Books by Charles Waterton

Wanderings in South America, the North-West of the United States, and the Antilles, in the years 1812, 1816, 1820 and 1824. Over 20 editions since its first publication by J. Mawman & Co. (London), in 1825. Reprinted four times by B. Fellowes (Mawman's successor) in 1828, 1836, 1839 and 1852. Several editions with an introduction by J. G. Wood, published by Macmillan between 1878 and 1893. 'Popular' editions published by Nelson & Sons (1891 & 1903) and by Blackie & Son (undated) in their School & Home Library and Library of Famous Books series. Twentieth century editions, with introductions by W. A. Harding, Edmund Selous, Gilbert Phelps, L. Harrison-Matthews and David Bellamy, published respectively by Hutchinson (1906), Dent's Everyman's Library (1925), Charles Knight (1973), Oxford University Press (1973) and Century (1984).

Essays on Natural History, Chiefly Ornithological, Series 1 (Longmans & Co., 1838); Series 2 (Longmans, 1844); and Series 3 (Longmans, 1857). Collected edition, with 'A Life of the Author' by Norman Moore (Frederick Warne, 1871).

Pamphlets by Charles Waterton

Letter to the Rev J. P. Simpson, Curate of All Saints Church, Wakefield (Richard Nichols, Wakefield, 1833).
A Letter to James (sic) Jameson, Esq (Richard Nichols, 1835). An open letter to Professor Robert Jameson, reprinted in *Fraser's Magazine*, Vol. XI (March, 1835).
A Second Letter to Robert Jameson, Esq (Richard Nichols, 1835).
To the Rev Disney Robinson (Richard Nichols, 1836).
An Ornithological Letter to William Swainson, Esq., F.R.S &c, &c (Richard Nichols, 1837).
A Letter on the Church of England by Law Established, Occasioned by Parson Gregg's Unprovoked Attack on the Catholic Religion (Richard Nichols, 1837).
The Law Church (Richard Nichols, 1837).
A Letter on the Reformation, Occasioned by the Attack of the Reformation Society of Wakefield (Richard Nichols, 1838).
Living in the Tropics (Unpublished), transcribed by Norman Moore, 1867.

Copies of all the above are in Wakefield Museum, except *Living in the Tropics* which is in the British Museum (Nat Hist).

Biographies

Four biographies and one book of reminiscences have been published since Charles Waterton's death. They range in quality from bad to imbecilic. In chronological order:

Hobson, Richard, *Charles Waterton, His Home, Habits and Handiwork.* (Longmans & Co., London, 1866. Second edition – enlarged – 1867). See text.

Gosse, Philip, *The Squire of Walton Hall.* (Cassell, 1940). Very limited research. Long quotations. Little else.

Aldington, Richard, *The Strange Life of Charles Waterton.* (Duell, Sloan & Pearce, New York, 1949. And Evans Bros., London, 1949). Sarcastic, opinionated, but curiously sympathetic towards the 'admirable' Richard Hobson. Alan Moore – son of Sir Norman Moore – took exception to Aldington's claim that his father had been professionally jealous of Hobson.

Phelps, Gilbert, *Squire Waterton.* (EP Publishing, Wakefield, 1973). Aldington abridged, and to some extent plagiarised, with a hint of BBC. 'A small episode took place which is typical of the man.' Then another one took place. Then, characteristically, another. Gilbert Phelps holds up the idiot cards and beams happily. Laugh. Applaud. Gasp.

Blackburn, Julia, *Charles Waterton, Traveller and Conservationist.* (Bodley Head, 1989). Authenticated tedium. Garbled fact and irritating little fictions presented as fact. All in the service of a drab pastiche of every biography ever written about anybody. It could be called an achievement to write a dull book about Charles Waterton.

Miscellanea

All of the following either feature long sections or chapters on Charles Waterton or refer to his life, directly or indirectly.

Aldington, Richard, *Four English Portraits, 1801-51* (Evans Bros., London, 1948). Disraeli, George IV, Dickens and Charles Waterton.

Allen, David Elliston, *The Naturalist in Britain: A Social History* (Allen-Lane, London, 1976).

Barber, Lyn, *The Heyday of Natural History* (Cape, London, 1982)

Barrington-Brown, Charles, *Canoe and Camp Life in British Guiana* (Edward Stanford, London, 1876). Good background material.

Bates, Henry Walter, *The Naturalist on the River Amazons* (John Murray, London, 1863. Everyman's, London, 1910).

Blunt, Wilfrid, *The Ark in the Park* (Hamish Hamilton, London, 1976).

Bunn D, Warburton A B & Wilson R D S, *The Barn Owl* (Poyser, London, 1982).

Burland, Cottie, *Men Without Machines* (Aldus Books, London, 1965).

Caufield, Catherine, *In the Rain Forest* (Heinemann, London, 1985).

Darwin, Charles, *Autobiography*, ed. Gavin de Beer (Oxford University Press, 1974).

Dickens, Charles, *Barnaby Rudge* 'Grip' the raven is based on Dickens' own pet ravens but influenced by Charles Waterton's *Essays.* CW mentioned in preface to second edition – 1867. (Reprint – Hazell, Watson & Viney Ltd., London).

Douglas, Norman, *Experiments* (Chapman & Hall, London,1925). Pp.183-198 – *A Mad Englishman.*

Fitzgerald, Percy, *Stonyhurst Memories* (Richard Bentley & Son, London, 1895).

Hatton, Joseph, *Old Lamps and New. An After-Dinner Chat* (Hutchinson, London,1889). Chapters on Henry Irving, Victor Hugo, William Blake, Lord Tennyson, Henry Longfellow, George Stephenson and Charles Waterton.

Head, Sir George, *A Home Tour through the Manufacturing Districts of England in the Summer of 1835* (Reprint – Frank Cass, London, 1968). Ch.XI – Walton Hall, Yorkshire.

Howe, Bea, *Lady with Green Fingers. The Life of Jane Loudon* (Country Life Ltd., London 1961).
Irwin, R.A., *Letters of Charles Waterton* (Rockliff Publishing Corporation Ltd., London, 1955).
Keay, John, *Eccentric Travellers* (John Murray and BBC Publications, 1982).
Loudon, Jane, *The Ladies Country Companion, or How to Enjoy a Country Life Rationally* (Longmans, London 1845).
Massingham, H. J., *Some Birds of the Countryside: The Art of Nature* (Unwin, London, 1921). Ch.VIII – Charles Waterton.
Mesick, Jane, *The English Traveller in America, 1785-1835* (New York, 1922).
Miller, Lillian B, *In Pursuit of Fame: Rembrandt Peale, 1778-1860* (University of Washington Press, Seattle, 1992).
More, Thomas, *Utopia* (Everyman, London).
Pitt-Byrne, Mrs W (Julia Clara Busk), *Gossip of the Century*, Vol IV, Ch 13. *Charles Waterton, the Wanderer* (Downey & Co., London, 1899).
Ralegh, Sir Walter, *The Discoverie of the Large, Rich and Bewtiful Empire of Guiana*, etc., (London, 1596.)
Roth, Vincent, *Notes and Observations on Animal Life in British Guiana* (Daily Chronicle Publications, Georgetown, 1947).
Roth, Vincent, *A Plea for the Better Conservation and Protection of Wild Life in British Guiana* (Daily Chronicle Publications, Georgetown, 1949).
Schomburgk, Sir Robert H, *Travels in British Guiana and the Orinoco, 1840-44,* ed. Walter E Roth. (Daily Chronicle Publications, Georgetown, 1922).
Sellers, Charles Coleman, *Mr Peale's Museum: Charles Willson Peale and the First Popular Museum of Natural Science and Art* (W W Norton, New York, 1980).
Simson, James, *Contributions to Natural History* (Privately printed, London & Edinburgh, 1875. New York, 1878).
Simson, James, *Charles Waterton, Naturalist* (Baillière, Tyndall & Co., London, 1880).
Sitwell, Edith, *The English Eccentrics* (Dobson, London, 1933).
Smith, Philip, *Arrows of Mercy* (Doubleday, Toronto & New York, 1969).
Spry, I.M. (ed)., *The Papers of the Palliser Expedition, 1857-60* (The Champlain Society, Toronto, 1968).
Stonyhurst Centenary Record (Stonyhurst College, 1894).
Swainson, William, *On the Natural History and Classification of Birds* (Lardner's Cabinet Cyclopaedia, Vol.I, London, 1836)
Swainson, William, *Taxidermy, Bibliography and Biography* (Lardner's Cabinet Cyclaepedia, Vol.XI, 1840).
Thomas, K Bryn, *Curare, Its History and Usage* (Pitman, London, 1964).
Tweedie, Mrs Alec, *George Harley, F.R.S. The Life of a London Physician* (The Scientific Press, London, 1899).
Wilson, Alexander, *American Ornithology*, ed. Robert Jameson. (Constable, Edinburgh, 1831).
Wood, G Bernard, *Yorkshire Villages* (Hale, London, 1971).
Wood, Rev J.G., *My Feathered Friends* (Routledge, London & New York, c1868).
Wood, Rev J.G., *Out of Doors* (Longmans, London, 1895). *The Home of a Naturalist.*

Serials

Hundreds of thousands of words have been written by or about Charles Waterton and his immediate concerns, in scores of periodicals, during the past 200 years. Almost all have been overlooked by previous biographers. Authors and article headings are in the relevant places in the Notes and References section.

American Midlands Naturalist, 66:200-05 (1961).
American Quarterly Review, 20:245-58 (Dec 1831).

Ampleforth Journal, June, 1959.
Anesthesiology (Toronto), 3:481 (1942).
Archives of Natural History, 15(1):77-88 (1988).
The Athenaeum, 549:321 (5/5/1838); 850:140 (10/2/1844); 856:276 23/3/1844); 874:692 (27/7/1844); Nos.1743-1784:passim (23/3/1861-4/1/1862).
Auk, 5:248 (1888); 6:51 (1889); 15:11-13 (1898); 22:31-4 (Jan 1905); 22:248-55 (Jul 1905); 34:275-82 (Jul 1917); 38:585 (Oct 1921); 56:472-4 (Oct 1939); 67:210-16 (Apr 1950).
Bentley's Miscellany, 60:263-71 (1866).
Bird Study, 3:42-73 (Mar 1956).
Blackwoods, June 1826; 58:289-300 (Sept 1845).
British Critic, Vol.II (Apr 1826).
British Journal of Anaesthesia, 55:221-33 (Mar 1983).
Canadian Anaesthetists' Society Journal, 29:195-202 (May 1982).
Chamber's Edinburgh Journal, 492:188-9 (3/7/1841); 158(4th ser):14-16 (5/1/1867).
Cornhill, 1023:200-208 (Spring 1960).
Country Life, 30/4/1981; 3/6/1982; 24/3/1988; 9/12/1993; 16/12/1993.
Dublin Review, 8:321-2 (May 1840); 446:55-70 (1949).
Dublin University Magazine, 25:305-18 (Mar 1845).
Edinburgh Journal of Natural History and Physical Sciences, 1:152 (July 1838); 2:14-15 (May 1839).
Edinburgh Review, 43:302-3 (Feb 1826).
The Examiner, 23:645-6 (10/10/1830).
Franklin Journal (Philadelphia), 2:144 (Aug 1828).
Fraser's Magazine for Town and Country, 11:325-9 (Mar 1835); 56:631-49 (Dec 1857).
Gardeners' Chronicle and Agricultural Gazette, 18/8/49; 15/2/1862; 22/2/1862.
Gardener's and Farmer's Journal, 1/9/1849.
Gentleman's Magazine, Pt.ii:716 & 843 (1783); 95:448 (May 1825); 163:526 (May 1838); 203:543-4 (Nov 1857).
The Guardian, 7/7/1989; 1/8/1989.
Guiana Morning Chronicle, 1/4/1809.
Ibis, 101:118 (1959); 126:67-9 (1984).
Illustrated London News, 24/8/1844; 17/6/1865.
The Leader, 388:836-7 (29/8/1857).
Leisure Hour, 1:282-6 (22/4/1852).
Literary Gazette and Journal of Belles Lettres, 468:4-8 (7/1/1826); 470:39-40 (21/1/1826); 471:56-7 (28/1/1826); 1109:244-46 (21/4/1838).
Living Age (Philadelphia), 44:649.
London Quarterly Review, 33:323 (Mar 1826); 53:382-420 (Jan 1880).
Loudon's Magazine of Natural History, Vols.I-IX (1828-36). Every issue. (Bi-monthly until Feb 1835. Then monthly until Sept 1836).
Macmillan's Magazine, 39:326-31 (Feb 1879).
Mechanics' Magazine, 447:403-5 (3/3/1832).
Morning Advertiser, 16/9/1861.
Monthly Review, 108:66-77 (Sept 1825).
National Review, 5:304-32 (Oct 1857).
The Naturalist, 2:34-5 (April 1837).
Nature, 24/4/1879; 188:548 (12/11/1960).
New Edinburgh Philosophical Journal, 2:172-84 (Oct 1826-Apr 1827); 3:21-30 (Apr-Jul 1827).
Nottingham Journal, 12/4/1839.
Once a Week, 13:48-52 (1/7/1865).
Pennsylvania Monthly, 10:443.
Philosophical Transactions, 78:Pt.2 (1788); 101:178 (1811).

Punch, 10:144 (Mar 1846).
St James's Chronicle, 23/2/1839.
Saturday Review of Politics, Literature and Art, 6:307-8 (3/10/1857).
Science, 130:630 (1959).
Selborne Magazine, Feb, Mar, Aug & Nov 1889.
The Spectator, 11:355 (14/4/1838); 83:490-1 (7/10/1899).
Timehri (Georgetown), May 1915:257-65.
The Times, 30/8/1825; 21/10/1852; 17/9/1861; 25/9/1861; 2/9/1949; 12/8/1995.
Times Literary Supplement, 29/12/1932; 26/8/1949.
Wakefield & West Riding Examiner, 14/8/1849; 1/9/1849; 12/10/1849.
Wakefield Express, 3/4/1992; 10/4/1992.
Westminster Gazette, Sept 1866.
Yorkshire Society of Anaesthetists Newsletter, Waterton bi-centenary issue, 1981.
Yorkshire Topic, 1989:12-13.

Ancestry

Charles Waterton can only be fully appreciated in the context of his family background. The Waterton ancestry is long and fascinating but has been almost completely ignored and misrepresented by previous biographers. Many primary sources are now available as collected volumes of close rolls, patent rolls, papal registers, society proceedings and manuscripts etc.

Bawden, Rev William, *Dom Boc: A Translation of the Records of Domesday so far as Relates to the County of York* etc. (London, 1809).
Blomefield, Francis, *An Essay Towards a Topographical History of the County of Norfolk.* (London, 1807). Vol.VI, Oxburgh.
Boothroyd, B. *The History of Pontefract* (Leeds, 1807).
Bossy, John, *The English Catholic Community* (Darton, Longman & Todd, London, 1975).
Boyle, J. R. *The Lost Towns of the Humber* (London, 1889).
Brooks, F. W. *The Council of the North* (Historical Association, London, 1953).
Burke's Landed Gentry and Burke's Peerage, Baronetage & Knightage (Harrison, London, various edns). Bedingfeld, Clifford, Edmonstone, Fairfax, Fleming, Northumberland, Waterton etc.
Darbyshire, Rev Hubert Stanley and Lumb, George Denison (eds), *The History of Methley* (Thoresby Society Publications, 1934).
Ellis, H. *Original Letters Illustrative of English History* (London, 1824).
Finch, H. *Nomotechnia* (London, 1627).
Foxe, John, *Book of Martyrs* (London, 1583).
Given-Wilson, Chris, *The Royal Household and the King's Affinity* (Yale University Press, New York, 1986).
Hailstone, E. (ed), *Portraits of Yorkshire Worthies* (2 vols., Leeds, 1869).
Hingston, Rev. F. C. (ed), *Royal and Historical Letters During the Reign of Henry IV* (Longmans, London, 1860).
Inquisitions Post Mortem Relating to Yorkshire, of the Reign of Henry IV and Henry V, Yorkshire Archaeological Society Record Series, Vol.LIX. (1918).
Kelly's Directory of the West Riding of Yorkshire (Kelly & Co., London, 1893 & 1936 edns).
Kirby, J. L. (ed), *Calendar of Signet Letters of Henry IV and Henry V (1399-1422)* (H.M Stationery Office, London, 1978).
Parliamentary Survey of Church Lands (Lambeth Palace). Vol.XVIII
Previta-Orton, C. W. *The Early History of the House of Savoy – 1000-1223* (Cambridge University Press, 1912).
Public Records Office, *Records of the Duchy of Lancaster, Lists and Indexes.*

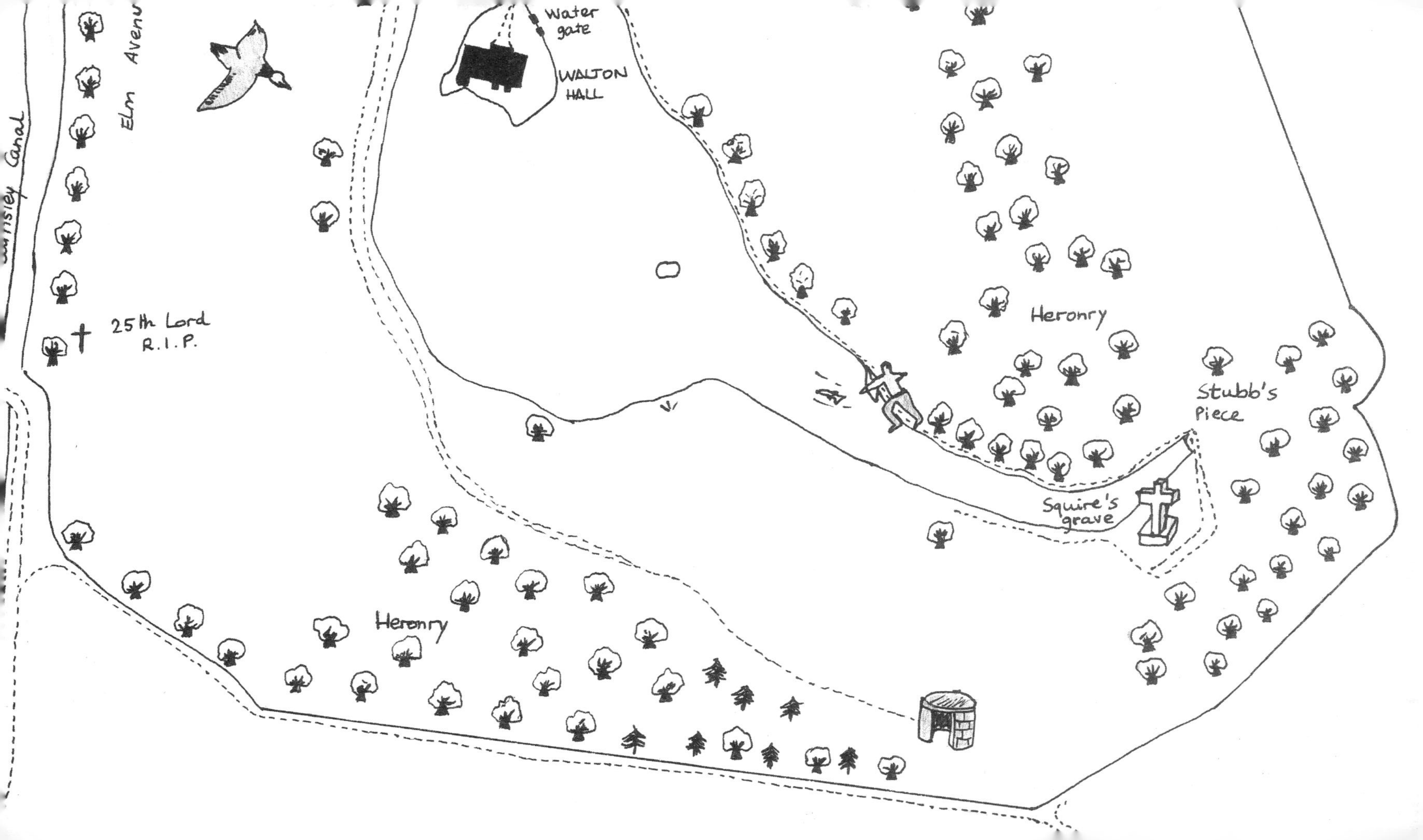

Canal
Elm Avenu
Water gate
WALTON HALL
25th Lord
R.I.P.
Heronry
Stubb's Piece
Squire's grave
Heronry